WATER AND THE ENVIRONMENT

WATER AND THE ENVIRONMENT

Editors:
J.C. CURRIE
Director, Tweed River Purification Board,

and

A.T. PEPPER
Independent Consultant

Institution of Water and Environmental Management

ELLIS HORWOOD
NEW YORK LONDON TORONTO SYDNEY TOKYO SINGAPORE

First published in 1993 by
ELLIS HORWOOD LIMITED
Market Cross House, Cooper Street,
Chichester, West Sussex, PO19 1EB, England

A division of
Simon & Schuster International Group
A Paramount Communications Company

Printed and bound in Great Britain
by Redwood Press, Melksham

British Library Cataloguing in Publication Data

A Catalogue Record for this book is available from the British Library

ISBN 0–13–950692–6

Library of Congress Cataloging-in-Publication Data

Available from the publishers

Table of contents

IWEM 91 — Water and the Environment foreword for conference proceedings

FOREWORD

There is growing public concern about environmental issues, and such topics as global warming and sustainable growth are frequently the subject of comment in the press and other media. Public pressure continues to grow for greater efforts to be made to clean up the environment and remedy the pollution and degradation that has arisen over the last hundred years.

Water makes up a major part of the environment, together with land and air, and is vital to the maintenance of life. Within the hydrological cycle, the rain falling from the skies, flowing over land, being used and discharged to rivers, and returning to the sea, affects both the air and land phases of the environment. The interaction between all three phases must be considered whenever pollution is being remedied, and 'integrated pollution control' requires this multi-media approach.

The Institution of Water and Environmental Management's IWEM 91 conference took as its title 'Water and the Environment'. The papers presented at the conference are published in these proceedings and, although primarily concerned with the water environment, there are papers on solid waste disposal and the interaction between all three phases of the environment. In setting out new developments in the treatment of potable water supplies, wastewater disposal and river and coastal water pollution prevention, particular emphasis is given to impacts on the environment and ways of minimizing adverse effects. It is now a statutory requirement for consideration to be given to ways of enhancing the environment, rather than just dealing with pollution and maintaining the status quo.

Changing public perception of environmental needs was highlighted at the conference with a series of papers dealing with the water industry in general, coastal development management and the scope for minimizing waste by recycling. Privatisation of the water industry created a need for regulation of standards and services and this has led to more press and other comments on potable water quality, sewerage and sewage treatment. In addition, there is the impact of the European Community directives.

Conference papers also dealt with water resources and water quality and the need for their protection. Surface water and groundwater are taken for supplies in many parts of the world and each demands its own type of protective measures and control to ensure that both quantity and quality are protected. The impacts on the environment of abstraction from underground sources, potentially drying up river headwaters, have to be taken into account. However, the creation of a new reservoir by impounding tributaries can create opportunities for enhancing nature conservation, as do works that are designed to prevent flooding from rivers or protect coastal areas from tidal inundation.

As an aid to the prediction of changes that might take place in the water environment, mathematical modelling has developed in recent years. Together with the rapid improvement in computer technology, modelling of river and coastal water quality, sediment loads, and tidal patterns of coastal discharges have been of major benefit to the water industry, as has the successful development of geographical information systems. The value of these techniques was recognized within the conference programme with the presentation of case histories.

Environmental impact assessment for major projects in the industry was also covered. It is not only in building a reservoir or changing the course of a river that the technique has to be considered. Growing concern about 'green' issues is leading to increased awareness and scrutiny of corporate policy in water and other industries. There is seen to be commercial advantage in being able to show that conservation and sustainability are an essential part of management decisions.

Conservation and consideration of environmental improvement is not only of importance in rural sites, but also in the development and redevelopment of urban areas. The conference included papers on the London Dockland redevelopment and canalside regeneration of Birmingham. The former has made a particular feature of the water environment, necessitating a higher water quality than earlier uses, and in Birmingham improvements to the canal system have been made at the same time as reconstruction of the city centre to improve amenities for local residents and to develop tourism.

Overall, the conference covered an impressive list of topics of concern to the water and environmental industry and it clearly demonstrated the wide range of professional disciplines required to determine the best environmental solutions to the many problems.

Ken Guiver, MBE
President

Part I Environmental policy

1

The water industry and changing environmental perceptions

P. G. Soulsby, BTech, MIBiol[†]

INTRODUCTION

The privatization of the water authorities occurred at a time of unprecedented concern for environmental issues. The process itself highlighted the role the industry played in environmental protection through the services it provided. The industry has had to adapt to a variety of cultural upheavals brought about by the change in ownership and in regulation. However, it is inevitably the views of the customers which will have the greatest long-term impact. The chapter assesses current perception on the environment and suggests possible cultural changes within the industry.

ASSESSING THE ENVIRONMENTAL MOVEMENT

Membership of organizations

An important barometer of the public's interest in environmental issues is the membership of environmental groups (Table 1). The identified figure of 4.5 million does not include the multitude of small local and single issue groups.

Many of the organizations have shown extraordinary growth in the last few years. Friends of the Earth (FoE) from 60 000 in 1985 to 200 000 in 1990; Greenpeace from

[†] Scientific Manager, Southern Water Services Ltd, Hampshire.

120 000 in 1988 to 385 000 in 1990 and Council for the Preservation of Rural England from 2000 in 1986 to 44 000 in 1990. The numbers may still represent only a small proportion of the population but are, nevertheless, likely to represent many heads of households and high disposable income groups. The figures can certainly be enhanced by those who support the aims of the various organizations, either continually or from time to time. The numbers also do not necessarily reflect their influence. Greenpeace membership advertisements are targeted to the quality newspapers and a recent breakdown of FoE membership showed the bulk of members to be well educated professionals with 70% having tertiary level education.

Table 1. Membership of environmental organizations – 1990

Organization	Membership
British Trust for Conservation Volunteers	11 000
Council for the Protection of Rural England	44 000
Friends of the Earth	200 000
Greenpeace	385 000
National Trust	1 500 000
Royal Society for the Protection of Birds	820 000
World Wide Fund for Nature	1 300 000
County Naturalists Trusts	213 000
Marine Conservation Society	5 000
Soil Association	6 000
Total	4 484 000

Market research

The importance of environmental attitudes in purchasing habits has prompted market research companies to undertake surveys independently and on behalf of clients. An ICM Research Survey for The Guardian newspaper (July 1990) of 1400 people showed: the environment was second only to the poll tax in serious issues facing the country; women were more concerned than men, and younger people more concerned than older; top of the list of environmental concerns was pollution of the seas and oceans followed by clearing the rainforests and global warming.

Such surveys also reveal the growing sophistication of the public with 69% preferring to buy products that have biodegradable or recyclable packaging, and growing cynicism with 49% not believing labels stating that the products were environmental friendly.

A Millward Brown Survey of 1000 people suggested nearly 70% of the population would accept zero economic growth to save the environment and 75% said they would buy environmentally friendly products, even if they cost more than their conventional counterparts. Four out of five people rejected the idea that the 'Green' movement is a passing phase. In the same survey, the water industry was seen as one of the least environmentally damaging.

Even single product analysis reveals environmental awareness as an important factor in purchasing preferences. A Mintel Survey of the cosmetics industry revealed 40% of consumers looking for environmentally friendly products and perhaps, more importantly, the image of the retailer influencing product choice.

Advertising

The hard evidence that environment sells is to be found in the use of environmental responsibility to boost sales. In the last few years, the weight of advertising material using 'green' associations has been remarkable. They have been used for many different purposes (Table 2) and many in a manner so irresponsible that it has lead to the Friends of the Earth 'Green Con Awards'. The Worldwide Fund for Nature benefited by £4 m last year as a result of licensing agreements with a variety of products.

Table 2. Marketing campaigns using environment associations

Company	Campaign aims	Technique
British Gas	Corporate image	Key environmental issues adverts
British Oxygen	Corporate image	Protecting habitats
Bayer Chemicals	Corporate image	Protecting the sea
Tipp-Ex	Brand enhancement	Tie up with WorldWide Fund for Nature
Persil	Brand enhancement	Tie up with BTCV
BP/Shell/Esso	Product sales	Unleaded petrol adverts
Peaudouce	Product sales	Environmentally friendly nappies

Nevertheless, while the marketing managers of companies engaged in furious competition see environment as a competitive edge, then one can be sure that customers have green aspirations which the industry needs to satisfy.

Sales of products

However, although advertising is important as a barometer, it is sales which are the real test. The most obvious success is unleaded petrol already claiming 35% of the market in three years although benefiting from a positive price differential. Recycled paper of one sort or another has seen impressive growth. 'Green' cleaning products have had a mixed reception with poor performance and price premiums causing negative consumer reactions. The major supermarket chains, however, all carry their own brands and Safeway's 'Ecologic' range captured 8% of the market within 16 weeks. In battery sales, Varta reached a 62% increase in market share after the launch of mercury-free 'green' batteries. Sales of organic food were forecast to treble from £40 m in 1988 to £120 m in 1990, with significant growth not only in vegetables but in dairy products, bread, meat and wine. They can even gain a premium of between 15% and 100% over non-organic foods (Coopers & Lybrand Deloitte).

THE ENVIRONMENTAL ISSUE – WILL IT LAST?

The important question for the water industry is whether the environmental issues will stay at the top of the agenda. Resistance to price rises will increase if there is not a perceived value for money. Satisfying the needs of environmental aspirations could be a significant part of the industry's value.

Theories abound to explain the sudden concern for 'green' issues. These range from the international issues of ozone depletion, acid rain and the greenhouse effect to Chernobyl, and the great storms. The media has clearly played an important role; it has been able to strike a chord and thus keep up or increase viewing and readership figures. Clearly there is a mixture of many factors, not least of which is the rise into positions of influence of a post-second world war concerned generation.

Perhaps the nearest equivalent boom to the environmental movement is the health foods fad of the 1970s. This centred on the question of food additives. Companies that ignored that issue went to the wall and health foods moved from specialist stores to mainstream supermarket shelves. The development into organic food is now spearheaded by the supermarkets themselves. That fad has now become an integral part of food retailing and the disastrous setback in sales of eggs and beef after their various scares shows how seriously consumers continue to view the issue.

All the evidence suggests that there is too much weight behind the environmental issue for it to disappear. Perhaps the most telling remark to support that came from Chris Patten at the last Conservative Party Conference:

'Every political instinct I have tells me that the environment is going to be a central issue, a crucial issue for as long as my generation is active in politics and far beyond. If you don't believe that – ask your children what they think!'

THE POSITION OF THE PLCs IN THE ENVIRONMENTAL DEBATE

The water industry lies at the centre of the environmental debate (Fig. 1). Not only are its own products affected by macro problems, global warming, acid rain, but also on a micro scale by individual discharges or spillages to watercourses or the diffuse actions of intensive agriculture, pesticides, nitrates, etc. Its own products are also issues – the effect on river flows of both groundwater and surface water abstractions or the problems of eutrophication or public health aspects of bathing and recreational waters.

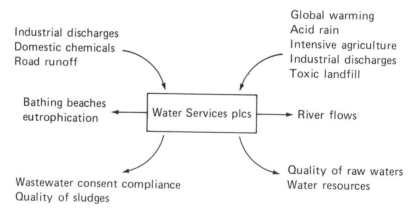

Fig. 1. Environmental issues and water service activities.

Protection of the environment is a core value for water services plcs. As a core value it should trigger a behavioural response which could go beyond obeying standards laid down by regulation. Successful business does not thrive on obeying regulations; they are but the framework on which customer-orientated products and attitudes are built.

Three possible responses can be identified: auditing of environmental impact within the company; auditing suppliers' environmental records, and evaluating attitudes to new standards.

Environmental audit

The impact of plcs goes far beyond effluent discharged or water supplied. Claiming environmental success for cleaning up a river is one thing, but that can be offset by: profligate use of energy; failure to encourage recycling or to conserve natural habitats. A logical process therefore is to encourage an environmental audit approach throughout a company in all its activities. Very successful companies engaged in a variety of activities have already imposed these management systems. Among them are Volvo, IBM, 3M, ICI and British Airways. They might not have solved all the problems but they are ongoing dynamic systems, part of a corporate policy running along the same lines as financial auditing or quality assurance. The EC is likely to impose environmental auditing on the water industry during the decade. It should therefore be in the forefront of developing and using such systems and selling its expertise to the rest of the industry.

Audit of suppliers

As well as looking inwards, companies can have a significant effect on the behaviour of others.

With four of the water service plcs being in the Footsie 100 and all with a massive annual spending power, it is the approach to suppliers which could establish the companies' environmental reputation. In a corporate survey of companies in Europe (but not the UK) by 3M in 1989, one in five respondents said that concern for the environment was the most important attribute they would look for in a prospective supplier. Such scrutiny is already commonplace in the food retailing sector but B & Q, for instance, now sends a 30-page questionnaire to each supplier asking questions on energy conservation policies, waste management, and emissions and discharges. Volvo has stated that the way suppliers approach environmental issues will be taken into account when supplier contracts are renewed. A continuation of this trend will see companies lose business not on a financial basis, but on an environmental attitude basis.

However, perhaps the most difficult area for the water industry is its relationship with industrial customers who discharge to sewer, particularly the majority who do not come under the Her Majesty's Inspectorate of Pollution (HMIP) umbrella.

At present, the consent standards for discharge to sewer – for example, metal levels – vary depending on the policy of the particular water services plc. Many industrialists meet these standards by dilution – a double advantage to a plc in terms of income. The forcing mechanisms to make the plc change its attitude are any threat to treatment works compliance or to disposal routes and costs for sludge. If these are not threatened, for instance, by using sludge incineration, then the status quo could continue. This is likely to exacerbate regional consent differences, particularly where plcs have opted for a sludge re-use strategy as a result of the closure of the sea route and are looking to reduce metal loads in their sludge. Plcs could therefore be in the position of encouraging some companies to continue wasteful practices which is not easily compatible with an environmentally responsible image. Nor will it find favour with the European single market philosophy which is unlikely to tolerate competitive distortion based on lowered waste treatment standards.

The attitude to standards

The third aspect of behavioural response is the attitude to standards and their role in the environmental debate.

The water industry needs to weigh carefully its resistance to standards based on investment costs against their customers' perception of the need for those standards and the value for money in achieving them. The Pesticide Standard in the Drinking Water Directive and the Virus Standard in the Bathing Water Directive are high-profile examples. Both have been described in the industry as being unrealistic in terms of achievement and in risk to public health. However, unless those also are the perceptions of customers, such views may be seen in a negative way and not as giving value for money.

On the other hand, attaining a standard does not necessarily satisfy the public's perception of risk. Modern long sea outfalls met all the current standards but not public approval. The industry needs to identify the standards their customers expect them to meet and, if necessary, lobby for their implementation.

A secondary effect of dismissing these standards is to lose a role in the environmental debate. This particularly applies to the pesticide question. Southern Water plc received excellent media response for its agreement with a county council to ban the use of triazine-based herbicides. This led to other users following suit and has locally opened up the debate on the question of pesticide application as a whole, the management of their use and the alternative approaches. The water industry, along with the regulators, can have an important role to play in these issues and can create a positive corporate image complementing the value for money of their charges.

THE EXPANDING MARKET AND PRICE RESISTANCE

The market for environmental technology

The worldwide annual expenditure for environmental technology has been estimated for 1989 as £135 billion, forecast to rise to more than £220 billion by the year 2000. Equivalent UK expenditure has been estimated at £6.1 billion and £11.3 billion respectively (Helmut-Kaiser). These costs are being forced by present and future regulations but will eventually have to be paid for by the final customer. The water industry is a significant contributor to that expenditure but is probably unique in that the costs represent a major proportion of the charges and are seen as a burden by the City concerned about cost recovery and maintenance of profits. The City, however, is not used to perceiving investment in environmental protection as an opportunity. However, with such large and growing markets to be tapped, this view will change. Certainly, the boards of several water service plcs have demonstrated their confidence by diversifying into these areas. Although still tiny, the UK ethical investment sector, now numbering 18 funds representing £200 m, will be increasingly attracted to the water industry if it shows real commitment. In America, where 10% of investments through Wall Street have ethical tags, there are now funds, currently at £250 m, which invest only in environmental service companies.

CONCLUSIONS

It can be argued that the industries' investment costs are safeguarded by the need to meet regulations enshrined in statutory law with consequent income to support it having to be made available from customers. Regulations tend to follow public or political opinion. However, the fate of the Control of Pollution Act or the EC Bathing Water Directives in the 1970s show that if the public pressure is not present, they can be postponed or interpreted in a less rigorous manner. The banning of sludge disposal to sea can be seen as a contract for the industry to produce a higher value product which customers will pay for. It is an opportunity brought about by environmental concern. Logically, if the concern reduces, then so will the opportunities and so will the willingness to pay.

Although the market research quoted and surveys on organic food purchase have demonstrated a willingness of a majority to pay a premium for perceived environmental protection in products or 'safe' food, there is little doubt that in the last few months the Gulf War and the growing effects of recession have had an effect. But no revolution could continue at the pace shown in the last few years. The evidence that the environment is slowly being degraded will not go away and the Gulf War is likely to strengthen, not weaken, that view. Privatization freed the water industry from the constraints of government spending but has made it more dependent on the customer's view of its worth. The premise of this chapter is that the water industry's role in the environment is the added value of its product.

2

Archaeology in the water industry

M. R. L. Hall, BA, MIFA, MIWEM[†]

INTRODUCTION

Archaeology is the detailed study of man's past, by the scientific examination of the physical remains which he leaves behind him. These remains can range from large built structures to the smallest artefacts relating to his everyday life, and from the effects of man's attempts to reshape his environment to the remains of man and those creatures which served him in both life and death.

Archaeology is now acknowledged to be a conservation discipline. Although excavation is the activity which is perhaps most readily associated with archaeology, it is not the main aspiration of today's professional archaeologist. There is still a great deal of excavation undertaken, mainly as a result of development projects, but the aim of archaeology is now to ensure that some representative aspects of the past are preserved *in situ*, for the education and enjoyment of future generations.

The discipline which has come to be known as industrial archaeology is also of great relevance to the water industry. The aim of industrial archaeology is to identify, record and, where possible, preserve and display industrial monuments and artefacts. In practice, the study of industrial archaeology in Britain tends to address the innovative period which commenced about two hundred years ago, at the time of the Industrial Revolution. The contribution of the water industry to the improvement of the human condition, and to the economic growth of the nation, needs no amplification within this chapter.

[†] Archaeologist, Thames Water Utilities Ltd.

The specialists who provide the needs of these disciplines are mostly professionals, employed by museums, universities and local and national government bodies. Although the role of the amateur has been and continues to be significant, the needs of an environmentally aware nation can only be fully satisfied by staff who have a professional place as advisers, enforcers and curators.

A recent development in the archaeological structure has been the emergence of a number of independent, self-financing individuals and contract-units. These concerns serve the various needs of archaeology in the development process, from initial desk-studies, through representation at planning inquiries, to the undertaking of major fieldwork projects.

Fig. 1. Excavation of a 9000-year-old mesolithic settlement at Thatcham, Berks, in advance of reconstruction of Newbury Sewage Treatment Works (Thames Water Utilities).

THE LEGAL CONSTRAINTS

Three current Acts of Parliament span and direct the water industry's activities as they impinge upon the historic landscape.

(a) The Water Act 1989 sets out the general duties of the water companies, and the National Rivers Authority, with regard to buildings or other objects of archaeological interest.

(b) The Ancient Monuments and Archaeological Areas Act 1979 (as amended by the National Heritage Act 1983) sets out the procedures for the protection of certain significant monuments, procedures for the carrying out of necessary works on or near such sites, and the penalties for damaging these 'scheduled monuments' in any way.

(c) The Planning (Listed Buildings and Conservation Areas) Act 1990 consolidates the safeguards and procedures for dealing with 'listed buildings', those buildings and structures which are protected for their architectural value or their historic associations. The water industry is rich in such monuments, many of which continue to perform their original function.

THE CODE OF PRACTICE

The Code of Practice on Conservation, Access and Recreation was published by the Secretary of State for the Environment in response to Section 10 of the 1989 Water Act, and aims to give the new water bodies practical guidance on how to carry out their statutory duties in a satisfactory and consistent manner.

The Code is not a legal document, and failure to comply with its guidelines is thus not an offence. It is not intended to be just another piece of law to add to the water managers' burden. The Code has been welcomed and adopted by the majority of the water organizations as a real aid to carrying out statutory functions in a hitherto ill-defined area. Its strength is that it does provide some insight into what is implicit in the bland wording 'to have regard to the desirability of protecting and conserving buildings, sites and objects of archaeological, architectural or historic interest'. (Section 8 of the 1989 Water Act).

In particular, the Code makes the point that archaeological sites are a unique and finite resource, which, if damaged, cannot be replaced. Unlike natural features, their destruction cannot be reversed, or their condition enhanced. In addition to the most obvious upstanding monuments, such as forts, burial mounds or stone circles, which are visible, and easy to avoid, the Code acts as a reminder that there are groups of more fragile sites, such as the waterlogged deposits of the Fens and the Levels, which are just as valuable, and worthy of protection. Such sites are often not even visible to the casual observer, but can be damaged beyond recovery by activities quite common within the industry, such as land-drainage and drawing down of standard water levels. On the topic of industrial archaeology, the Code draws attention to the water industry's rich past, urging that details be recorded of machinery, equipment and documentation identified as being of historic interest, and that interested bodies be notified when such items are earmarked for disposal.

The Code also addresses the subject of pipelaying, which is one of the industry's most invasive activities, and urges the undertakers to plan sewers and trunk mains along routes which will by-pass, and avoid damage to, sites of importance for conservation.

The final, but perhaps the most fundamental, recommendation of the Code is that the water organizations should ensure that they are fully informed about all the archaeological and historical sites in their care, and have access to the best available data thereon, in order to formulate robust management plans.

THE INDUSTRY'S ASSETS AND ACTIVITIES

The various elements of the water industry are inextricably bound up with archaeology, by the nature of their functions.

First, the water companies are sizeable landowners. Inevitably, within these holdings are a certain number of listed buildings, and, in the case of companies with large catchment areas, a considerable number of scheduled monuments. In most cases the company's activities will have no effect on the monuments. In other instances where, for instance, the monument comprises a set of totally buried features in arable land, a change of agricultural regime involving deep-ploughing, land-drainage or sludge injection may have a serious implication for the archaeology of the site, and may thus become a consentable activity.

The case of listed buildings is particularly relevant to the industry. It has already been stated that many such structures continue to perform their original function. As such, they are likely to attract adequate finance for their upkeep. The managers' problems begin when a building is no longer capable of serving that initial function. If the building can be re-used for a suitable purpose (subject to Listed Building Consent if necessary) then its survival can be financially justified. If, however, a listed building can no longer perform a useful function, the supply of operational money ceases. However, the law states that the building must continue to be maintained in its original condition, at the expense of the owner. What budget code does a hard-pressed manager invoke to guarantee such expenditure?

Fig. 2. Abbey Mills Sewage Pumping Station, West Ham, London. Built by Sir Joseph Bazalgette in 1860, it still pumps sewage to Beckton Sewage Treatment Works in East London (Thames Water Utilities).

THE RESPONSE

There are three basic levels at which the industry can respond to its historical and archaeological obligations.

(a) Voluntary initiatives, whereby the water companies and the National Rivers Authority take into account the implications of their proposals, and take spontaneous steps to put in place the necessary mitigating measures.

(b) Response to advice or conditions supplied by a local authority planning department. (It must be borne in mind that archaeology is now acknowledged to be a material consideration in the determination of planning applications.)

(c) Response to legal constraints imposed by the Secretary of State for the

Environment – this will apply particularly in the case of Scheduled Ancient Monuments and Grade 1 Listed Buildings.

However, it is now recognized that there exists a significant number of monuments which, although they carry no statutory protection, will nevertheless be taken into account in the determination of a planning application. These sites may outnumber the scheduled monuments in a given county by a factor of twenty. It is important that water engineers should be aware of the existence of such sites at the design stage of projects, and it is recommended that county archaeologists should be consulted for the relevant information. Experience has shown that there is no way in which all of this information can be incorporated into in-house databases, nor could it be used, if there are no staff trained in its interpretation. However, the basic conservation database of any water company or National Rivers Authority region must contain information about all Scheduled Ancient Monuments and Areas of Archaeological Importance within its operational area.

Implicit in most of the routes above is the presumption that where the water industry finds itself involved, voluntarily or otherwise, in the promotion of archaeological projects, it will also be the major paymaster for such projects. The availability of external funding for archaeology arising from development proposals by the water industry is minimal.

CONSULTATION PROCEDURE

Fundamentally, it is important to know the archaeological potential of a site or area in which development is proposed, in order to: (a) avoid those sites and monuments which are formally protected from development pressure; (b) make arrangements for survey, watching-brief or excavation on such sites and monuments as are known to be affected; (c) make arrangements for notification, examination and safeguarding of such other sites or artefacts as may be newly located in the course of the work.

As has been suggested, archaeological information is available from county archaeologists, and is incorporated into the county Sites and Monuments Record (SMR). The SMRs comprise annotated maps containing all relevant information on sites and finds, backed up by descriptive text on computer or index card, amplifying the cartographic details. The SMRs are available to all *bona fide* users, although commercial interests may now have to pay for their use. It must be reiterated that the interpretation of such information by non-specialist staff may be fruitless, or even misleading.

Thames Water has employed a professional archaeologist since 1973, and has thus been able to build up databases, internal appraisal procedures for capital works, and management plans for land and buildings. It has been clearly demonstrated that the sooner archaeological input can be made the less risk there is of burdening essential schemes with extra costs and delays.

COSTS AND BENEFITS

It would be totally irrelevant for this chapter to synthesize a profit and loss account which could financially validate the water industry's involvement in the promotion of archaeology and related disciplines. While archaeology itself is undertaken as a scientific process, the circumstances under which it becomes a part of the water engineering and management cycle can sometimes be less systematic, driven by variable responses from government bodies, planning authorities and pressure groups.

It is to be hoped that the recent Planning Policy Guidance Document on Archaeology & Planning (PPG 16) will serve to standardize the requisite levels of response, and to underline the financial implications of archaeology in a development context, however unpalatable they may be to the unwilling supporter of archaeological endeavour.

In terms of real costs, it has been found by experience that where full-scale archaeological excavations have been funded, they have averaged out, over 10 years, to cost only half to one quarter per cent of the capital schemes to which they are adjunct, whether reservoirs, pipelines or sewage treatment works modifications. In financial terms, this is not a very severe burden to a project.

By far the most serious implications of archaeological involvement are related to the uncertainty which can arise, as a result of three factors:

Delay: It is fair to say that if archaeologists knew exactly what they were going to find, they probably would not need to dig. Thus, the risk of delay always hangs over a scheme.

Fig. 3. Roman pottery excavated from Lower Farm, Sandford-on-Thames, Oxon, in advance of mainlaying. The site proved to containa major pottery production site (Thames Water Utilities).

Alteration: As a result of archaeological remains being discovered, it is not uncommon for the regulatory bodies to require the relocation of pipelines or control structures, or the modification of building foundations.

Refusal: Where archaeological work is requested by a local authority as part of the planning determination process, it is possible that the developer may ultimately be financing his own frustration, if the work produces remains of a quality which are deemed worthy of preservation. This is a rare occurrence, but that comes as little consolation to the developer involved in such a situation.

The benefits to the national heritage of a positive response from the water industry are incalculable.

The benefits to the industry itself are:

Compliance: In many cases, the water undertakings will be required to fulfil certain obligations towards archaeology and the historic landscape, in order to comply with the legal duties imposed by the Water Act (and the advisory measures laid out in the Code of Practice).

Promotability: As many of the industry's activities are 'plannable works', it is an increasing requirement that local planning authorities should be satisfied that adequate provision has been made for archaeological input to new schemes.

Credibility: A company's response to archaeology and other environmental issues is closely observed in the present 'green' climate. In addition to smoothing its own path by gaining a good environmental record, the company can often make use of the results of archaeological work undertaken in its name to create good positive public images through popular publications, exhibitions and media presentations. The additional cost of such initiatives will often justify the initial outlay, even when planning conditions are not paramount.

Developer funding has been the main source of archaeological money for the last decade. During this time, many major discoveries have been made, and fully exploited, which would not have been maximized if they had relied on government funding. In this way, even the most reluctant developers have become benefactors of our heritage. The policies of English Heritage, which advises the government on such matters, are set to increase the emphasis on developer funding even more in the foreseeable future, and this will affect the water industry in a major way.

However, the water industry in this country comprises a proud and competent group of experts who are, at all levels, concerned about their own origins and those of the industry itself. Often, the involvement of archaeology in a scheme adds a degree of interest to the day to day work.

Conflicts are rare, and can stem from the imposition of unreasonable conditions on either side. In the majority of cases, archaeology and engineering can be integrated into time-tables. In the minority of cases, where something is discovered which is so unexpected, and so important, that work has to stop, and priorities have to be redefined, the chances are that public opinion will influence the subsequent decision-making process.

CONCLUSIONS

Because of its past achievements and its future aspirations, the British water industry will continue to attract the attentions of the archaeologist and the historian.

Sometimes this relationship will be merely academic. Increasingly, however, the interest will take the form of positive intervention, backed up by the requirements of the law. In this case, those who represent the archaeological viewpoint will themselves be professionals.

Past co-operation between the industry and archaeologists has produced some exciting results, typified by recent work at Roadford Reservoir, Flag Fen, Gatehampton and Newbury Sewage Treatment Works.

With the guidance provided by the Code of Practice on Conservation, Access and Recreation, the industry must expect more demands from the archaeological regulators. It is certain, however, that such demands will lead to exciting conclusions.

DISCLAIMER

The views expressed in this chapter are those of the author, and not necessarily those of Thames Water.

3

Changing objectives in coastal management

D. J. Parker, BA, PhD[†]

INTRODUCTION

The integrated management of coasts is an objective which presents a growing challenge to the water industry. The finite nature of our coasts is increasingly being recognized. Coasts are being heavily pressured by economic development and other demands of an affluent, technologically advanced society. The coastal environment is under great stress from effluents, from commercial developments and from erosion. Our coasts have many conflicting uses and sharing the coast is increasingly important, making necessary a high degree of integrated or harmonious management. Institutional arrangements for coastal management are complex, responsibilities are fragmented and the community's needs are rapidly changing. Water and environment managers must discover ways of successfully binding together the various threads of coastal management for which they are responsible, and of exerting greater influence for integrated coastal management.

Policies and practices which are conflicting or not mutually reinforcing lead to unharmonious or non-integrated management. Thus, integrated coastal management involves developing policies and practices which are positively reinforcing and conflict-reducing, both between different functions within the water industry, and between water industry functions and the coastal management functions of other

[†] Head of the School of Geography and Planning and member of the Flood Hazard Research Centre, Middlesex Polytechnic.

public and private agencies and groups (Fig. 1). A wide range of 'integrating instruments' is available, including statutory or legislative, administrative (including planning liaison; and research), regulatory and economic or market-based instruments.

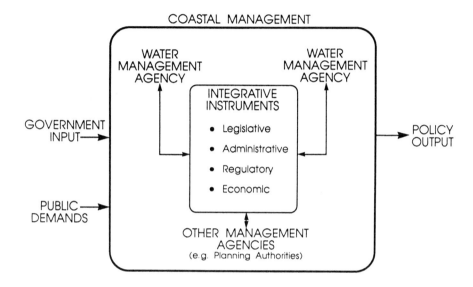

Fig. 1. Integrated coastal management.

COASTAL MANAGEMENT FAILURES

Coastal zones are natural resource systems with the potential to generate major flows of economic, social and environmental benefits for society. Inevitably, because coastal zones are a major attractor of economic and social activity, as well as an important source of environmental and scenic assets, problems and conflicts stem from multiple use pressures. Unfortunately an array of problems arises which reduces the flow of benefits to society. Unmanaged conflicts erode the value of coastal resources to one or more sets of users. The efficient use of coastal resources is often significantly reduced by market failures associated with common property resources. Thus, for example, the full socio-economic and environmental costs of sewage disposal may be inadequately reflected in volume and strength related charges, and therefore incentives to control discharges to coastal waters may be inadequate. Often, without recognizing that they do so, customers therefore 'externalize' their costs onto the coastal environement with negative effects on coastal water quality. Other problems include over-use and degradation of open access resources. Finally, unless institutional arrangements exist which permit successful intervention through management plans the flow of benefits to society from coastal resource systems will often be much less than is potentially possible [1].

COASTAL MANAGEMENT CASES

Three rather different cases taken from the south coast of England illustrate the
wide scope and complexity of problems faced by coastal managers (Fig. 2). These
cases also demonstrate the growing need for integrated management; how integrated
management has been achieved using 'integrating instruments'; and some of the
problems which remain.

LOCATION OF THREE COASTAL MANAGEMENT CASES

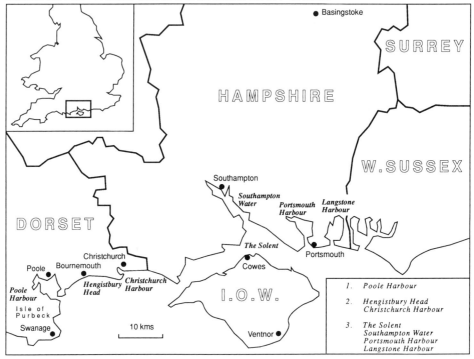

Fig. 2. Location of three coastal management cases.

Coastal erosion near Bournemouth and Christchurch

The coastline west of Hengistbury Head is eroding at an average rate of 1.125 m per
annum (Fig. 3). East of the Long Groyne cliff erosion was estimated in the 1980s
at 0.8 m per annum (Fig. 4). A severe storm overtopping the cliffs at Double Dykes
could erode a channel into Christchurch Harbour – currently protected from the
open sea. This channel would become a new permanent entrance to Christchurch
Harbour triggering a sequence of damaging consequences [2]. An archaeological site
would be lost; wave action in the Harbour would reduce the safe water area; and
within the Harbour higher tides would increase the flood risk. The Headland, which
is a major amenity, would become an island. The environmental and ecological
character of the Harbour-Head complex, which includes a local nature reserve,
would be degraded and some valuable areas lost. East of the Long Groyne cliff
erosion is progressively revealing geological sections of national importance but
also causing the loss of ecologically important heathland [3].

Thus unchecked erosion had created risks of severe economic and environmental loss, yet a coast protection scheme promoted by Bournemouth Borough Council (BBC) could also be economically wasteful as well as costly on environmental grounds. Several 'integrating instruments' were combined to generate a solution. However, these were not always consciously used and there was a degree of adversarial agency conflict before a satisfactory solution was adopted.

Fig. 3. Erosion of coastline west of Hengistbury Head.

Of principal importance were economic instruments in the form of a large central government grant-aid subsidy for coast protection works, without which the coastal authority might not have raised the capital required. Secondly, the project's economic viability was investigated using benefit-cost analysis, extended by an assessment of environmental consequences [3,4]. Thirdly, a mix of regulatory and legislative instruments ensured that some of the environmental outcomes of the project were satisfactory; although at first these outcomes had threatened to be unsatisfactory.

The Nature Conservancy Council (NCC) initially objected to BBC's project because it would prevent active erosion at the geological Site of Special Scientific Interest (SSSI). The threat of a public inquiry caused BBC to agree the signing of a Section 15 agreement under the Countryside Act 1968. This enabled the NCC to manage the SSSI with BBC as their agents to regularly clear eroding material from the base of the cliff, thus permitting erosion to continue and the cliffs to remain in pristine condition [1].

   The BBC has now implemented an economically efficient coast protection project comprising rock and timber groynes and other works costing £3.71 million at 1986 prices. The project now protects the Harbour-Head complex including environmentally important zones and related economic activities, and yet still provides for active erosion of the geological SSSI.

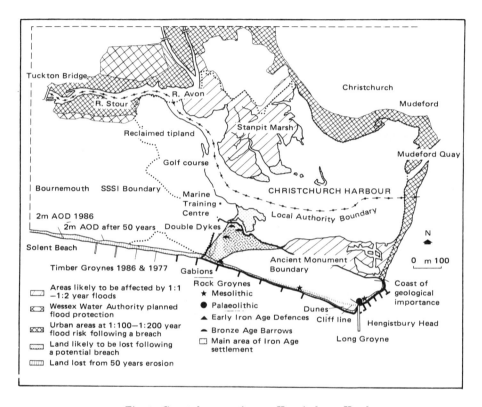

Fig. 4. Coastal protection at Hengistbury Head.

   Whilst the outcome of BBC's project was largely satisfactory, the administrative arrangements for coast protection and sea defence suffer from major weaknesses and intervention failures. Processes for consultation between coast protection agencies and others over ecological impacts need improving. During 1990 such a consultation procedure was agreed at a seminar at Middlesex Polytechnic attended by representatives of the Ministry of Agriculture, Fisheries and Food, coast protection agencies and environmental bodies (Fig. 5.) [5]. Coast protection is currently a district council responsibility leading to a piecemeal approach. The interventionary actions of one district council may have adverse impacts on a neighbouring authority. The administration of development control and statutory plans is often poorly co-ordinated with coast protection, and unfortunately developments have been permitted too close to eroding coastlines. On the south coast the need for an integrated approach to coastal management is now recognized by the Standing Conference on the Problems Associated whith the Coastline (SCOPAC) – a non-statutory forum comprising representatives from many statutory bodies [6].

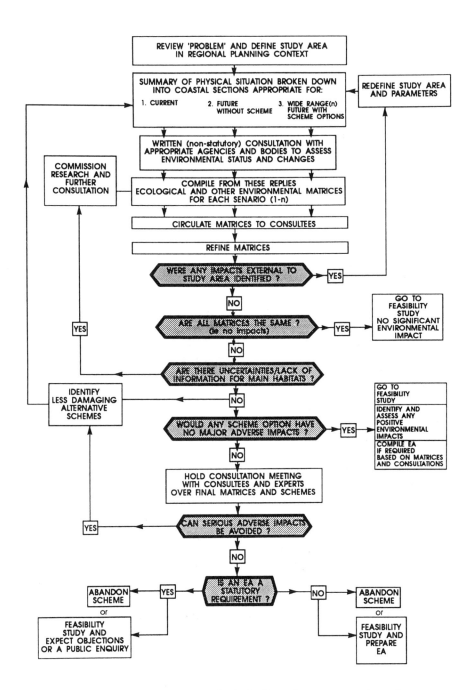

Fig. 5 Consultation procedure on ecological effect of coast protection schemes.

Coastal water quality improvement in the Solent area

The Solent, Southampton Water and Portsmouth, Langstone and Chichester Harbours are important receivers of wastes from coastal urban and industrial sources, including the expanding Southampton-Portsmouth conurbation and the Isle of Wight (Fig. 6). By the early 1970s approximately 50 per cent of sewage generated in the Hampshire coastal zone was discharged to the Solent, much of it through unsatisfactory short sea outfalls, and many sewage treatment works did not provide sufficient treatment to prevent pollution. The Solent has an immensely rich variety of navigable estuaries, salt marsh, sand and shingle beaches, impressive cliffs and a host of historical and geographical features making it an Area of Outstanding Natural Beauty (AONB). The Solent coastline is important for tourism and the Solent is an international sailing centre. The recreational resources are major contributors to the local economy. Shellfisheries have been developed in the Solent and Southampton Water. The potential size of the clam fishery is reduced by pollution, and levels of bacteria in oysters are higher than permitted levels, making cleansing essential [7].

Coastal managers must develop an approach which ensures that urban and economic development is unconstrained by inadequate sewage disposal infrastructure, whilst also maintaining and improving coastal water quality to ensure that recreational, amenity and environmental needs are not harmed, and that the area's overall development is sustainable. Recently public interest in coastal water quality has increased, generating growing pressures to alleviate pollution. Contact sports, such as boardsailing, have grown rapidly, necessitating environmental improvements, for example in Langstone Harbour. The public now expects bathing waters to comply with European Community (EC) standards. Pollution has caused major concern to the growing yachting interests at Cowes.

Whilst some problems remain to be overcome, a number of 'integrating instruments' have contributed to harmonious development. Since the early 1970s the integrated planning of population growth and sewage disposal infrastructure in South Hampshire has ensured that structure plan population growth forecasts can be accommodated by the development of facilities. Planning authorities recognized the need to protect the Rivers Itchen and Test from pollution because they are water sources; the unsatisfactory nature of discharges into Portsmouth and Langstone Harbours and the Hamble estuary; and the potential of the Solent as a natural receiving water for effluents from inland sources. Southern Water Authority developed a new sewage treatment works at Peel Common (Fig. 6) bringing substantial improvements to water quality in Fareham Creek, in Portsmouth Harbour and in Southampton Water.

Legislation provided an impetus for substantial coastal water quality improvements. Laws include the Clean Rivers (Estuaries and Tidal Waters) Act 1960 followed by the Control of Pollution Act 1974 (Part II). By the 1980s consent standards could be applied to all industrial discharges. Unsatisfactory sewage treatment works and outfalls have been closed or upgraded, and levels of sewage treatment increased. Aesthetic conditions have improved in important conservation and amenity areas. In addition, three European Community Directives (the Bathing Water Directive of 1976, the Shellfish Waters Directive of 1979, and the Directive on pollution caused by certain dangerous substances discharged to the aquatic environment of 1976) have required further coastal water quality improvements. A range of improvements including new sea outfalls has been implemented (Fig. 6).

The Southern Water Authority employed research to further integrated management by undertaking investigations into: the behaviour of sewage when discharged to the sea (including death rates of bacteria); the construction of a hydrodynamic model of the Solent; development of computer models of water movement to predict movement of sewage plumes; and a mathematical model of tidal currents in the Solent and Harbour areas which simulates the dispersion of diluted effluent and the mortality of faecal coliform bacteria.

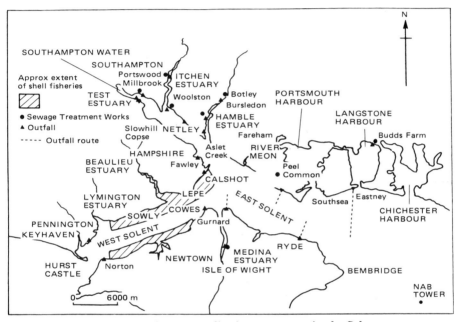

Fig. 6. Coastal water quality improvements in the Solent.

Coastal managers continue to face considerable challenges regarding coastal water quality management in the Solent area, not least because of changing attitudes to sea outfalls and the government's timetable for discontinuing sewage sludge disposal at sea. The population projections of the 1960s proved to be incorrect, but plans for sewage treatment capacity relied upon them. The UK government broadly accepts the polluter-pays philosophy but currently policy instruments which support volume-related charges are not strongly developed, such that there remains an element of 'market failure'.

Coastal oilfield development at Poole Harbour

The final case shows how heavy industrial development has been permitted within a fragile coastal environment of national and international importance. The Poole Harbour-Isle of Purbeck area is one of great beauty and is designated a Heritage Coast and an Area of Outstanding Natural Beauty. It contains three National Nature Reserves and 20 SSSIs (Fig. 7). British Petroleum's (BP) oilfield development shows that commercial and enviromental objectives may be successfully integrated. Here water agencies have not been amongst the principal actors but the case is highly relevant to water and environment managers because it further demonstrates the effective use of 'integrating instruments' in coastal management.

The Wytch Farm oilfield became the first major on-shore oilfield to go into production in 1979. Since then exploration and extraction have expanded [8]. In 1985 BP were granted planning permission by Dorset County Council (DCC), against objections from statutory and non-statutory environmental bodies, to establish oil exploration facilities at Furzey Island. The number of oilwells within the Wytch Farm oilfield is 67.

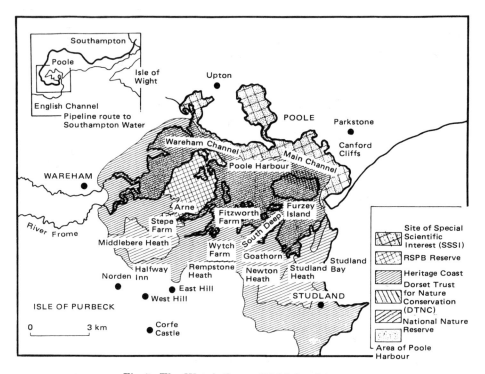

Fig. 7. The Wytch Farm oilfield development.

The stimulus for integrated management was the statutory regulatory instruments, including the planning application process, and the existence of strong environmental pressure groups. From this came the need for BP to satisfy DCC that further commercial operations were acceptable within this fragile coastal zone. The administrative planning system was employed to develop a staged process of extensive negotiations between BP and DCC, also involving other interests, in which a large number of options were examined. British Petroleum was required to undertake an environmental impact assessment to which DCC contributed and which was made publicly available. Through this process DCC managed to steer BP away from developing the most fragile environments, and towards adopting a highly sensitive approach to their works. The County Council reserved its right to refuse development, and some options were rejected as being environmentally unacceptable. A total of 338 planning conditions were set for works on Furzey Island as part of the final planning permission. Specialized research sponsored by BP is contributing to knowledge of the area's environment and how best to manage it, and the construction phase is being monitored by a consultative committee [9].

The Poole Harbour oilfield developments are a showpiece of integrated management and BP have taken great care to conserve the ecology and landscape of the area. Both BP and DCC have received awards for their efforts. Some limited environmental damage followed early developments, but so far the visual impact of the latest developments is negligible. The possibility of further oilfield development, this time off-shore, remains a challenge.

CONCLUSIONS

The coasts are under great pressure and currently the nation is not making the best use of them – the flow of benefits to society is not being maximized. Water and environment managers are faced with particularly challenging problems and responsibilities in the fragmented area of coastal management. A central problem is that they individually represent but one actor, albeit a most important actor, within the coastal management sphere. A central objective should be the explicit development of integrated or harmonious management policies. Water managers must persuade others of the need for integrated or harmonious management. The cases cited above demonstrate the range and potential effectiveness of 'integrating instruments'. These instruments, including legislative, administrative, regulatory and economic instruments, deserve more explicit focus and development.

ACKNOWLEDGEMENTS

Parts of this study were undertaken for the Organisation for Economic Cooperation and Development. The Hengistbury Head coast protection scheme appraisal was commissioned by Mr R. E. L. Lelliot of Bournemouth Borough Council's Engineer's Department. Fig. 5 was prepared by Annabel Coker, Cathy Richards and Sylvia Tunstall. Mr K. Guiver, Southern Water Services Ltd, provided information on water quality improvements in the Solent area and other useful comments, and Mr A. Price, Dorset County Council, assisted with information on the Poole Harbour oilfield developments.

REFERENCES

[1] Parker, D. J. (1990) *Coastal zone protection in east Dorset and Hampshire, United Kingdom.* Report to the Organisation for Economic Cooperation and Development, Paris.

[2] Parker, D. J. and Thompson, P. M. (1988) An 'extended' economic appraisal of coast protection works: a case study of Hengistbury Head, England, *Ocean and Shoreline Management*, **11**, 45–72.

[3] Thompson, P. M. and Parker, D. J. (1986) *Hengistbury Head coast protection proposals: assessment of potential benefits and costs.* Summary and Supplementary Reports, Middlesex Polytechnic Flood Hazard Research Centre, Enfield.

[4] Penning-Rowsell, E. C., Coker, A., N'Jai, A., Parker, D. J. and Tunstall, S. M. (1989) Scheme worthwhileness, In: *Coastal Management*, Institution of Civil Engineers, Thomas Telford, London, pp. 227–41.

[5] Coker, A. and Tunstall, S. M. (1990) Survey-based valuation methods. In: Coker, A. and Richards, C., (Eds) *Ecological evaluation and economic evaluation: proceedings of a workshop*. Flood Hazard Research Centre, Middlesex Polytechnic, Enfield.

[6] Court, C. D. (1989) *The Standing Conference on Problems Associated with the Coastline (SCOPAC)*. Paper presented to the Conference of River and Coastal Engineers, Loughborough University, 11–13 July.

[7] Southern Water, *Sewage disposal to the Solent: An overview*. Southern Water, Hampshire Division, Otterbourne. Undated.

[8] Tree, I. (1989) Defining Dorset's Gold, *Geog. Mag.*, **61**, 1, 18–22.

[9] British Petroleum (1988) *Wytch Farm Oilfield: Caring for the environment*, Wareham.

4

Coastal defence: the retreat option

Jan S. Brooke, BSc, MSc, MIWEM[†]

INTRODUCTION

In a recent report, the Intergovernmental Panel on Climate Change predicted an average rise in global sea levels due to global warming of 180 mm by the year 2030, and 440 mm by 2070 (Fig. 1). These increases could potentially compound a problem which is already appearing on the UK flood-defence agenda with increasing frequency, that is whether the country should continue to defend areas of agricultural land which currently have little or no national economic value.

This chapter explores the opportunities for nature conservation which might arise if 'retreat' from the existing line of defence is accepted as an option. The current situation in the UK and the United States is reviewed, and some possible responses are discussed.

BACKGROUND

Some of the defences protecting the UK's smaller areas of low-lying agricultural land against salt-water inundation are reaching the end of their design life. Many currently provide little more than the minimum standard of defence required to sustain viable agricultural production. An acceleration in the rate of sea-level rise could make the situation in these areas untenable, because the economic viability of any improvement works must usually be demonstrated if grant aid is to be attracted from central government.

[†]Manager, Posford Duvivier Environment, Peterborough.

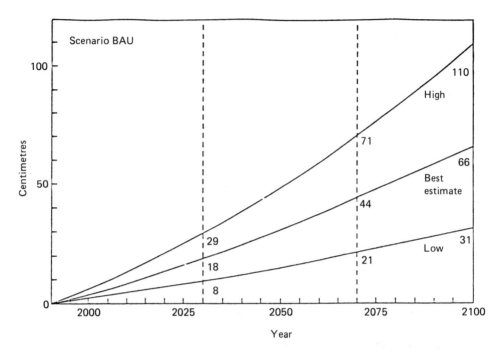

Fig. 1. Global sea-level rise, 1990–2100, for policy scenario Business-as-Usual.
(Source: Climate Change, The IPCC Scientific Assessment, 1990).

Until recently the national value of agricultural production was relatively high, and the economic justification for flood defences protecting agricultural land was usually straightforward. In 1985, however, the Ministry of Agriculture, Fisheries and Food advised that the output value of cereals, oilseed rape, beef and sheep production, used in the economic benefits assessment process, should be reduced by 20% in acknowledgement of the cost of support paid to farmers under the terms of the European Community's Common Agricultural Policy.

Even though yields in many cases have continued to improve, demand has remained relatively stable, and the national economic value of production has been reduced to the point where economic net margins may be negligible or even negative.

NATURE CONSERVATION ISSUES

Agricultural practices have changed considerably since the beginning of the twentieth century and, during the 1970s and early 1980s in particular, farmers were encouraged to intensify many types of agricultural production. These agricultural changes, together with developments undertaken by the port and harbour industry, marina developers, and others, have led to the UK losing much of its remaining wetland, marsh and intertidal environmental resource. The Nature Conservancy Council's report 'Nature Conservation in Great Britain', [1] demonstrates, for example, that prior to 1967 less than 1000 ha out of 20 000 ha in the Yare Valley,

Norfolk, were in arable production. By 1982, however, an additional 6000 ha of grazing marsh had been converted to arable land, and the grassland regime over a further (unspecified) area had been intensified. Since 1935 the extent of permanent pasture in the North Kent marshes has similarly been reduced by more than 50% as a result of agricultural intensification. The Royal Society for the Protection of Birds (RSPB) reports that, of 123 estuarine sites surveyed in a recent nationwide review, 80 were found to be 'at risk' with 30 being in danger of permanent damage [2]. The intertidal area of the Tees estuary, for instance, has been reduced by about 90% as a result of land claimed for industrial development during the last 100 years.

If rates of sea-level rise increase significantly as a result of global warming [3], further habitat losses are anticipated, particularly if this change is accompanied by an increase in storms. A reduction in the area of intertidal mud and sandflats, marsh, and seasonal wetlands is likely to occur, and this would be exacerbated by the maintenance of hard tidal defences. Salt marsh, reed beds and other vegetation fronting flood defences could also be lost, not only through submergence but also through erosion.

RETREAT OPTION

Under a scenario of global warming and sea-level rise, the UK has a number of choices. One option is to defend all agricultural land, irrespective of whether or not the required engineering works can be justified in economic terms. If this policy is selected, the UK will almost certainly experience a net loss of some of its most valuable coastal wildlife habitats. An alternative is to accept that it is not realistic to 'protect at any cost', particularly where the land area in question is small. In these cases it may be prudent to accept a retreat from the existing defence line.

The retreat option is unlikely to be appropriate in urban or industrial areas or in areas like the Fens where many thousands of hectares of high-grade agricultural land would be lost. Elsewhere, however, a tremendous opportunity exists to compensate for the recent degradation of coastal environmental resource by 'managing' retreat in a way which creates or restores desirable habitats.

Natural or unmanaged retreat is not unknown in the UK. In parts of East Anglia, for example, flood defences have failed and land has effectively been abandoned to agricultural production. In some cases, natural processes have created a habitat of substantial environmental interest. At Walberswick in Suffolk, the progressive failure of flood defences since 1902 has led to a substantial area of grazing marsh reverting to intertidal mudflats fringed by salt marsh and reed beds. The site has become extremely important for nature conservation, being notified as a Site of Special Scientific Interest (SSSI) and designated under the terms of the Ramsar Convention. The defences protecting Skippers Island in Hamford Water were breached in 1953, and the diverse habitat which subsequently developed supports a range of breeding birds and unusual plants. This site is now an Essex Naturalists Trust Reserve.

The development of interesting sites such as these should not, however, be regarded as 'automatic'. Cases also exist where embankments have failed with no real ecological gain. For example, if the level of the land relative to sea is too low, a subtidal habitat might develop where a salt marsh or mudflat would be considered to be of greater ecological value. Elevation is therefore one of a number of important parameters which play a role in determining whether or not the resource, which develops following failure, is likely to be of particular significance. If the creation of an environmentally desirable habitat is to be promoted, these parameters must be identified and carefully controlled. Similarly, the economic, social, and legal implications of this type of habitat creation must be defined properly.

The concept of managed retreat therefore raises a number of questions which must be answered if habitat creation is to become an established engineering option where it is no longer economically viable to maintain tidal flood defences.

REQUIREMENTS FOR SUCCESSFUL HABITAT CREATION

A number of physical, chemical and biological factors combine to influence both the character and the sustainability of the habitat which will develop in a given situation. Site elevation and area, soil and sediment characteristics, tidal hydraulics, wave exposure, and climate/weather patterns are among the physical parameters of greatest importance. Water quality, salinity, vegetation establishment options (i.e. natural colonization from adjacent areas versus planting) and the composition of sediments, represent the primary chemical and biological controls. Many examples of wetland habitat creation in the US involve the use of dredged material to raise the site elevation to a suitable level relative to the tide. The characteristics of such material are therefore of equal significance. Finally, the design should also take account of likely changes in sea level and the implications for sediment supply and demand.

A review of fourteen marsh creation and/or restoration projects in the San Francisco Bay area [4] reported that six sites had succeeded in meeting their stated objectives. Of the remainder, five had partially succeeded and three had failed. Four factors were of particular importance in cases where marsh restoration was judged to be successful:

(1) the site elevations achieved were suitable for the desired habitat;
(2) the site was adjacent to an existing 'source' of flora, fauna, etc;
(3) water circulation objectives were achieved and water quality was satisfactory; and
(4) the sites had been the subject of careful planning and detailed design by suitably qualified personnel.

Suitable soil conditions, successful planting, and a (permit) requirement for maintenance were also cited as contributing to successful projects.

Among the reasons given for the total or partial failure of projects were the following:

(a) restoration had simply not been completed;
(b) problems with soil chemical composition or with soil structure (e.g. soil had been compacted during construction);
(c) poor planning or unauthorized modification of the approved plan;
(d) the site elevation was not suitable for the desired habitat; and
(e) adverse impacts of man (i.e. disturbance by all-terrain-vehicles, pets, etc.).

The report therefore highlights another vital prerequisite of successful habitat restoration/creation schemes; that is the need for careful planning. The authors of the report discuss the problems which arose when the goals of the project, particularly those relating to the type of habitat to be created, had not been carefully defined. Thorough planning by qualified and experienced specialists and the execution of the project in accordance with the plan approved at the permit stage proved, with hindsight, to have been key factors in determining the success or failure of many of the schemes which were assessed.

Some examples of both marsh and other types of habitat which have been restored or created in the US and the UK, with varying degrees of success are listed in Table 1. In each case, limited details about creation methods and objectives have been tabulated. One example from the UK is illustrated in Fig. 2.

STRATEGIC RESPONSES TO SEA LEVEL RISE

Titus [5] discusses a number of possible strategic responses to global warming. However, he warns that in most cases the response will involve a fundamental choice between maintaining economic activities in their current locations and protecting the environment. The maintenance and/or improvement of coastal defences (which may include hard defences, beach nourishment, or works offshore to reduce wave attack) represents one possible strategic option, but it should be noted that coastal ecosystems cannot migrate inland if their path is blocked by any kind of structure, including hard flood defences. An alternative approach [5] is therefore physically to remove such structures and accept the retreat of the shoreline. A further proactive option is to prohibit new development in the coastal strip.

In common with many countries, several US states have already introduced regulations which include limiting development within a defined set-back zone, prohibiting any new hard defences, and insisting that any structure (property, defence, etc.) should be demolished if the cost of repairing storm damage exceeds a certain percentage of its value. If global-warming-induced sea-level rise significantly increases existing erosion rates and/or storm damage, however, such policies may become increasingly difficult to sustain.

COMPENSATION ISSUE

A recent position paper [6] in the US examined the viability of the retreat option and concluded that:

'Strategic retreat, whether on the beach or in war, has often been the key to ultimate self-preservation and victory. The greatest resistance comes from a misplaced sense of pride and from the very real possibility of short-term but large private economic setbacks... the interests of private property owners are important and politically powerful. The wisdom of strategic retreat will not be accepted, emotionally or legally, unless the needs of property owners are adequately addressed'.

Under a retreat scenario, the issue of compensation will be of comparable importance in the UK. This is particularly relevant if the notion of managed retreat is to gain widespread support outside the conservation lobby.

Table 1. Examples of habitat creation/restoration projects in US and UK

Habitat	Location	Details
Marsh	North Carolina, US	Several sites; planting of vegetation on dredged material on Barrier Islands.
	California, US	Many examples; some use dredged material, some natural siltation; vegetation planting and natural colonization.
	Lousiana, US	Marsh creation at rear of Barrier Islands.
	Texas, US	Dredged material study site.
	Georgia, US	Planting of vegetation on dredged material.
	New York, US	Dredged material placed to create wildlife refuge.
	Hampshire, UK	Marsh restoration as sea defence.
	Essex, UK	Salt-marsh restoration using dredged material.
	Lancashire, UK	Stabilization of intertidal flats.
Other intertidal habitats	California, US	Lagoon habitat for birds.
	New England, US	Clam and worm beds on mudflats.
	Oregon, US	Clam flats.
	Essex, UK	Artificial rock pools.
	Cardiff Bay, UK	Proposed mudflat creation as mitigation for barrage construction.
Sand dunes	Louisiana, US	Sand fencing, planting of vegetation and use of dredged material to increase elevation.
	Lincolnshire, UK	Foredune creation by kidding.
	Merseyside, UK	Dune stabilization, planting of vegetation.
	Northern Ireland, UK	Foredune restoration, planting of vegetation.
Shingle features	California, US	Sand/shingle lagoon habitat.
	Kent, UK	Shingle redistribution.
Artifical islands	Florida, US	Dredged material placed to create wildlife refuge.
	Alaska, US	Gravel islands.
	Alabama, US	Dredged material for bird refuge.
Marine habitat (sub-tidal)	Florida, US	Sea-grass planting.
	New York, US	Artificial reef creation.
	Maryland, US	Sea-grass beds.
	Maryland, US	Oyster reef.
	Alabama, US	Artificial reef for fish habitat.
	Dorset, UK	Artificial reef.
Seasonal wetland	Connecticut, US	Feeding area for Canada geese.
	California, US	Winter waterfowl habitat.

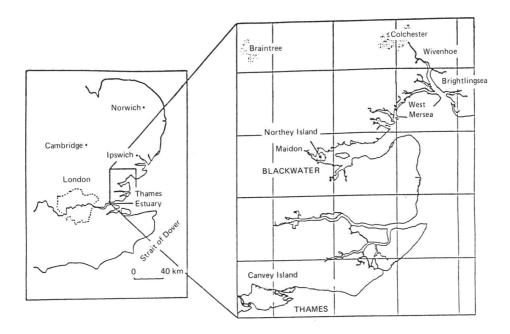

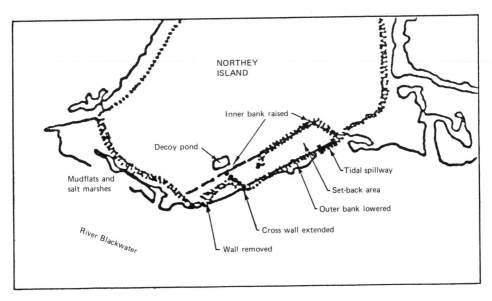

Fig. 2. Northey Island in Essex, where an experimental managed retreat project to create an area of salt marsh is underway. (Source: The Institute of Estuarine and Coastal Studies, 1991. A report to English Nature.)

In England and Wales, the National Rivers Authority's flood-defence powers are permissive. Once a flood defence fails, the Authority is not generally under any obligation to reinstate that defence. However, some habitat creation options will require engineering works in advance of the failure of the flood-defence structure. Works might include, for example, the placing and grading of material to raise the elevation of the land and/or the digging of watercourses to ensure good tidal circulation when the structures do fail (Fig. 3). It is therefore likely, and quite reasonable, that the landowner will expect to be compensated if the residual productive life of his land is deliberately terminated in order to develop an environmentally desirable resource.

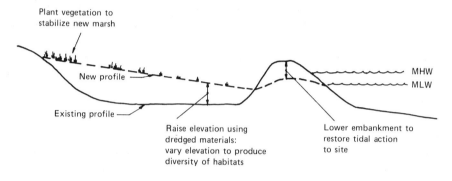

Raising elevation to produce intertidal areas

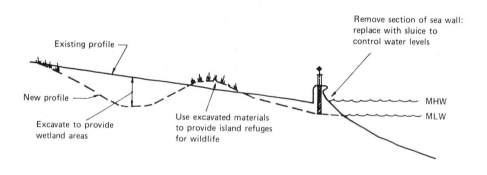

Lowering elevation to produce intertidal areas

Fig. 3. Examples illustrating the principle of retreat option (not to scale).

In some parts of the US local, state, or federal government is currently purchasing areas of land with the objective of protecting or restoring important ecosystems. In San Francisco, for example, the state is buying agricultural land for future wetland restoration. Voluntary groups both in the US and UK are also buying land for nature conservation purposes. As a general solution to the compensation issue, however, the cost of land purchase by public agencies is likely to prove prohibitive. Alternative management options will therefore need to be explored if environmental enhancement opportunities are to be fully exploited.

A variety of options exist, both for meeting the capital and long-term management costs of habitat creation/restoration projects and for compensating landowners. Two such options are discussed below.

MITIGATION

In the US, the protection of wetlands is a high-profile issue. Mitigation of wetland loss is a requirement of the federal development review process under section 404 of the 1972 Clean Water Act. Only 5% of California's 1.6 million ha of wetlands, for example, now remain intact; the rest have been filled and developed. Since the early 1970s, conservation agencies have therefore been working to improve wetland restoration and creation techniques, and many would-be developers have been required to undertake such projects as mitigation for habitat losses likely to result from their proposals.

Section 404 requires that steps are first taken to see if the proposed project can be relocated, or if damage can be minimized to an acceptable level. If this fails, but it is thought that habitat creation would represent an acceptable alternative, compensation may be required to take one of the following forms: (1) in-kind (i.e. similar habitat to that being lost), on-site; (2) in-kind, off-site (i.e. elsewhere); (3) out-of-kind (i.e. alternative habitat), on-site; or (4) out-of-kind, off-site.

A fifth option has also emerged more recently, that is conservation 'banking'. This has occurred, in California for example, when no suitable site for immediate compensatory habitat creation can be located either on or off-site, and involves the developer banking a sum of money to pay for future works when an appropriate site is found.

At present in the UK there is no real precedent for environmental mitigation as such. If both the loss of some agricultural land and the principles of managed retreat for habitat creation are accepted, it must also be acknowledged that someone has to pay. Meanwhile, the UK might begin to recognize mitigation as an option where development in the coastal strip is likely to detrimentally affect the environmental resource, and where relocation or damage minimization options are not viable. If there are no opportunities available for on-site mitigation in such cases, maybe it is possible to set up a conservation banking system to enable the developers of ports, harbours, and marinas to fund off-site habitat creation projects in other coastal areas by way of compensation.

MANAGEMENT AGREEMENTS

The purchase of land, either using public money or as part of a mitigation deal, requires the landowner to accept a one-off payment and hence the loss of any controlling interest in the land. A management agreement, however, represents an option whereby the landowner might maintain some degree of control, possibly in return for an annual payment in the form of compensation or rent.

Already there are many examples of this type of arrangement in the UK. Management agreements are negotiated by the Nature Conservancy Council under the terms of section 15 of the 1968 Countryside Act on SSSIs. Payments are made to farmers by the Countryside Commission as part of the recently introduced Countryside Stewardship Scheme, and sites of nature conservation interest are leased from private landowners by voluntary environmental agencies such as the RSPB.

The success of many initiatives of this type suggests that national and/or local nature conservation bodies might be keen to make full use of the tremendous opportunities on offer if a managed retreat policy is accepted. Some funding to meet the costs of a management agreement might be available from donations and subscriptions, but if the capital costs of development are high, the agencies could look to the public or private purse for contributions.

ECONOMIC JUSTIFICATION OF HABITAT CREATION OPTIONS

As discussed earlier, coastal defence works in the UK should be economically viable if they are to attract public funding. It is possible that a similar argument could be applied to the creation of environmentally desirable habitats. The true 'do-nothing' option costs nothing. The value of a created resource, over and above that which would develop as a result of natural processes, might therefore have to be quantified. Environmental economics offer a number of techniques which could be applied to value the creation of such sites, but a full review of techniques such as the 'contingent valuation methodology' would be the subject of a paper in itself. It is nevertheless interesting to note that some evaluation techniques being used in the US [7] are resulting in values of £18 500–£26 000/ha being placed on the services and functions provided by natural wetlands. On the costs side of the equation, however, as scientists' and engineers' understanding of the complex physical, chemical, and biological requirements of natural habitats improves, the cost of creating a sustainable environmental resource increases. In 1989 the US Army Corps of Engineers was reporting costs in the range of £3200–£44 500/ha for tidal marsh creation.

Whether or not economic justification is needed for environmental enhancement projects will be governed to some extent by the funding agency. In both the US and Canada, the federal government has allocated a budget for habitat creation. In San Francisco, for example, the US Fish and Wildlife Service is spending money to acquire and restore habitats in the Bay area. There is no formal requirement for benefit : cost assessment on these projects, but there is a national recognition (public and political) of the importance of natural habitats. In Canada, £2.4 million of the £52 million budget allocated for the environmental management of the St Lawrence River is destined for habitat creation initiatives. The UK Government could therefore choose to allocate a similar budget for projects designed to restore environmental resources in low-lying coastal areas.

CONCLUSIONS

1. The low national value of agricultural production is increasingly resulting in rural flood-defence schemes showing marginal or negative economic benefits. Meanwhile, many coastal and tidal defences are continuing to deteriorate.

2. Retreat from the existing defence line is therefore not an option which will have to be considered only if rates of sea-level rise increase. Such a rise would simply compound an existing problem.

3. Unless there is either a major change in grant-aiding policy or a significant upturn in agricultural fortunes in the near future, the decision will not be 'should we abandon this land', but how can we retreat in a way that makes maximum use of the environmental, engineering and economic opportunities?

ACKNOWLEDGEMENTS

The author is grateful to the National Rivers Authority, Department of Environment, Nature Conservancy Council, and Countryside Commission for allowing information collected under the terms of the Research Project 'Environmental Opportunities in Low Lying Coastal Areas Under a Scenario of Climate Change' to be used in the preparation of this chapter. This research project assessed the extent and nature of possible habitat creation opportunities associated with managed retreat throughout England and Wales, as well as investigating many of the other issues discussed in this chapter.

REFERENCES

[1] Nature Conservancy Council (1984) *Nature conservation in Great Britain.* UK.

[2] Royal Society for the Protection of Birds. (1990) *Turning the tide, a future for estuaries.*

[3] World Meteorological Organization/United Nations Environment Programme. Intergovernmental Panel on Climate Changes. (1990) *Climate Change: The IPCC scientific assessment.* Cambridge University Press, Cambridge.

[4] San Francisco Bay Conservation and Development Commission. (1988) *Mitigation: an analysis of tideland restoration projects in San Francisco Bay.* Staff Report. SFBCDC, San Francisco.

[5] Titus, J. G. (1990) Strategies for adapting to the greenhouse effect. *J. Am. Planning Assoc.*, 1990, p. 311.

[6] Second Skidaway Institute of Oceanography Conference on America's Eroding Shoreline. (1985) *National strategy for beach preservation.* Skidaway Institute of Oceanography, Savannah, USA.

[7] Personal communication.

Part II Water engineering and the environment

5

The benefits of environmental assessment

M. W. Child, BSc (Hons), CEng, MICE, MIWEM[†], and A. M. Mills, BSc, CBiol, MIBiol[‡]

INTRODUCTION

The role of the NRA

The National Rivers Authority (NRA) is one of the strongest environmental protection agencies in Europe, and therefore has a major role to play in the protection and enhancement of the water environment. The 1989 Water Act [1] imposes on the NRA a general duty of care in respect of conservation, public access and recreation. The Code of Practice on Conservation, Access and Recreation [2] gives guidance on the discharge of these responsibilities.

The responsibilities of the NRA are far reaching. Statutory duties and powers relate to water resources, pollution control, flood defence, fisheries and navigation. Flood defence includes both freshwater and sea and tidal protection. A declared objective of the NRA is to protect people and property from flooding by rivers and the sea. Property includes homes and houses, roads and utility services, agricultural land and sites of heritage and environmental value. The NRA spends £400 million each year on flood defence. Some £135 million of this expenditure relates to new construction projects to replace, repair and renew flood defences.

[†] Engineering Manager, NRA (Anglian Region)
[‡] Consultant, Environmental Assessment Services Ltd.

N

.0   25   50
Scale (km)

Fig. 1. Tidal flood risk area (NRA – Anglian Region).

The Anglian region

Much of the region is flat, low lying and below maximum recorded sea level (Fig. 1). It is therefore very sensitive to flooding both from freshwater runoff and tidal storm surges. The region covers one of the most vulnerable and variable coastlines in Britain, stretching from the Humber in the north to the Thames in the south: vulnerable from the flooding risk, and variable in its coastal geomorphology and the many conflicting interests to be taken into account. The flood risk area, which is roughly equivalent to the size of Essex, has a population of over 750 000 who are protected by 1500 km of defences. Without the embankments, dunes, beaches, flood walls and groynes much of the area would have been lost to the sea long ago.

The region spends some £40 million annually on new construction projects to replace, repair and renew flood defences, although an increase in investment is likely as a result of the forecast climatic changes and sea level rise. To maintain the standard of protection the defences will need to be higher and more substantial, which will inevitably increase the environmental impacts.

Flood defence construction and the environment

New construction for flood defence is very often carried out in areas of natural beauty, of special scientific interest and of high heritage value. Flood defences prevent economic, environmental, social and psychological damage arising from flooding and provide an improved and safer environment, although there is a risk that the building of flood defences may in itself cause environmental damage both during and after construction. River and coastal engineers have in the past developed informal procedures to ensure consultations with affected parties, and to minimize possible impacts. A formal procedure of data collection, analysis and presentation is now required under The Land Drainage Improvement Works (Assessment of Environmental Effects) Regulations 1988 [3] for improvement works (Statutory Instrument 1217), and The Town and Country Planning (Assessment of Environmental Effects) Regulations 1988 [4] for construction requiring planning approval (Statutory Instrument 1199).

All projects in the region undergo an environmental assessment (EA) procedure. In addition a number of larger projects, particularly those in sensitive environmental areas, require an environmental statement (ES) in accordance with the statutory procedure for assessment of environmental effects [3,4].

ENVIRONMENTAL ASSESSMENT IN THE REGION

Project appraisal

The project development procedure, which commences with the production of a project appraisal, includes EA as an integral part of the procedure. On most projects a multi-discipline project group is formed, often including conservation staff, to produce the appraisal. The appraisal describes the problem, the need, the options, the disadvantages and advantages, the costs and the economic justification. A key requirement of the appraisal is to detail potential environmental impacts and interests. Likely significant impacts are described, and the measures envisaged to avoid, reduce or remedy these effects are outlined.

The details of any elements of the scheme which positively further conservation or enhance the landscape are also included, as are details of all consultations with affected parties and environmental bodies. The Project group also recommends whether or not an ES will be required. Depending on the scale, the impact and the complexities of the project, the ES is developed either in parallel with the appraisal, or subsequently.

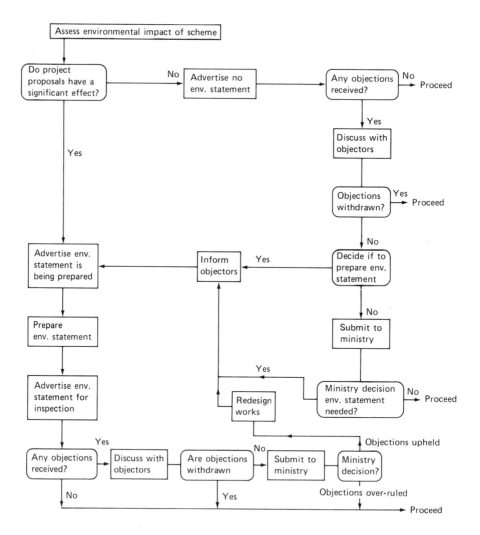

Fig. 2. Outline procedures to comply with Statutory Instrument 1217.

Statutory procedures

The flow chart in Fig. 2 describes in outline the activities, decisions and advertising necessary to comply with Statutory Instrument 1217 [3]. The procedure identifies two courses of action, which result in the decision to produce an ES, or alternatively not to produce an ES. In either case it is necessary to advertise the decision. In Anglian region a total of 86 projects had been advertised to the end of 1991, of which only 16 had an ES produced. The remaining 70 were advertised as having 'no ES', yet these produced not a single objection.

Environmental statements procedures

If it is decided that an ES is necessary the procedures include:

    (a) early identification of the consultees to (i) establish fields of interest; (ii) obtain views;

    (b) reviewing impacts both during construction and on completion for all options, including 'do nothing';

    (c) identifying environmental opportunities for enhancement and mitigating factors;

    (d) organizing public exhibitions and/or questionnaires and consulting widely with interested parties;

    (e) producing an ES that (i) states clearly the preferred option and the supporting reasons; (ii) maximizes use of colour maps, simple diagrams, photographs and non technical language; (iii) is as far as practicable a report independent and separate from the project promotion, although it must include supporting technical appendices.

The costs

Table 1 sets out a number of actual costs for producing an ES.

Table 1. Environmental statement costs

Project title	Project cost £ 000s	ES cost £ 000s	ES cost as % of project cost	Project type
Lincolnshire Coast Strategic Works	45 000	45	0.10	Coastal
Broadland	46 000	25	0.05	Fluvial & tidal
Dovercourt	2 350	20	0.85	Coastal
Woodbridge	3 600	5	0.14	Tidal
Colne Barrier	12 500	25	0.20	Tidal
Ouse Washes	16 000	49	0.31	Fluvial & tidal
Aldeburgh	4 900	8	0.16	Coastal
Happisburgh & Winterton	49 000	20	0.04	Coastal
Hunstanton & Heacham	7 100	5	0.07	Coastal

The costs in Table 1 include consultations and the production of the ES documents by independent consultants. Excluded are costs incurred on: EA prior to ES; design development; NRA consultations and liaison costs.

Most of these costs were being incurred before the 1988 legislation and Table 1 therefore broadly represents the increased costs arising from the 1988 legislation.

COLNE ESTUARY DEFENCES: THE NEED AND BACKGROUND

The Colne Estuary is the navigable tidal channel to the port of Colchester in Essex (Fig. 3). Some 240 ha of land adjacent to the estuary are at risk of flooding. The estuary is important for commercial shipping, a base for inshore and estuary fishing vessels, an oyster lay and an area actively used for boating and sailing. Conservation and environmental interests include flora and fauna, landscape and archaeological sites. The area contains several nature reserves, Sites of Special Scientific Interest (SSSIs) and conservation areas.

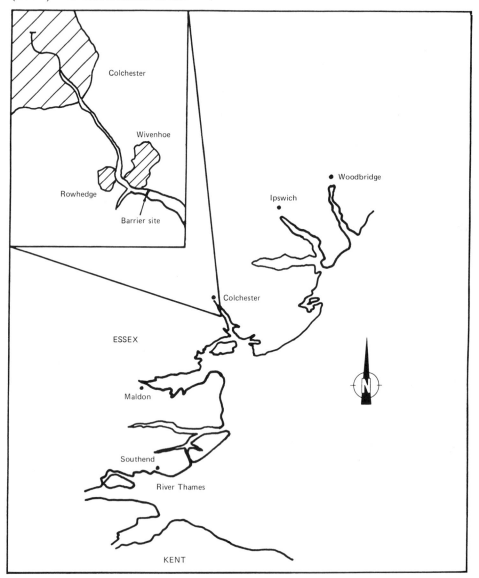

Fig. 3. Location plan of Colne Barrier.

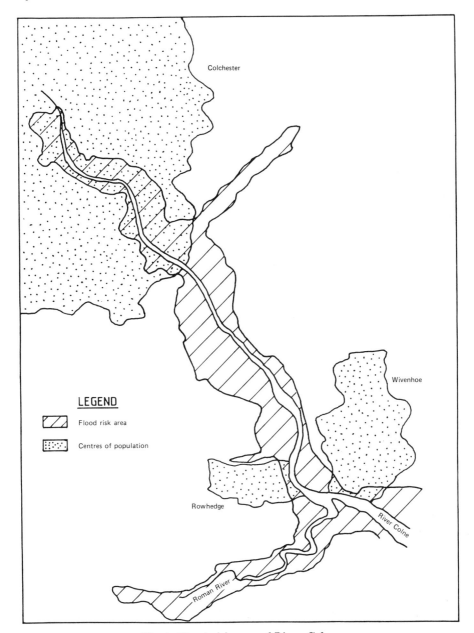

Fig. 4. Flood risk area of River Colne.

Area protected

Since Roman times the low-lying land bordering the estuary has been protected from flooding. Existing flood walls built after the 1953 floods are obsolete and cannot provide an effective long-term flood defence without further investment. Some 240 hectares of land including 100 ha of industrial land and 500 houses require protection (Fig. 4).

A preliminary examination of the area in 1986 indicated that the costs of flood protection would be significantly lower than the damages of any possible flooding and therefore a sound investment for public money. The need for careful attention to landscape, conservation, the environment and the boating and sailing interests was also highlighted.

COLNE ESTUARY DEFENCES: ENVIRONMENTAL ASSESSMENT

Flood protection options

Flood protection options, costs and benefits were produced in 1988 as part of a project appraisal which included an EA. Three options were evaluated in outline: (1) do nothing; (2) flood walls and banks; (3) construction of a flood barrier.

Consultations

Early consultation took place on these options with statutory bodies, the relevant planning authorities and selected non-statutory organizations. Interviews with local groups and individuals provided further basic information, options and concerns. General conclusions were drawn and attention focused on those aspects of the environment which would require further detailed consideration and sympathetic management, should the project proceed.

Outline options only were available at this stage and for this reason only selected authorities and bodies were consulted, in order to obtain an overall opinion on the best available option. Further detailed consultation at the local level was limited by the nature of the outlines and the preliminary EA, and could not have been successfully undertaken until more detailed design information had become available.

This highlights a problem as in the early stages of development it is not economic to prepare and present detailed design information on each and every option, neither is it feasible to embark on a widespread consultation programme without detailed design information. The consequence of this dilemma was that some objectors at the subsequent public inquiry were of the opinion that, at a local level in particular, the opportunity for external influence on the choice of design option had been too restricted too early.

Impacts and benefits

As part of the EA a simplified preliminary assessment of impacts was drawn up as illustrated in Table 2.

According to the severity of the impact, numerical values were assigned to the environmental aspects under consideration, on a scale of 0 – 4, where: 0 = severe adverse impacts; 1 = moderate adverse impacts; 2 = minor adverse impacts; 3 = no change; 4 = beneficial effects.

The lower the score the greater the likelihood of highest impacts and lowest benefit.

Table 2. Preliminary assessment of impacts

Environmental Aspect	Option		
	1	2	3
	Do nothing	Banks	Barrier
Economic	0	1	4
Visual	3	0	1
Social	0	0/1	2
Ecological	3	0/1	2
Archaeological/historical	0/1	0/1	2
Total score	7	4	11

Do nothing option

Although the 'do nothing' option did not, with 7 points, have the lowest score, it was discounted from further analysis because flooding would cause severe and unacceptable social and economic effects. Consultations at this stage also confirmed widespread support for effective and improved flood defences rather than 'do nothing'.

Bank raising option

The bank raising option was considered acceptable, but the cost was greater than the barrier option. Loss of amenity at quayside areas because of new and higher walls and the potential for widespread ecological and archaeological disturbance by further bank building was not in its favour, and hence the score of 4.

Barrier option

The barrier scored 11 points and was therefore likely to have the least impact and the most benefit. It also had the least cost of the options identified. Some adverse effects were identified from the social, ecological and archaeological/historical interests in the study area, although it was predicted that these would be of the order of minor local disturbances.

Explanation of allocation of scores to the economic aspect

Do nothing: To maintain the status quo would subject commercial areas below Colchester to an increasing threat of flooding, and as existing flood defence works were in danger of collapse, this option scored zero.

Banks: The proposed walls and banks would reduce flood risk but would result in particularly restricted access to the many active quays and wharves in the area, hence a score of 1.

Barrier: The barrier would reduce the risk of flooding without interfering with the economic use of the quays and would remove the blight of flood risk from some commercial areas, hence the score of 4.

EA conclusions

The EA identified at an early stage the aspects which could give rise to undesirable environmental effects. It included recommendations, to be integrated with the outline design, to avoid or mitigate adverse effects. This aided the streamlining of the development process and allowed the outlined design to proceed with confidence. A comprehensive list was compiled and appended to the report, to indicate those groups and organizations which should be formally consulted as part of any future ES.

The option achieving the highest overall score (Table 2) indicated the most acceptable option in environmental terms which included the economic aspect.

The barrier option was therefore identified for detailed investigation as part of an ES.

COLNE BARRIER: ENVIRONMENTAL STATEMENT

Introduction

Basic information was already available from the EA completed as part of the detailed project appraisal in 1988. An extensive consultation programme was initiated to discuss the proposal and its implications in detail with statutory consultees and other organizations and parties, in order to determine their views and concerns. The information gained from the consultations was studied in conjunction with site visits, surveys, and various technical reports on the barrier option. The results of the consultation were analysed and possible impacts, both adverse and beneficial, were identified. Measures to mitigate or avoid significant adverse impacts were proposed and recommendations were made for a monitoring programme. The ES for the barrier option was completed in August 1989.

Environmental effects

The barrier option was examined in terms of: (a) planning and policy; (b) visual and landscape; (c) ecology; (d) social and amenity; (e) economics, (including navigation and commercial); (f) physical and geomorphology; (g) air and water quality; (h) noise and vibration; (i) archaeology and heritage; (j) health and safety.

The effects were assessed separately for the construction stage and the completed barrier. It was concluded that the most significant adverse effect would be the visual intrusion of the structure, and measures to reduce the severity of this impact were proposed. During construction the major impact identified was traffic movement on the adjacent riverside town of Wivenhoe. Measures to reduce and control this impact were proposed.

Other adverse effects of the barrier were deemed less serious, principally affecting social and amenity interests (e.g. boating and sailing) and methods of ameliorating these concerns were discussed.

Model studies of flow in the estuary predicted that the effects on the ecology of the area should be insignificant. To confirm the predictions it was recommended that a programme to monitor the relevant parameters be undertaken.

Planning approval

The completed ES was submitted in support of the planning application and the relevant planning authorities subsequently granted planning approval for the scheme, subject to schedules of conditions. In November 1990 a public inquiry into the Barrier Order was held and the ES was submitted as part of the formal documentation.

THE PUBLIC INQUIRY

The ES was one of 63 documents submitted by the NRA to the inquiry. The ES was a key document and assisted the NRA in establishing thoroughness of the technical, environmental and economic aspects of the study. The NRA confirmed that it would implement all recommendations of the ES, particularly: (a) attention to visual design; (b) reduction of traffic movement by undertaking to bring 95% of materials by river; (c) alternative arrangements for boating and sailing; (d) a monitoring programme for ecology.

In June 1991 the Minister confirmed the Barrier Order and work commenced in October 1991. The barrier was due for completion in March 1993.

BENEFITS OF EA

The major benefits of EA are more effective linkages between design, decision making and the environment. Some examples are:

(a) improvements of siting and design;
(b) availability of information on environmental benefits which may be included in the benefit cost analysis;
(c) availability of additional information for engineering design and decision making;
(d) streamlining of the consultation processes;
(e) early indication of possible mitigating measures;
(f) improved co-ordination and understanding between environmental bodies;
(g) the ability to trade-off impacts and benefits in a more meaningful way;
(h) improved public acceptance of new projects;
(i) improved awareness and understanding of the importance of flood defence (the benefits, the impacts, the cost and who pays);
(j) improved public relations.

KEY PRACTICAL ISSUES

The aim of EA is to achieve a systematic and objective account of the environmental effects as an aid to decision making. In practice achieving this aim is complex, requiring independent, timely and adequate consultations. Some of the key issues are: (a) timing of consultations; (b) quality of the EA; (c) options considered in the ES; (d) independence; (e) monitoring.

Timing of consultations

Successful involvement of consultees can be affected by the timing of consultation.
(a) If commenced too early, without adequate definition of the project, misunderstandings arise and constructive comment will not be achieved.
(b) If commenced too late, after detailed project development, consultees feel that they have little opportunity to influence.
(c) If commenced too late inadequate data and information will be available for objective analysis in the EA and any subsequent ES.

Quality of the EA

The quality of the EA is dependent on:

(a) adequate lead in time to acquire necessary data, information and viewpoints;
(b) objective rather than subjective analysis;
(c) targeting of breadth and depth of content to the real environmental effects rather than the perceived need of individuals or organizations.

Options considered in the ES

For practical and financial reasons it is necessary to reduce the number of options to be included in the ES. This means that parties who perceive a need for detailed environmental examination of all options, irrespective of cost or feasibility, may object. To avoid such problems early consultation at feasibility stage on a full range of options is essential even though full design details may not be available.

Independence

The EA process needs to be seen to be independent, although there is an equal need to ensure that the aims and concerns of all, including the promoting organization, are properly understood. The EA process therefore needs to be independent but without remoteness and isolation from the design and project development.

Monitoring

Monitoring requirements need careful control, specification and management to ensure meaningful data is available for future action and post project appraisal.

CONCLUSIONS

1. The EA procedure requires a decision on the need for an ES, depending on the type of project, its complexity and likely environmental impact. If a project advertised as not requiring an ES subsequently raises objections then objectors may be pursuaded to withdraw objections, although failing this the Minister may make a final decision. Out of 86 projects in the Anglian region 70 projects have been advertised as not requiring an ES and no objections have been received. This would indicate that the procedure and criteria used for making the decision are adequate.
2. A total of 16 environmental statements have been produced to comply with SI 1217 and SI 1199. The cost of producing these statements, which broadly represents the increased cost of compliance with the 1988 legislation change, ranges between 0.04% and 0.85% of the project cost.
3. To achieve a systematic and objective EA, and any necessary ES, early consultation at feasibility stage on a range of options including 'do nothing' is essential. Consultation needs to be comprehensive, wide ranging and fully recorded. The procedure needs close integration with design decisions but at the same time needs to maintain a degree of independence.
4. EA can benefit all stages of a project from early justification through design decisions to final approval, and, in some cases where monitoring programmes are approved, through construction and for the life of the project. In the

Colne barrier project the ES was a key document at the public inquiry and assisted the NRA in establishing thoroughness of technical, environmental and economic justification.

REFERENCES

[1] Her Majesty's Stationery Office. *Water Act 1989.*
[2] Department of Environment/Ministry of Agriculture, Fisheries & Food/Welsh Office. *Code of Practice on Conservation Access and Recreation July 1989.*
[3] Ministry of Agriculture, Fisheries and Food/Welsh Office. *Land Drainage Improvement Works (Assessment of Environmental Effects) Regulations 1988.* Statutory Instrument 1988 No.1217 (6th July 1988).
[4] Department of Environment. *The Town and Country Planning Regulations (Assessment of Environmental Effects) Regulations 1988.* Statutory Instrument 1988 No.1199.

6

Managing environmental impact – experience in the private sector

D. R. Edge, BSc, DMS, DWM, MIWEM[†]

INTRODUCTION

The environmental movement and growing concern about 'green' issues is leading to increased awareness and scrutiny of corporate behaviour across the full range of commercial and industrial activities. The legislative environment created by both the United Kingdom Parliament and the European Community is becoming more robust, and pressure groups more vocal. Organizations which can respond effectively and efficiently to these challenges and demands will benefit and enhance their reputations. Anglian Water plc is committed to environmental improvement and a corporate environmental policy is being translated into specific action programmes. This chapter describes the policies and procedures which have been adopted by Anglian Water and details experience of their implementation to effect environmental improvement.

CHANGING ENVIRONMENTAL STANDARDS

The attention given to environmental standards has derived much impetus from a groundswell of public opinion, driving and driven by effective pressure groups and by European Community legislation. The public perception of the need for action to protect the environment has its roots in the post war environmental

---

[†] Environmental Impact Manager, Anglian Water Services Ltd.

movements that have from time to time gained more or less publicity and support. The recent momentum of this movement has been provided by several factors. Two of the significant influences are the series of environmental disasters such as Bhopal, Chernobyl and the Exxon Valdez which highlighted the environmental agenda, and the global issues of apparently irreversible change but which can be influenced by individual behaviour.

Chlorofluorocarbons (CFCs) are an example where the capacity to contribute towards an environmental improvement lay in the hands of the consumer. This demonstrated the power of individual and of co-operative action to effect meaningful results. The lesson has application in the European arena where the Commission is responding to the complaints of individuals or groups on environmental issues and using them to effect change in the conduct of environmental affairs in Member States [1].

The European dimension

Environmental policy within the Community has been implemented in a series of 5-year action programmes. These have resulted in the various environmental directives many of which impact on the water industry. The Fourth Action Programme (1987–1992) places the emphasis on establishing strict environmental standards, on the practical implementation of Community directives, and on the use of education and information to accompany and increase the effectiveness of the legislative approach. It develops the principles that appear in earlier programmes and seeks integration with other policy areas and the establishment of a global and transfrontier approach to pollution control. Finally it provides for initiatives in the areas of biotechnology and the management of natural resources [1].

The Commission is responsible for ensuring that Community law is properly implemented and pays particular attention to its integration into national law. It also intends to increase public awareness, to organize seminars and training, and to encourage private individuals and associations to bring to the Commission's attention cases of failures to respect Community legislation [1]. The Water Act 1989 and associated regulations are being used to implement Community law within British legislation replacing the administrative procedures that previously gave them effect.

CORPORATE ENVIRONMENTAL STRATEGY

The water service companies (WSCs) are in an environmental business. They carry with them the traditions of public health engineering and are able to embrace the current trend toward 'green' policies, indeed they should be at the forefront of that movement. They must also recognize the influences to which they should respond if their environmental policies are to be effective. The key influences operating on the WSCs are customer reaction, financial performance and environmental regulation. Anglian Water's customers will receive larger bills but might perceive little in the way of tangible benefits. To ensure that they understand what they are paying for requires that they recognize the cost of higher standards and that they perceive the company as one concerned with their protection and with the protection of their environment. The shareholders or City institutions may want an ethical investment or may wish to be reassured that their investment is in capable hands. This would not be the case if a company falls foul of the regulators or fails in any of its environmental obligations. A positive stance towards environmental issues recognized in the corporate policy is essential not only to promote a positive corporate image but also to ensure the operations, procedures and practices of the core business are focused on this key area.

Environmental influences on company performance

The financial success of the core businesses depends on secure performance of the water supply and sewage disposal operations and on the successful completion of the K investment programmes over the next nine years. To promote the timely completion of that programme will require that the investment in treatment plant, the locating of these plants and the obtaining of the necessary licences and consents is carried through effectively. A positive environmental policy will help to reduce, though not eliminate, the potential for delay.

There is a trend towards ethical investment both as formal investment portfolios and in the behaviour of some City institutions in seeking to influence those organizations in which they invest towards more environmentally friendly practices. A green image and a robust environmental policy will enable companies to attract investment of this sort and maintain a management tone appropriate for an environmental business.

The core business is closely prescribed by regulations and by regulators and it is essential for the company to have systems and procedures that ensure that regulations are recognized, monitored and reported on. Failures to meet the required standards will always reflect badly upon the company.

ANGLIAN WATER

Anglian Water plc provides the water and sewerage services to eastern England serving the largest area of the ten WSCs. The structure of the group of companies is broadly similar to that of the other plcs, the core business being carried out by Anglian Water Services Ltd (AWS) as the licence holder. Other companies within Anglian Water group provide engineering, financial and computer services to the core company and beyond. Within AWS the Directorate of Quality provides laboratory services, process science and quality standards support to four operational divisions. The Quality Standards Department is the focus within the company for all environmental issues and provides the point of contact for the regulators on all but day to day operational issues.

Anglian Water plc has adopted a Group Environmental Policy which recognizes not only the need to comply fully with the statutory requirements but also a duty of care for the environment within which it operates. It intends to execute its utility services and commercial interests in such ways as to limit or avoid all impacts on the environment and to minimize the consumption of non-renewable resources in doing so, within practicable constraints. This policy is being put into effect throughout the companies of the group. Its implementation is promoted both within and outside the company and its success is assessed and reported in terms of progress towards measurable objectives identified for individual policy items. The policy brings together existing practices and, in certain areas, less familiar approaches to produce a coherent strategy for the environmental operations of the company.

MANAGING ENVIRONMENTAL IMPACT

The approach to the management of environmental impact in Anglian is one which integrates vigorous audit monitoring of water and sewage operations with a proactive environmental emphasis on in-house assessments of conservation value and environmental impact. Anglian Water's first priority is to get its actions right in relation to the regulations, that is, to ensure that all activities and all environmental investment is consistent with the requirements of the regulations and the regulators. Compliance with consents and regulations is an area of work which receives greater attention and scrutiny in the post privatization industry. The next priority is to ensure that AWS operates in ways which, though not regulated, are nevertheless environmentally friendly. Anglian Water Services supplies water to 70% of the population of the region and operates sewage disposal sites in every almost every community. These operations and their associated sites are the sharp end of the potential and actual impacts on the environment and on the perceptions of its customers.

Environmental impact encompasses not only sewage disposal operations but also water supply operations including disposal of water treatment effluents and the effect of abstractions on groundwaters. However, the quality of the product that can have the most direct impact on customers is drinking water. In this area where the company is rightly most closely regulated, the emphasis of regulation is on effective self-monitoring with periodic audits of Anglian Water's practices and procedures by the Drinking Water Inspectorate. In the area of sewage treatment the emphasis of the regulator is on external monitoring. Although the NRA will visit sites regularly an extensive programme of audit monitoring is continued in-house. This supplements the data obtained from the NRA and provides a valuable input to the management networks on plant performance and on compliance with regulations and consents. To supplement and illuminate the results of this monitoring there are two further tools, Environmental Assessments which concentrate on the impacts of disposal activities on the receiving environment, and Wildlife and Landscape Surveys which assess the conservation value of our sites.

Existing operations

To monitor the impacts of its operations and operational sites, Anglian Water undertakes Local Environmental Assessments (LEAs) [2]. These are carried out by a team of field scientists dedicated to this task. They complement the routine quality monitoring which is carried out by laboratory and operational samplers. The purposes of LEAs are:

(i) to ensure that all Anglian Water's operations are consistent with the relevant consents or licences;

(ii) to describe and quantify the impact of discharges on the environment;

(iii) to assess the assimilative capacity of the receiving environment; and

(iv) to provide information to assist in discussions with the regulator about the appropriate conditions for consents or licences.

Clearly the principal targets of the LEA programme are the sewage treatment works. Anglian Water operates some 1080 sewage treatment works of which two thirds serve populations of less than 1000. Anglian Water does not attempt to emulate the degree of monitoring of the environment that the NRA might be expected to carry out, but by targeting works and being selective about priorities it has been able to assess the more significant sites. Selection criteria for inclusion in the LEA programme include size, risk of non-compliance, sensitivity of receiving environment and capital expenditure plans. In addition to the LEA work the field scientists also undertake the in-house monitoring of those small works which have descriptive consents.

The LEA provides input to Anglian Water's assessment of options for sewage treatment schemes. Although AWS must already meet some of the tightest standards for effluent quality [3] the arrival of Statutory Water Quality Objectives will result in the need for very high quality effluents in certain environments. Therefore particular attention must be paid to the location of the discharge point and possible alternative locations. The field scientist investigates the receiving water to identify other discharges, downstream sites of special scientific interest (SSSI) and other factors that might have a bearing upon the effluent standard. For although AWS will invest to achieve the required standards, it must also seek out the most cost effective option and the LEA provides input into that process.

The LEA comprises four components.

(1) An assessment of the data held about the site. This data, on the flow and quality of the receiving water, on the design and operation of the site and on the sources of the sewage flows provides the context for the LEA and must be verified or gathered from other sources.

(2) A site inspection, provides the opportunity to check that the works is being operated in accordance with its consents and to check the accuracy of the information held about the site.

(3) The local environmental monitoring, in particular the receiving environment is sampled chemically and biologically in order to monitor impact. Other potential impacts are also assessed, for example smell, noise and fly nuisance.

(4) The final stage is the preparation and circulation of the report to operational managers.

The report is the means by which the assessment results are communicated to operational managers. It comprises sections as follows.

Management headlines
Context data

(a) The consent conditions that apply to the works and associated discharges.
(b) Design flows, populations and processes.
(c) Trade discharges, contributing pumping stations.
(d) Environmental data, river flows, and quality objectives in the immediate and main receiving water.

Site inspection report
  (a) Commentary on the design and operation of treatment units and how they might impact on performance.
  (b) Conspicuous health and safety issues.
  (c) The management of on site conservation areas.

Operational report
  (a) Feedback from local operational managers on the site inspection report.

Environmental sampling results
  (a) Commentary on the status of the receiving environment.
  (b) Final effluent results and previous half dozen results.
  (c) Results of chemical samples upstream and downstream of the discharge.
  (d) Results of biological sampling upstream and downstream of the site, generally using macroinvertebrate samples.

Data presentation
  (a) Graphs of final effluent performance including a projection of current performance trends.
  (b) Maps and plans of receiving waters and site layout.
  (c) Photographs of the site and of any items noted in the reports.

Impact report
  (a) Summary Report.
  (b) Recommendations.

These assessments are carried out principally at sewage treatment works, but have also involved water treatment works and storm sewage overflows. More specialist assessments of, for example, marine outfalls have so far been undertaken by external consultants as would formal Environmental Impact Assessments should they prove to be required.

To date 30% of the works within Anglian Water Services with numeric consents have been assessed. Those have generally been works at risk of non-compliance or where capital investment is required.

Descriptive consents
Some 30% of AWS's sewage treatment works (STWs) serve a population of less than 250 and are subject to descriptive consents. These consents describe the plant on site and require that it is operated correctly, that the discharge is not having any significant impact on the receiving water and that there is no carry over suspended solids. The in-house assessment of compliance is carried out by the team of field scientists. The reporting is not simply in terms of pass or fail but, like the LEA, includes context information as well as comment on the operation of the site and secondary impacts.

Wildlife and landscape surveys
In addition to assessing the impact of operations on the environment Anglian Water

is conscious of the intrinsic conservation value of many of its operational sites. Some of these are well known, in particular major reservoir sites, and some are local botanical or ornithological havens. However it is recognized that many sites would benefit from improved landscaping and more sensitive management. To maximize, within the constraints of the operational needs, the conservation value of all sites, a series of Wildlife and Landscape Surveys have been undertaken to assess existing conservation value. From the surveys site management plans are developed within which it is often possible to provide opportunities to enhance the conservation value of the sites and to limit their impact within the landscape. This work is being carried out both by external contractors and by Anglian Water's own team of conservation scientists.

A dedicated budget is available to fund these environmental initiatives, which include tree and shrub planting, the creation of wildflower meadows, ponds and wetlands. While the site maintenance costs fall to operational budgets it can be the less expensive option to manage the site so as to enhance its wildlife value. The use of signs which indicate those areas of the site which are being managed for conservation provides a useful reminder to operators and managers.

Capital schemes

In addition to the LEAs prepared for STWs or similar activities, Anglian Water's procedures require that for all capital schemes a conservation statements is prepared by the conservation team for inclusion in the project appraisal report. The statements are prepared following site visits which identify anything of conservation value, whether these be formal sites of interest, for example, SSSIs, archaeological sites, trees with preservation orders, or informal sites of local value. These are, as far as is practical, protected by adjusting the proposals to avoid impact. An essential component of this process is full and effective consultation with both statutory bodies like the Nature Conservancy Council, County Archaeologists or Countryside Commission but also with local wildlife trusts and other local or national groups who might have a interest in the scheme.

Many of the schemes involve pipelines and as a result of the surveys routes have been adjusted to avoid not only major sites of interest but also features of local importance. The statement also identifies ways in which the impact of schemes can be mitigated through appropriate landscaping or by the creation of new features of conservation interest. This is known in Anglian as the Added Conservation Value and is normally funded out of the capital scheme's budget.

ENVIRONMENTAL STATEMENTS

The conservation statement and LEA can provide valuable input to the many improvement schemes that will be completed over the next few years, however major schemes to provide new STWs, or major water resources require a fuller assessment. When required Anglian Water will prepare an Environmental Impact Assessment. Even if the planning authorities do not require a full EIA the Environmental Policy requires the preparation of an environmental statement for large STWs and other major schemes. The environmental statement will normally be prepared by external consultants who provide not only the resources but also the independence that will aid the presentation of the case to local interests.

The area of coastal sewage disposal illustrates the need for proactive environmental management. The impact of the Urban Waste Water Directive on investment plans in estuarial and coastal areas will be significant in terms of the treatment options in that a site must be found for the treatment works. The selection of the best discharge option is aided by the application of dispersion models, enabling the Anglian to demonstrate that the design fully protects the uses of the marine environment. There is a need to quantify fully the status of marine ecosystems work undertaken, both before and after the discharge commences, by specialist consultants. The challenge in meeting the required timetable is in ensuring that the needs of the regulators and of local community are addressed early and effectively. The preparation of an environmental statement will be an important component of this process.

CONCLUSION

The water industry and the context in which it operates has undergone a change that requires the development of positive environmental policies. This involves the bringing together of existing practices and procedures and their integration with new policies which recognize the need to be seen to be environmentally friendly. The application of this policy must be wholehearted and must not be seen simply as a public relations exercise. The benefit to the company is in being environmentally responsible, a position from which PR benefits will follow. In Anglian the implementation of the Group Environmental Policy is influencing attitudes and approaches within the company and will lead to benefits among staff, customers and financial stakeholders.

ACKNOWLEDGEMENTS

The author wishes to thank the Director of Quality for his permission to publish this chapter.

REFERENCES

[1] Jonson, S. and Corcelle, G. (1989) *The environmental policy of the European Communities.* Graham and Trotman.
[2] Price, D. R. H. (1990) Environmental impact assessment as practised by Anglian Water Services Ltd. Paper presented at IWEM Symposium, Churchill College.
[3] National Rivers Authority (1990) *Discharge consent and compliance policy: a blueprint for the future.* Water Quality Series No.1. July.

7

Inverness main drainage scheme: a case study for engineering options and environmental assessment

M. Betts, C. Cowley and D. Siddy[†]

INTRODUCTION

The Environmental Impact Assessment (EIA) for the proposed Inverness Main Drainage Scheme is an important case study for both the theory and practice of environmental assessment, and for national and international developments in coastal sewage disposal.

The work is also a test case for the interpretation and implementation of several recent or impending developments in legislation and policy, including the EC Directive on Environmental Impact Assessment; the draft EC Directive on Municipal Waste Water Treatment; the conclusions of the Third North Sea Conference, and the interpretation of the Environment Minister's statements in 1990 on sewage and sewage sludge disposal.

The Scheme was called in by the Secretary of State for Scotland and was brought to the attention of the European Parliament. The Scheme (and the associated EIA) received considerable public attention and media coverage, due to its emotive and political nature. This was largely because the receiving waters of the Inner Moray Firth contain the last remaining resident population of bottlenose dolphins around the North Sea.

[†] Respectively Managing Director, Associate, and Senior Researcher, Environmental Management Limited.

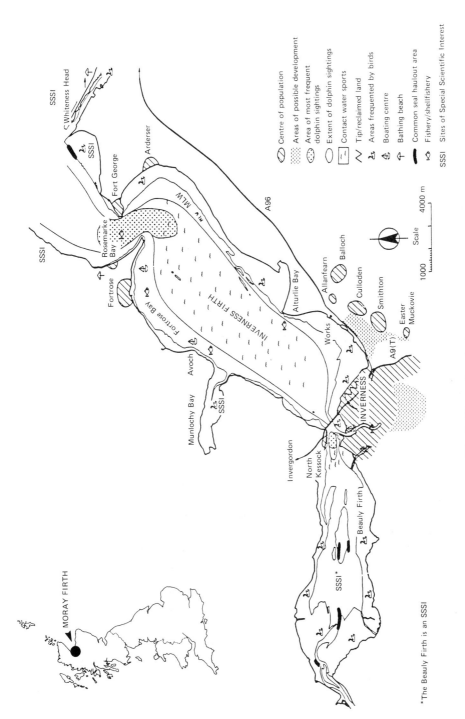

Fig. 1. Plan of study area.

MORAY FIRTH

Whiteness Head
SSSI
SSSI
Fort George
Arderser
A96
Rosemarke Bay
MHW MLW
Fortrose
SSSI
Fortrose Bay
Avoch
Munlochy Bay
SSSI
Alturlie Bay
INVERNESS FIRTH
Allanfearn
Balloch
Culloden
Works
Smithton
Easter
Muckovie
A9(T)
INVERNESS
North Kessock
Invergordon
Beauly Firth
SSSI*

Centre of population
Areas of possible development
Area of most frequent dolphin sightings
Extent of dolphin sightings
Contact water sports
Tip/reclaimed land
Areas frequented by birds
Boating centre
Bathing beach
Common seal haulout area
Fishery/shellfishery
SSSI Sites of Special Scientific Interest

Scale
1000   4000 m

*The Beauly Firth is an SSSI

BACKGROUND

Environmental Management Limited, in association with Watson Hawksley Limited and the University of Strathclyde, was appointed by the Highland Regional Council in September 1989 to undertake an environmental impact assessment (EIA) of the Inverness Main Drainage Scheme and to prepare an environmental statement (ES) in accordance with The Environmental Assessment (Scotland) Regulations 1988. A Final Study Report on the EIA was presented to the Highland Regional Council in November 1990.

Events preceding the EIA have a major bearing on this case study. In April 1987, the Highland Regional Council received from its engineering consultants, Mott MacDonald, a report which considered the design, environmental impacts and costs of possible schemes for the collection, treatment and disposal of sewage from Inverness. Inverness is a coastal town on the Inner Moray Firth, NE Scotland, whose population of over 40,000 was forecast to increase substantially over the following few decades. The existing arrangements for sewage disposal (unscreened discharge through a short sea outfall) were recognized as unsatisfactory even for existing loads and flows. There was thus a very widely accepted need for a new system of main drainage for Inverness.

Based on the 1987 report and further discussions with Mott MacDonald, the Council proceeded to approve plans for the construction of a long sea outfall discharging into an area of the Inner Moray Firth known as the Meikle Mee Deep, with associated headworks incorporating preliminary treatment located in the Longman Industrial Estate, Inverness. Discharge consent was granted by the regulatory authority, the Highland River Purification Board.

However, after opposition from local district councils, environmental pressure groups and the local community, the Secretary of State for Scotland called in the planning application and instructed the Highland Regional Council to prepare an environmental statement for the proposed scheme.

The consultants' report was prepared following twelve months of detailed study and investigation by a team of leading environmental scientists and other specialists. The study represented the most comprehensive and thorough examination of the environmental status of the Inner Moray Firth and the largest assessment at that time of a scheme of its kind to be carried out under the Environmental Assessment (Scotland) Regulations 1988.

The final study report was produced with the following aims.

- To summarize the findings of the study and discuss the suitability of the existing environmental quality objectives and standards.
- To outline further options for the scheme and discuss the compliance of the proposed scheme with the current objectives, and any alternative objectives that could be considered.
- To describe the different options for treatment and outfall location and give a preliminary economic analysis of these options.
- To present the arguments for and against the various feasible options and assist Highland Regional Council in deciding on a preferred option.

The significance of the study's findings was discussed in full to provide Highland Regional Council with the necessary information to decide upon a preferred option, for which engineering designs and an environmental statement to accompany a planning application could then be prepared.

WHEN IS AN EIA/ES NEEDED?

The Environmental Assessment Regulations apply to two distinct categories of developments:
- 'Schedule 1 developments' for which an ES is obligatory in every case; and
- 'Schedule 2 developments' for which an ES may be required if significant environmental effects are anticipated.

The proposed Inverness Main Drainage Scheme falls under the Schedule 2 category; an environmental impact assessment was not required until the Scottish Secretary intervened in the planning process. His intervention was a direct result of local opposition to the Scheme. The opposition arose because the objectors were not convinced that the Scheme would have minimal environmental effects. This illustrates three important points.
- A voluntary EIA can often make a useful contribution towards the planning and design of a development.
- Interactive consideration of the environmental and engineering aspects of major projects is frequently essential.
- Public as well as statutory consultations are an important feature of the EIA/ planning process.

CONSULTATIONS

Consultations should not be limited to statutory authorities, but should also include other interested parties, including possible objectors to a scheme. Furthermore, it is important that consultations begin at an early stage in the EIA.

Public perceptions of environmental effects were particularly important in a study such as this, where there existed a significant degree of concern combined with lack of trust in official procedures. Highland Regional Council continued to involve interested or affected parties in the development of the scheme, and to keep the public informed about proposals as these were developed. Amongst other things, this helped to prevent false rumours from developing.

SCOPING

'Scoping' is a word that is often used in EIA; scoping is essential.

The essential nature of this exercise is to undertake an initial identification of key environmental issues and to evaluate the adequacy of existing data and knowledge. Scoping plays a vital role before the commencement of the assessment, in order to place the significant issues in an appropriate context, and to ascertain whether additional data collection and fieldwork are required. It is important at this stage to consider the construction as well as the long term operating aspects of the development.

The focus for a coastal sewage scheme is usually the offshore environmental impacts, especially in a case such as Inverness where the resident dolphin population is a central and emotive issue. Care should be taken, however, not to ignore terrestrial impacts; noise or odour problems, for example, could be of considerable significance to the local human population and to the final choice of scheme.

DATA BASE

The establishment of an adequate data base is vital for the proper assessment of possible significant impacts.

When considering data requirements with regard to fauna and flora, it should be recognized that the range of potential impacts will vary with the species involved. For example, migratory species such as birds and dolphins can require appraisal over a larger area than the vegetation along a pipeline route. Care must be taken to think this through early on in the environmental assessment.

Similarly, there are seasonal considerations for elements such as ecology and hydrography. This in turn means that a thorough work schedule should be prepared at an early stage in the EIA.

In the case of the Inverness EIA, it was necessary to build a data base including detailed information on hydrography, fisheries, benthic communities, marine mammals, birds, water quality and microbiology.

UNDERTAKING ENGINEERING WORK
IN PARALLEL WITH ENVIRONMENTAL ASSESSMENT

As stated earlier, it is normally desirable for environmental and engineering aspects of a project to be considered simultaneously and interactively. If this is not done, it is possible for extensive engineering design work to be undertaken on aspects of a scheme which otherwise would have been identified at an early stage as having major environmental (and hence planning) implications.

On the other hand, if an interactive, multi-disciplinary approach is adopted, it is more likely that the scheme will be acceptable to conservation bodies, planning authorities, and other organizations and that lengthy delays, such as public inquiries, can be avoided.

It is sometimes preferable if the environmental assessors are independent from, though co-operating closely with, the engineers designing the scheme and the scheme proposers. This gives interested parties and possible objectors a channel to express their views at an early stage in the scheme design, and the confidence that any appropriate points raised will be considered seriously.

Effective communications go hand-in-hand with environmental assessment and engineering. It is often difficult for developers to explain to all interested parties that they are not opting for the 'cheap and nasty' solution. How is the public presented with a relatively cheap option that may actually provide considerable environmental benefits? Similar problems can arise in the situation where it is pragmatic to proceed initially with a partial solution, with an assurance that additional treatment or other works will be implemented later, as required. It can be difficult to convince the public that this is an acceptable environmental solution, even though it may be the only way that a scheme can actually receive adequate funding and be commenced as soon as possible.

Finally, the interactive approach in the context of EIAs of this kind allows the environmental implications (positive or adverse) of innovative technologies for sewage treatment or disinfection to be assessed at an early stage.

LEGISLATIVE REQUIREMENTS

Legislative requirements are of major importance as scheme design criteria and must be related to local conditions and perceptions. Such legislation is changing and becoming increasingly strict, and brings with it new challenges.

Fig. 2. Bottlenose dolphins in the Moray Firth.

For instance, what is a 'sensitive' environment? The precise definitions of a sensitive environment under the draft Directive on Municipal Waste Water Treatment and the proceedings of the Third North Sea Conference were still under debate in 1991. For the Inner Moray Firth, (which comprises the Beauly and Inverness Firths,) 'sensitivity' concerns the presence of one of the UK's last resident populations of bottlenose dolphins, the semi-enclosed nature of the Inner Moray Firth, and any potential for eutrophication the upper Beauly Firth. Until the national authorities provide guidance on such definitions there will continue to be disagreements between developers, regulatory authorities and conservation bodies on such issues.

It has often been said over recent years that 'the goalposts have changed'. What does this really mean for a developer who will need to meet the draft Municipal Waste Water Directive? Early agreement should be reached between the developer and the regulatory authorities on the timescale for which a scheme is being planned, and whether additional treatment can be phased in later.

The Third North Sea Conference agreed that where there is an element of doubt concerning the effects of a discharge on the marine environment, or the mechanism by which it is dispersed, 'the precautionary principle' should be followed. The question of how to implement this principle is debatable. However, the point raised above relating to voluntary EIAs reveals that the absence of conclusive evidence of any significant environmental impact is often not acceptable as an argument in support of a proposed development; rather, the developer must increasingly demonstrate that there is no significant environmental impact.

Table 1. Environmental quality objectives required
to maintain existing potential beneficial uses

Beneficial Use \ Objective	Aesthetic	Bacteria viruses	Dissolved oxygen	pH	Nutrients	Slick	Suspended solids	Dangerous substances
Commercial fisheries	X	✓	✓ (critical)	✓ (critical)	✓	✓	✓	✓
Intertidal & subtidal invertebrates	X	✓	✓	✓ (critical)	X	X	✓	✓
Bathing	✓	✓ (critical)	X	✓	X	✓ (critical)	✓	✓
Secondary contact recreation	✓	✓ (critical)	X	✓	X	✓ (critical)	✓	✓
Navigation & shipping	X	X	X	X	X	X	X	X
Aesthetics: land based recreation	✓	✓	X	X	X	✓ (critical)	X	X
Marine mammals & birds	X	✓ (critical)	✓	✓ (critical)	✓	✓	✓	✓
Shellfisheries	X	✓	✓	✓	✓	X	✓ (critical)	✓

✓ Objective required   X No objective required

▨ Critical EQSs required

Similarly, there is confusion over how to interpret the need to provide 'secondary or equivalent' treatment as required under the draft EC Directive. This has been better explained in the previous draft revision, but a national consensus for this, to allow the best solution for individual coastal locations, should be agreed. For Inverness, the provision of conventional secondary treatment in addition to primary settlement was unlikely to have much benefit for the environment of the Inner Moray Firth because the reduction of the organic load in the sewage was not considered to be a critical factor. The option of lime-assisted primary treatment was recommended, because it conferred particular microbiological advantages.

THE SIGNIFICANCE OF IMPACTS

The current environmental situation must be fully appraised in order to assess any potentially significant environmental impacts of the proposed scheme, whether beneficial or adverse, temporary or permanent, direct or secondary or cumulative.

Having identified the likely significant impacts, the assessment basically encompasses the following sequence of steps: (1) prediction of impacts; (2) evaluation of impacts; (3) a comparison of alternatives within the context of the specific scheme being assessed; (4) organization and presentation of the information on impacts.

Table 2. Possible environmental quality standards for the Inner Moray Firth

EQOs	Objective/Extent of Treatment	Critical EQS Requirements
(a)	Current HRPB objectives	ES mandatory faecal coliform limit 2000/100ml (95%ile) at shoreline, solids removal for no visible slick
(b)	Marine recreation	EC mandatory faecal coliform limit 2000/100ml (95%ile) outside pool above outfall, solids removal for no visible slick
(c)	Cetaceans	ES mandatory faecal coliform limit 2000/100ml (95%ile) and 0.1 enterovirus/10 litres (50%ile) in pool above outfall, solids removal for no visible slick
(d)	EC Municipal Wastewater Treatment Directive	Effluent quality from preliminary, primary and secondary treatment only. Assumes also compliance with (a)
(e)	EC Municipal Wastewater Treatment Directive ('sensitive water')	Effluent quality from preliminary, primary and secondary treatment with nutrient removal of total N to 10mg/l and total P to 1mg/l. Assumes also compliance with (a)

For the Inverness study, a major aspect of the EIA involved a thorough study and consideration of the definition and application of Environmental Quality Objectives (EQOs), and Standards (EQSs), and relevant emission/discharge standards, as a precursor to predicting and evaluating the significant impacts. Tables 1 and 2 summarize respectively the EQOs which the study identified as necessary to protect/maintain existing or potential beneficial uses of the Inner Moray Firth, and the possible EQSs which were considered.

Determining the significance of an environmental impact is inevitably mainly subjective, and critically depends upon the definition of significance. Some impacts of the various scheme options would be detectable but not significant. The study adopted the criteria for significance contained in the HMSO publication Environmental assessment; A guide to the procedures (1989). These criteria provided general guidance in trying to determine whether or not a potential impact should be regarded as significant but the assessment, nevertheless, still relied on the collective experience and judgement of the study team in reaching conclusions on the significance of particular impacts.

EXTENT OF STUDY REMIT

Determining the best solution for a problem as rapidly as possible sometimes requires that an environmental assessment is extended beyond the standard remit for an environmental statement, as set out in the European Community (and implementing UK) legislation. Work of direct environmental, and so engineering, relevance to an outfall scheme is often essential to the developer's decision-making process.

This can be presented as part of an associated study report for a client, to be followed by an environmental statement for submission with a full or outline planning application, once the preferred option has been agreed. This approach should help to avoid planning inquiries and other subsequent delays.

Such an associated study report sets out and evaluates information of relevance to the scheme, that would normally fall outside the remit of a formal environmental statement.

The terms of reference for the Inverness EIA were unusually wide, but further opportunities to amend them in the light of changing legislative requirements or increased understanding of the local environment would still have been welcomed. This is of relevance to most coastal discharge EIAs where the question of sewage sludge disposal from an extended treatment scheme must be addressed in conjunction with works and pipeline designs, and also where associated trade effluents would benefit from a separate control policy.

USE OF ENVIRONMENTAL BALANCE SHEETS

The use of environmental balance sheets or matrices as part of an integrated decision analysis framework allows detailed environmental comparison of engineering options in as simple way as possible.

There are some dangers inherent in such matrices, as they may tend to create a false impression of scientific objectivity about essentially subjective environmental decisions.

Nevertheless, a matrix comparing environmental impacts, employing a very simple scale, can greatly facilitate the appraisal of engineering options. In this case the classification of environmental effects (considered separately for the construction and operation of the scheme) was restricted to 'significantly adverse', 'significantly positive' or 'not significant or of unknown effect'. Introducing weightings to try and take account of the relative magnitude or importance of environmental effects can lead rapidly to controversy about a matrix, making it of little further use in the decision process. Any such matrix must be the result of extensive discussions and consensus between the relevant experts for all aspects of a scheme.

The extension of an environmental matrix to form part of a wider decision model, which tries to take account of other factors such as social impacts, engineering feasibility, costs, and other factors, becomes even more controversial. This requires impact weighting and ranking, on which there is unlikely ever to be full agreement from all parties. Although of some value for in-house decision-making, further methodological development is needed before such extended decision models can be employed effectively in reaching key decisions.

LONG-TERM MONITORING

Even when the environmental data base is considered to be adequate for the assessment of any significant impacts of a scheme, there is a need for on-going monitoring.

For the Inner Moray Firth, further fieldwork and analysis relating to the salinity distributions, particularly at low river flows, was required. Associated work would enable the creation of a mathematical model for long-term planning of the nutrient budget of the Beauly Firth. This would be particularly relevant to evaluating the potential for eutrophication and hence to determining the need for designation of the waters of the Firth as 'sensitive' under the draft EC Municipal Waste Water Treatment Directive.

A comprehensive programme of continued monitoring was also recommended, to be developed and agreed with the Highland River Purification Board, for implementation as soon as possible. This was to provide further information on background environmental conditions, and would allow proper assessment of the impacts of the construction of the scheme, of impacts after an initial stage was completed, and indicate any need for additional treatment or other environmental mitigation measures.

SOME LESSONS FROM INVERNESS

All the lessons which have been learned from the Inverness study are too numerous to list here.

However, some of the key lessons of relevance to future EIAs of such schemes are:

- EIAs should not just be seen as a statutory requisite after the planning and design concept of a scheme have been finalized. Even where a statutory ES is not obligatory, an EIA conducted in parallel with the initial planning and conceptual design will usually result in an environmentally more robust and acceptable scheme.
- Public as well as statutory consultations are a vital feature of the EIA/planning process, particularly if the nature or location of the scheme is likely to be sensitive. Apart from perhaps generating information of value to the EIA and planning process, public consultations help to build confidence that fears and concerns are being taken seriously.
- 'Scoping' prior to embarking on the EIA itself is essential.
- The establishment of an adequate data base for the EIA is fundamental. Where this has to be developed, the associated time and costs can be considerable, as was the case in Inverness. However, this needs to be viewed against the possible delays and ensuing costs of *not* carrying out the EIA in sufficient depth.

- Presenting the results and conclusions of a complex EIA in a form which can be readily absorbed and understood by decision-makers is a difficult task. Techniques for presenting and communicating such information continue to be developed and refined.

8

Environmental implications of treatment of coastal sewage discharges

J. Gay, BSc, MS, CEng, MICE, FIWEM[†], R. Webster, BA, MSc, MIWEM[‡], D. Roberts, BA, MSc[§], and M. Trett, BSc, FLS[¶]

INTRODUCTION

In August 1989 the Department of the Environment (DoE) commissioned a wide-ranging study of coastal sewage discharges. The study included an assessment, in environmental and financial terms, of costs and benefits of a range of treatment options.

The study was prompted (a) by growing concern about the long-term effectiveness and acceptability of marine treatment systems, and (b) in anticipation of an EC draft Directive on urban wastewater treatment which was subsequently published in November 1990. The minimum level of treatment required by this Directive is secondary treatment for discharges to estuarine waters serving more than 2000 population equivalent and primary treatment for discharges to coastal waters from greater than 10 000 population equivalent.

This chapter concentrates primarily on the environmental aspects of the study. The chapter details the current situation in terms of treatment installed, then the potential impact of sewage discharges on the marine environment is assessed. A

[†] Director, Consultants in Environmental Sciences Ltd.
[‡] Environmental Scientist, Consultants in Environmental Sciences Ltd.
[§] Head of Water Environment Division B, Department of the Environment.
[¶] Marine Biologist, Centre for Research in Aquatic Biology, Queen Mary and Westfield College, University of London.

brief review of treatment options is included, followed by a section on the possible land-based impact of installing treatment plants at coastal discharges. A rationale that could be used to assign priority to the installation of treatment at coastal locations is presented, and possible approaches to the objective assessment of net environmental impact are discussed.

PRESENT SITUATION

National questionnaire surveys of water plcs, Scottish regional councils and the DoE (Northern Ireland) were carried out for outfalls included in the draft Directive. Tables 1 and 2 present a summary of the data which were collated. As the precise definition of estuarine waters in the draft Directive is still not clear, the definitions included in the revelant water quality surveys conducted by the DoE and the then Scottish Development Department, (now the Scottish Office Environment Department (SOEnD)), were used. Two surveys were carried out, one in advance of the draft Directive and based on data relating to summer resident population, and a second after publication of the draft Directive and based on population equivalent.

Table 1. National estuarine outfall data (>2000 population equivalent)

Type of treatment	Number of outfalls	Population equivalent
No treatment	83	1 797 000
Preliminary treatment only	57	2 489 000
Preliminary and primary treatment	81	5 062 000
Secondary treatment	120	9 435 000
Total	341	18 783 000

Table 2. National coastal outfall data (>10 000 population equivalent)

Type of treatment	Number of outfalls	Population equivalent
No treatment	49	1 645 000
Preliminary treatment only	54	3 359 000
Preliminary and primary treatment	11	395 000
Secondary treatment	3	43 000
Total	117	5 442 000

The data show that most of the sewage discharged to estuarine waters is given secondary treatment, while that discharged to open coastal waters normally receives only preliminary treatment, reflecting the emphasis on marine treatment in these locations.

POTENTIAL IMPACT OF SEWAGE DISCHARGES
IN MARINE ENVIRONMENT

The composition of sewage is variable, as are the length, location and condition of individual outfalls and the receiving waters into which they discharge; consequently it is difficult to draw definitive conclusions from studies at individual sites. It becomes increasingly difficult as pollutants dilute and disperse to trace the pollutant source. Therefore, where primary impacts are not wholly in the vicinity of the outfall, an assignation of the effects from individual outfalls may become problematic. A growing body of evidence indicates that synergistic effects may exist between components of effluents from different outfalls discharging to coastal and estuarine waters; the sum of the effects of individual components may be greater than both of the individual effects in isolation. In this section, potential impacts are considered in relation to pathogens, trace chemicals of concern, organic enrichment and biochemical oxygen demand (BOD), and particulate material. In the last section three case studies are cited to give examples of impacts which have been reported.

Pathogens

Most marine treatment systems have been designed to allow designated bathing areas to meet the coliform standards contained in the EC Directive on bathing water quality. They are normally designed on the basis of an assumed or experimentally derived half-life for bacteria in the particular marine environment and modelling of the hydrographic situation. Calibration studies are important to verify model parameters. There has recently been some evidence from Colwell [1] that reliance on the culturing of samples in the laboratory has led to an underestimate of bacterial survival in coastal waters. However, provided that this is taken into account when marine treatment system are designed, it does not invalidate the design technique.

Viral survival periods and viral standards are still a matter of debate. Goyle and Adams [2] demonstrated that viruses could still be detected in sewage sludge eighteen months after the cessation of dumping, although these results are not necessarily applicable to the conditions prevailing at outfall diffuser. The risk to public health from direct exposure to viral contamination is not clear, and although isolated cases of poliomyelitis have been contracted after swimming Waldichuck [3] was unable to associate them unequivocally with sewage in bathing water. In the summer of 1989 a DoE funded pilot study of the relationship between microbiological quality of bathing water and health risks was carried out at Langland Bay, near Swansea [4]. The water was well within EC bathing water standards and results from both study methods employed indicated that, although the rate of gastroenteritis did not differ significantly between swimmers and non-swimmers, those entering the water reported more symptoms, including irritation of the ears, eyes, nose and throat. Additional studies of a similar nature are currently in progress. It should be noted that the long incubation periods of certain viral diseases may make correlation with exposure to sewage effluents difficult to establish.

Particulate material

Sewage-derived particulates range from coarse litter that has escaped any screening processes to fine suspended solids (SS). Gerlach [5] noted that the introduction of low salinity effluents with high SS loadings leads to localized flocculation and precipitation. Where this material is deposited onto the sea bed, often within 50 m of the outfall, physical 'smothering' of the sediments may occur, reducing the penetration of oxygenated seawater into the sediment. In combination with a high BOD this can cause anoxic conditions in the sediments.

Solids that remain suspended in the water column or are re-suspended may exert other deleterious effects. Suspended particles clog the tentacles, fine filters and gills of suspension feeders, and may lead to the localized extinction of these species up to a few hundred metres from the point of discharge depending on the dispersion achieved and the pollution load. A further effect is increased turbidity, which causes attenuation of light, affecting the rate of photosynthesis by macroalgae and phytoplankton. Larger solids such as plastics, when ingested by birds, can accumulate in the proventriculus and gizzard, which will impair digestive efficiency and may have acute and lethal effects.

Organic enrichment and biochemical oxygen demand
The introduction of sewage-derived organic material can lead to depletion of oxygen in sediments and overlying water. An increase in sediment loadings of organic material leads to changes in species number, biomass and abundance of macrofauna. As distance from the point of discharge increases and concentrations of organic material decline, opportunistic species which can tolerate relatively low oxygen levels (such as certain polychaetes) reach high densities, and the biomass increases to abnormally high values. Further along the organic gradient, polychaete numbers decline and the number of species increases as the biomass declines to normal levels and the effect of the discharge diminishes.

Substances of concern
Studies at individual locations indicate that the effects of BOD and SS can be mitigated and the effects confined to a small mixing zone in the immediate vicinity of the outfall, provided that outfalls are well-designed so that large dilutions are achieved and the screening plant is effective. However, in a submission during the DoE study, Greenpeace highlighted concerns about eutrophication and the potential for bioaccumulation of synthetic organic compounds and certain heavy metals, and stressed their support for the 'precautionary approach'.

Nutrients
It has been estimated that in the UK 12% of nitrate and 25% of phosphorus in estuaries and coastal waters are contributed by sewage discharges. In the UK no clear evidence of marine eutrophication caused by man has been observed, but the south eastern North Sea and the Adriatic have been affected by algal blooms thought to result from increased nutrient concentrations due to anthropogenic inputs. In parts of the Mediterranean and Adriatic, algal blooms may be seen in the vicinity of outfalls, and sewage inputs may trigger blooms, but high background concentrations caused by other inputs are often an essential prerequisite in these instances.

Some evidence of eutrophication has been found in certain UK estuaries. Raffaelli *et al.* [6] reported a two- to threefold increase in nitrogen concentrations in river water entering the River Ythan over a 25-year period, leading to an increase in the abundance of certain weeds in the estuary. The effects of eutrophication on salt marshes have been observed by a number of observers, for example Ranwell [7].

Organic compounds

Many synthetic complex organic compounds such as PCBs, DDT and hexachloro-cyclohexane (HCH) have very slow or non-existent biochemical breakdown mechanisms. Consequently the possibility exists for bioaccumulation in organisms higher up the food chain and those which filter large volumes of water. DDT concentrations in seawater are normally in the ng/l range and in the mussel *Mytilus* are generally below 20 ng/l, while predatory fish may have up to 1900 ng/g in their liver oils. Law *et al.* [8] observed mean levels in the subcutaneous fat of seals of 3000 ng/g.

Polycyclic aromatic hydrocarbons (PAHs) may also accumulate up the food chain [9]. Laboratory studies by the same author demonstrated sublethal and lethal carcinogenesis and mutagenesis in aquatic organisms. These compounds are present in effluents from processes involving combustion, road runoff and household sewage [10].

In estuarine environments, sediment transport processes often deposit certain sizes of particles in particular areas. Thus micropollutants adsorbed onto particulate material may be selectively deposited, giving rise to the possibility for disproportionate exposure of benthic communities in certain areas of estuaries.

Heavy metals

At low concentrations many heavy metals are essential to life, but in high concentrations they can be toxic. However, unlike many organic compounds, they occur naturally in seawater; consequently marine organisms have developed regulatory control mechanisms. Most metals adsorb onto particulate material, and therefore organically rich sediments at discharge sites may contain high concentrations of metals. There is evidence that bivalve molluscs can accumulate high levels of heavy metals, but there is no strong evidence of significant transfer to demersal fish of commercial importance (e.g. sole, plaice and halibut).

Case studies

The number of detailed studies of the effects of sewage discharges in the UK is relatively small, but the results have not given rise to immediate concern. A few examples are given in this section to illustrate the effects that have been noted. Kayralla and Jones [11] surveyed the benthos of the Tay Estuary and found no evidence of gross pollution except in the immediate vicinity of sewage outfalls, despite the fact that 80% of the sewage from Dundee was untreated. This was attributed to the very high rate of flushing of the estuary.

Read *et al.* [12] assessed the benthic impact of the replacement of short outfalls in Edinburgh with primary treatment and a long outfall into the Firth of Forth. After the new outfall became operational, the populations of characteristic species of polychaetes declined and were replaced by other species found on nearby clean beaches. A similar increase in the diversity of meiofauna was also observed.

Cooper and Thompson [13] surveyed macrobenthic communities surrounding two major outfalls at Weymouth and Tenby, which respectively discharge 20 000 m^3/d

and $17\,500$ m^{3}d of fine screened sewage. At Weymouth the only effect observed was an increase in concentration of thermotolerant coliforms (TTC), coprostanol, metals and an increase in toxicity, as measured by the Microtox test, within 50 m of the outfall. No detrimental effects on the macrobenthic community could be detected. At Tenby no effect on the concentration of any of the parameters could be detected, and the macrobenthic community was apparently unaffected by the outfall.

The need for further studies enabling long-term impacts of sewage effluents on marine benthic ecosystems cannot be over-emphasized. Recent studies have indicated the value of meiofauna as sensitive indicators of the impacts of pipe-borne effluents [14]. Such studies should also provide important verification data for dispersion models.

REVIEW OF TREATMENT OPTIONS

Table 3 presents a review of the major treatment options currently available for sewage treatment. Membrane filtration and aerated biological filtration are also considered, since both systems appear to have some significant potential benefits, although these are not yet proven operationally on full scale. It should be noted that conventional treatment systems are designed primarily to remove BOD and SS. The removal of other parameters has normally been considered to be an incidental benefit. The chemical treatment in Table 3 is primary sedimentation assisted by the addition of lime and possibly flocculating agents.

Primary treatment is able to remove about one-third of the BOD from sewage, and is therefore beneficial where deoxygenation is a problem. Typically up to two-thirds of the SS may be removed, which can be advantageous where physical smothering of the sea or estuary bed would occur due to insufficient dispersion; otherwise the levels of removal achieved for other parameters are unlikely to be of major significance in terms of improving water quality.

Chemically assisted primary treatment is able to produce enhanced pollutant removal, notably of pathogens, phosphorus and heavy metals. The removal of complex organic compounds is also improved, but this is very compound specific. The major drawback is the large volume of sludge which is produced and the high energy cost associated with the power requirement in chemical manufacture.

Biological secondary treatment effects high removals of BOD and SS and achieves a significant reduction in a wide range of parameters. However, the 90–99% removal of pathogens has to be compared to the 99.999% (5 log unit) reduction required to meet the EC bathing water quality guideline standard. Consequently, secondary treatment alone without a marine treatment element cannot achieve bathing water standards close to the discharge. Two biological secondary treatment methods which offer similar performance in a more compact unit are deep-shaft aeration and aerated filters. These could find application where land availability is difficult.

Membrane filtration offers the potential for compact plants achieving high removals of pathogens, BOD and SS. This process is currently under development, and therefore detailed information is not yet available for full-scale plants.

Marine treatment is designed to utilize natural degradation processes that occur

Table 3. Comparison of sewage treatment processes operating on a (mainly) domestic sewage

Parameter	Primary treatment	Conv. activated sludge	Extended aeration	Deep shaft	Chemical treatment high dose	Chemical treatment low dose	Long sea outfall with maceration
SS (%)	50–65	90–95	90–95	90–95	85–90	60–65	0
BOD (%)	30–40	90–95	90–95	90–95	55–60	0–50	0
COD (%)	30–40	80–90	80–90	80–90			
Total N (%)	10–20	25–45	30–50	69% NH_3¶ 100% NO_3¶	10–20	10–20	0
Total P (%)	5–15	20–40	20–40		85–95	75–85	0
Total coli-forms (%)	30–60	90–99	90–99	90–99	99–99.9 (Faecal coli)	75–96 (Faecal coli)	0
Viruses (%)	0–50	75–99	75–99	—	90–99	80–96	0
Heavy metals‡ (%)	30–50	50–70	50–70	—	85–95	Good	0
Complex organics§ (%)	20–30	50–70	50–70	—	60–90♭	—	0
Dry solids (kg/hd.d)	0.04	0.07	0.07	—	0.125	0.064	0
Power required (W/hd)	0.2	2.4	4.0	—	9.7♮	3.2♮	Pump power
Approx. capital cost†(£/hd)	60	115	125	120	70	70	Low
Operating cost (p/m³)	4	9	9	—	13	7	Low
Land required (m²/m³.d DWF)	1	1.8	1.8	1.2	1.1	1.1	Lowest
Land environ. impact ranking	2	6	5	4	4	3	1

This table is designed to give an overall indication of the relative costs and benefits of various sewage-treatment options. In practice, costs and performance may vary considerably depending on the site and the operational regime. Costs are based on a plant sized for a domestic population of 45 000 with a dry weather flow (DWF) of 7000 m³/d which can treat up to 3DWF, and is provided with storm balancing tanks for flows in excess of 3DWF. Sludge disposal is by digestion and spreading to land.

% denotes percentage removal.

† Excluding cost of outfall, which is assumed to be similar in each case, and cost of screening, which is assumed to be required whatever treatment is given.

‡ Typical figures only, varies according to specific metal. Figures exclude Nickel.

§ Typical figures only, varies according to specific compound.

¶ Manufacturers' figures.

♭ Poor for chlorinated pesticides.

♮ Including power required for lime manufacture.

in the sea. The emphasis has been on achieving large initial dilutions rather than a reduction of pollutant inputs prior to discharge. The mitigation of adverse marine impact is achieved by a high degree of dilution and dispersion, and the visual impact on land is small because only a small pumphouse is required.

LAND-BASED IMPACT OF TREATMENT SYSTEMS

The major permanent impacts of implementation of a requirement for primary treatment for coastal outfalls and secondary treatment for estuarine outfalls are summarized below in terms of land-take, air quality, noise, visual impacts, sludge disposal and secondary impact. Additional temporary impacts and disruption would result during the construction of sewage-treatment works and sewers.

Land-take
It is estimated that the land-take for a medium-sized primary treatment works typically requires 0.15–0.20 m^2 *per capita*, while conventional secondary treatment requires 0.30–0.35 m^2 *per capita*. Many existing outfalls project from the base of protruding headlands which are often conservation features and may be designated as such. Some outfalls project from the base of sea walls or promenades, and the siting of treatment works in such locations would probably be impractical and would certainly be unpopular among local residents and tourists. Consequently it is likely that in some cases treatment works would need to be sited a considerable distance from the original outfall headworks.

Air quality
The major impact will arise from odours which may be produced in the sewage-treatment works, particularly at the inlet works and around sludge treatment units, and can be detected 200–300 m downwind. If it is necessary to site a treatment plant a considerable distance from the original inlet, particular problems may arise with septicity of the sewage arriving at the plant. In addition, if sulphate levels in the sewage are elevated, possibly due to seawater infiltration in seafront areas, the potential for hydrogen sulphide formation will be increased.

Noise impact
Noise will be generated by mechanical equipment and vehicular traffic. Impacts can be minimized by careful design, operation, equipment selection and timing of road traffic to avoid sensitive periods such as nights and weekends.

Visual impact
Visual impact arises from the siting of structures required for a sewage-treatment works and might be increased by the sensitive nature of many coastal areas. The greatest visual impact generally results from anaerobic digesters (which can be over 10 m high) and inlet works (which may be 4–7 m above ground level). The use of covered works, underground units, earth embankments and landscape screening hedges as well as high-quality building materials can considerably reduce visual impact.

Sludge disposal impact

It is estimated that implementation of the minimum provision draft Directive would lead to approximately 5.5 million m^3 of additional wet sludge (0.25 m tonnes DS) on a national basis. In some local areas, which have relied heavily on marine treatment, the amount of sludge for disposal will increase disproportionately. In conjunction with the phasing out of disposal of sludge to sea, it is clear that any implementation of the treatment requirement of the draft Directive is likely to increase existing pressure on sludge disposal routes.

The use of sewage sludge in agriculture is generally beneficial, but applications must be carefully controlled to minimize the risks to workers and livestock, and the danger of contamination of the soil and groundwater by heavy metals and other chemicals. If the total additional sludge were transported for disposal in liquid form it is estimated that about 4 million l/year of fuel would be consumed, producing about 100 000 kg/year of carbon monoxide and 120 000 kg/year of nitrogen oxides.

The option of dewatering sludge and disposal to landfill results in increased landfill gas production and the possibility of increased contamination of leachate by ammonia and other chemicals. Incineration is increasingly being considered as a sludge disposal option at works serving large conurbations. However, the process has potential air-quality impacts in terms of emission of residual sewage sludge components and combustion products.

Secondary impact

The major secondary impacts from the installation of treatment plants would be increased energy usage at the plants and consumption of energy and raw materials and the production of waste in the manufacture of equipment. It is estimated that implementation of the minimum treatment requirement in the draft Directive would result in an additional 200 million kWh/year of electricity being consumed for treatment, leading to the emission of about 200 000 t/year of carbon dioxide.

POSSIBLE PRIORITIZATION OF INSTALLATION OF TREATMENT

The cost of implementing the minimum treatment requirement is estimated at £2.2 billion, and in order to spread the cost of installing treatment some phasing of expenditure will be required. This section outlines a simple rational basis for deciding on the prioritization of outfalls.

The costs and impacts resulting from the installation of treatment are dependent on (i) the nature of the receiving water, (ii) the site of the treatment works, and (iii) the sludge disposal route. With regard to receiving waters, net environmental impacts will be considered in relation to whether the receiving waters are (a) estuaries and enclosed bays with limited dilution and mixing, or (b) open coastal waters. Fig. 1 presents a simplified approach in the form of a flow chart for the assignment of priority to individual discharges.

Estuaries or enclosed bays

(i) Where sewage contains industrial discharges with high concentrations of heavy metals and complex organic compounds, the installation of treatment plant is assigned the highest priority in the chart (priority 1), since estuaries have been shown to be sensitive to pollution by these substances. This is due to many reasons including relatively poor dilution and dispersion, and use of estuaries as breeding grounds by a large number of vertebrate and invertebrate species.

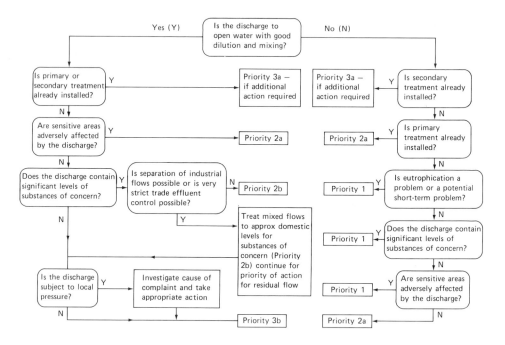

Fig. 1 Decision tree for priority assignment

(ii) Where eutrophication is a problem or potential problem, it is considered that the installation of treatment plant should be assigned priority 1 if phosphate is the limiting nutrient, since sewage can make a major contribution to phosphorus input. However, where nitrogen is the limiting nutrient, sewage discharges normally make a less significant contribution. Installation of treatment plant, while helpful, may not have a major effect in these circumstances.

(iii) Action at other estuarine discharges is assigned medium priority (2a) because of the more sensitive nature of the receiving waters.

Open coastal waters

(i) In general, lower priorities are assigned to discharges to open coastal waters through well-designed outfalls because of the greater assimilative capacities of these waters. Where largely domestic sewage is discharged via well-designed long outfalls in favourable circumstances, little marine environmental impact has been found beyond an immediate mixing zone of 50–100 m. Therefore little reduction in measurable marine environmental impact would occur if treatment plant was installed. In contrast, a clear impact on the land environment would result. These discharges are therefore assigned the lowest priority (3b).

(ii) When sewage discharged via long outfalls contains significant quantities of substances of concern from industrial inputs, the situation is more difficult. The effect of these substances is not well-defined, due to limitations of analytical techniques and the difficulty of collecting data; therefore the full assessment of marine environmental impact is not possible. Primary treatment would only reduce the concentrations of substances of concern by about 40%, while secondary treatment would effect greater levels of removal, but would also have greater adverse land impact. Action for these discharges is assigned medium priority (2b).

FUTURE APPROACHES TO ASSESSMENT OF NET ENVIRONMENTAL IMPACT

1. In general, most pollution control measures have a favourable impact in relation to some environmental media and a detrimental impact on others. In the case of sewage discharges to the marine environment, treatment will produce an improvement in the quality of the aqueous environment, while having a detrimental effect on the air and land. If the aim of environmental policy making is to minimize the environmental effects of essential human activities, then a means of comparing the cross-media impacts is required. At present this is limited to the use of subjective judgement, but the process might not be adequate to account fully for the complex nature of the problem.

2. It is considered that further studies, incorporating a technical environmental and an economic dimension, to assess the relative impacts in different media and then to evaluate the relative importance of the various factors would be valuable. The difficulty of evaluating weighting factors for all substances introduced to every class of environmental medium is recognized, but even a partial objective assessment of the net environmental impact would assist the decision-making process.

REFERENCES

[1] Colwell, R. R. (1988) Microbiological effects of ocean pollution. In: *Environmental Protection of the North Sea*. Newman, P. J., and Agg, A. R. (eds). Heineman Professional Publishing, 375–389.

[2] Goyal, S. M and Adams, W. N. (1984) Drug resistant bacteria in continental shelf sediments. *Appl. Environ. Microbiol.*, **48**, 861–862.

[3] Waldichuck, M. (1985) Sewage for ocean disposal. To treat or not to treat? *Mar. Poll. Bull.*, **16**, 41–43.

[4] Pike, E. (1990) Sea bathing – the search for seaside standards. *Water Bulletin*, **430**, 10–11.

[5] Gerlach, S. A. (1981) *Marine Pollution, Diagnosis and Therapy*. Springer-Verlag: Berlin.

[6] Raffaelli, D., Hull, S. and Milne, H. (1989) Long-term changes in nutrients, weed mats and shore birds in estuarine system. *Cah. Biol. Mar.*, **30**, 259–270.

[7] Ranwell, D. (1972) *Ecology of Salt Marshes and Sand Dunes.* Chapman and Hall: London.

[8] Law, R. J., Allchin, C. R. and Harwood, J. (1988) Concentration of organochlorine compounds in the blubber of seals from eastern and north-eastern England. *Mar. Poll. Bull.*, **20**, 110–115.

[9] Neff, J. (1979) *Polycyclic Aromatic Hydrocarbons in the Aquatic Environment, Sources, Fates and Biological Effects.* Applied Science Publishers Ltd, London.

[10] Green, J. and Trett, M. W. (1989) *The Fate and Effects of Oil in Freshwater.* Elsevier Science Publishers, London.

[11] Kayralla, N. and Jones, A. M. A survey of the benthos of the Tay Estuary. *Proc. of Royal Society of Edinburgh (B)*, (1974–75), **75**, 113–135.

[12] Read, P. A., Anderson, K. J., Matthews, J. E., Watson, P. G., Halliday, M. C. and Shiells, G. M. (1983) Effects of pollution on the benthos of the Firth of Forth. *Mar. Poll. Bull.*, **14**, (1), 12–16.

[13] Cooper, V. A. and Thompson, M. J. (1989) *Effects of Sea Outfalls on the Environment*, FWE Report DE 0031, Foundation for Water Research, Marlow.

[14] Newell, R. C., Newell, P. F. and Trett, M. W. (1990) Assessment of the impacts of liquid wastes on benthic communities. *Science for the Total Environment*, **56**, 1–13.

9

Hazard assessment for the Sha Tin Water Treatment Works

W. C. G. Ko, MSc, BSc, MICE, MHKIE, MIWEM, MIEAust, CEng[†]
and **W. S. Chan**, BSc, MICE, MHKIE, CEng[‡]

INTRODUCTION

In recent years, there has been a growing concern about the potential hazard to people living in the vicinity of installations storing and handling hazardous materials. In 1986, a Co-ordinating Commitee on Potentially Hazardous Installations (CCPHI) was formed within the Hong Kong Government. A potentially hazardous installation (PHI) is defined as an installation which stores materials in quantities exceeding the specified limit. For chlorine, the limit is 10 tonnes. The CCPHI also issued interim guidelines on acceptable individual risk and societal risk for hazard assessment of PHI as follows: (a) the maximum involuntary individual risk of death associated with incidents arising from PHI should not exceed 1×10^{-5} per year (1 in 100 000 /year); (b) the acceptability for societal risk associated with PHI is given in the form of a frequency/number of fatalities (FN) diagram in Fig. 1.

[†] Assistant Director, Water Supplies Department, Hong Kong Government.
[‡] Chief Engineer, Water Supplies Department, Hong Kong Government.

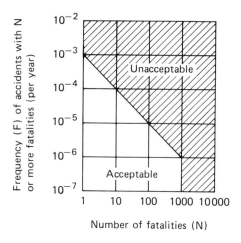

Fig. 1. Interim guideline for acceptable societal risks.

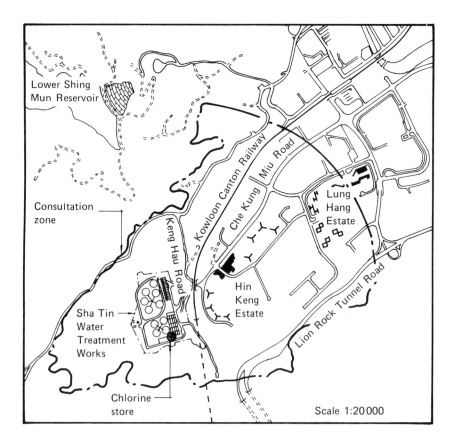

Fig. 2. Location of the Sha Tin Water Treatment Works.

SHA TIN WATER TREATMENT WORKS

Background

Sha Tin Water Treatment Works (STWTW) was first established in 1964 and was expanded in several stages. It is now the largest water treatment works in Hong Kong having an output of 1 227 000 cubic metres per day and is responsible for approximately 50% of the current potable water supply of the territory.

It is located about 30 metres above sea level at the head of Sha Tin Valley at the junction of Keng Hau Road and Che Kung Miu Road as shown on the map in Fig. 2. To the north, west and south are steep hills rising to about 300 m. To the east the land slopes downwards towards sea level and the Sha Tin New Town (population in 1987 about 415 000 and 1997 about 507 000). In close proximity to the works are low-density private housing and high-density public housing.

The chlorine storage system

Four bulk tanks of nominal capacity 125 tonnes each are employed to store liquid chlorine for the purpose of disinfection. The chlorine usage runs at about 2.3 tonnes per day.

Chlorine is delivered in 1 tonne drums to the site. Chlorine drums are unloaded into the new drum store, which can accomodate 10 drums. Transfer of chlorine from the drums into the bulk storage tanks is conducted one at a time using compressed air which takes about one hour per drum.

A range of safety measures are available and the more salient ones are:

(a) one bulk tank is always kept in reserve;
(b) the bulk tanks are fitted with bursting disc assemblies and a 12.5 tonnes pressure relief vessel;
(c) chlorine leak detectors and alarms are provided in all chlorine handling areas;
(d) although the capacity of each bulk storage tank is 125 tonnes, the maximum inventory is limited to 60 tonnes per tank;
(e) at least two people should be present when entering a chlorine store or whenever dealing with a chlorine leak or when working on any apparatus involving liquid or gaseous chlorine. Both people will be equipped with compressed air breathing apparatus, protective clothing and gloves with one person performing the work and the other monitoring and assisting should a difficulty arise;
(f) when dealing with a minor liquid leak from a drum, rotate the drum so that the leak becomes a gas leak and reduces the leakage;
(g) maintenance work of any kind requiring internal access to the bulk tank or breaking into pipelines must be undertaken only after a permit-to-work is obtained;
(h) the bulk tanks have to be examined and pressure tested at intervals of not greater than 26 months; and
(i) the chlorine plant emergency procedures should be followed when dealing with an emergency which may result in discharge of chlorine gas to outside the chemical house.

THE HAZARD ASSESSMENT

Objectives

Having chlorine storage in excess of 100 tonnes, STWTW is hence classified as a PHI. Accordingly, hazard assessment, planning study and action plans have to be made. Specialist consultants were employed to carry out a comparative hazard analysis, including the quantification of risk resulting from the use, storage and transport of chlorine, for the three options, which include bulk storage, on-site direct generation of gas/liquid chlorine, and drum storage; to recommend requirements for further improvement of the existing bulk storage system where the risk levels are high; to evaluate the consequences of such risks on the present and future development in the vicinity of STWTW; to provide a comparative hazard analysis of the various options for disposal of chlorine vapour under emergency conditions, including quantification of the reduction in risks achievable by these means and estimates of their costs; and to advise on the special planning considerations which should apply, and on the development controls necessary in the neighbourhood of the installation for the various options.

Individuals at risk

Two areas surrounding the works were identified to provide a reasonable indication of the level of individual risk in the risk assessment. They are:

(a) Person A lives in the lower floor of Hin Keng Estate about 350 m east-north-east of the STWTW chemical house. A spends 80% of the time at home.

(b) Person B lives at Lung Hang Estate about 1.1 km east-north-east of the STWTW chemical house. B spends 80% of the time at home.

Table 1. Distance from the chemical house of Sha Tin WTW

	0–250 m		250–500 m		500–750 m		750–1000 m		1000–2000 m	
Year:	87	97	87	97	87	97	87	97	87	97
Sector†										
0–30	—	—	20	20	60	50	100	100	3 600	3 850
30–60	—	—	145	2 237	1 635	2 195	4 642	4 987	30 800	39 450
60–90	—	—	20	3 175	100	2 730	—	1 638	1 450	1 800
90–300	—	—	—	—	—	—	—	—	—	—
300–330	—	—	2	2	—	—	—	—	—	—
330–360	—	—	6	6	—	—	78	81	600	350
Total	—	—	193	5 440	1 795	4 975	4 820	6 806	36 450	45 450

† 0 degree is the direction of North and 90 degrees the direction of East and so on.

Society at risk

Chlorine is a greenish-yellow gas with a vapour density of about 2.5 times that of air, and tends to disperse relatively close to the ground. The majority of the population in the vicinity of STWTW is housed in high rise blocks. Excluding the residents of the upper floors that are unlikely to be threatened by a chlorine release, the population at risk in the assessment of societal risks are given in Table 1 overleaf.

Risk assessment of bulk storage of chlorine

Cases for consideration

In assessing the risk associated with bulk storage of chlorine, the following four cases were considered: (a) failure of a drum to give 1 tonne release; (b) failure of a bulk storage tank to give 60 tonnes release; (c) failure of a line carrying liquid chlorine; and (d) failure of a line carrying gaseous chlorine.

Risk to an individual

The risk to an individual is assessed as follows.

Release	Risk of an individual becoming a fatality/year	
	A	B
1 tonne	1.2×10^{-6}	—
60 tonnes	3.7×10^{-6}	1.0×10^{-6}
0.8 kg/sec	0.2×10^{-6}	—
0.35 kg/sec	0.3×10^{-6}	—
0.2 kg/sec	—	—
Total	5.4×10^{-6}	1.0×10^{-6}

Societal risk

The societal risk is assessed as follows and shown in an FN diagram at Fig. 3.

Fatality	1987 Population	1997 Population
> 1	170×10^{-6}	180×10^{-6}
> 10	33×10^{-6}	130×10^{-6}
> 30	33×10^{-6}	130×10^{-6}
> 100	26×10^{-6}	39×10^{-6}
> 300	8.6×10^{-6}	23×10^{-6}
> 1000	7.2×10^{-6}	9×10^{-6}

Risk assessment of on-site generation of chlorine

Cases for consideration

In assuming a buffer store of 20 tonnes of liquid chlorine in four bulk tanks of 5 tonnes each, the following three cases were considered: (a) failure of a bulk tank to give 5 tonnes release; (b) failure of a line carrying liquid chlorine; and (c) failure of a line carrying gaseous chlorine.

Risk to an individual

The risk to an individual is assessed as follows.

Release	Risk of an individual becoming a fatality/year	
	A	B
5 tonnes	1.5×10^{-6}	0.1×10^{-6}
0.8 kg/sec	0.4×10^{-6}	—
0.2 kg/sec	—	—
Total	1.9×10^{-6}	0.1×10^{-6}

Societal risk

The societal risk is assessed as follows and shown in an FN diagram in Fig. 4.

Fatality	1987 Population	1997 Population
> 1	35×10^{-6}	35×10^{-6}
> 10	17×10^{-6}	20×10^{-6}
> 30	17×10^{-6}	19×10^{-6}
> 100	9.6×10^{-6}	19×10^{-6}
> 300	1.6×10^{-6}	10×10^{-6}
> 1000	1.6×10^{-6}	2×10^{-6}

Risk assessment of using 1 tonne chlorine drums

Cases for consideration

In assessing the risk associated with the use of 1 tonne chlorine drums, the following three cases were considered: (a) failure of a drum to give 1 tonne release; (b) failure of a line carrying liquid chlorine; and (c) failure of a line carrying gaseous chlorine.

Risk to an individual

The risk to an individual is assessed as follows.

Release	Risk of an individual becoming a fatality/year	
	A	B
1 tonne	1.0×10^{-6}	—
0.8 kg/sec	1.0×10^{-6}	—
0.35 kg/sec	2.7×10^{-6}	—
0.2 kg/sec	—	—
Total	4.7×10^{-6}	—

Societal risk

The societal risk is assessed as follows and shown in an FN diagram in Fig. 5.

Fatality	1987 population	1997 population
> 1	220×10^{-6}	230×10^{-6}
> 10	23×10^{-6}	140×10^{-6}
> 30	23×10^{-6}	140×10^{-6}
> 100	8×10^{-6}	83×10^{-6}
> 300	—	10×10^{-6}
> 1000	—	—

MITIGATION MEASURES

The 1-tonne uncontained release contributes substantially to the unacceptable societal risk of the chlorine drum option without using bulk tank. It was assessed that by converting the existing bulk storage to 1-tonne drums storage with building containment and installation of chlorine scrubber, the societal risk would be reduced to figures as follows and shown in an FN diagram in Fig. 6.

Reduced societal risks (per year) for drums
storage with contain-and-absorb system

Fatalities in	1997 population
1	14×10^{-6}
100	8×10^{-6}
300	1×10^{-6}

The societal risk for the 1997 population after provision of contain-and-absorb facilities will be acceptable. The mitigation measures for converting to 1-tonne drums storage is estimated at HK $29 million, which is much cheaper than providing an on-site generation plant.

RECOMMENDATIONS FOR FUTURE ACTION

Recommendations for future action were, inter alia:
- (a) in order to ensure no delay in reducing societal risk, funds should be immediately applied and design should immediately start on converting the storage of chlorine from bulk tanks to 1-tonne drums with contain-and-absorb facilities;
- (b) measures should be taken to review the possibility of reducing the quantity of chlorine in bulk storage tanks for the interim period;
- (c) in order to minimize the number of fatalities in the events of an accident, an effective emergency/evacuation plan should be drawn up; and
- (d) it is recommended that, before remedial works are actually completed, developments within 1000 m from Sha Tin Water Treatment Works should be subject to control.

UP-TO-DATE POSITION

A decision has been taken to construct a new building for new plant and storage facilities for 1-tonne drums with contain-and-absorb facilities to minimize risk due to potential chlorine leakage. As at October 1990, tenders for construction of the building were invited. The tender document for the mechanical and electrical plant contract is being finalized. It is estimated that the new 1-tonne drum storage and dosing plant will be commissioned by October 1992. The existing bulk storage tanks will be evacuated by December 1992 in time for the Hin Tin Development. As an interim measure, the chlorine inventory has been reduced to 25 tonnes in each of the three tanks in use. Together with the 40 tonnes in 1-tonne drums, total chlorine on site is limited to 115 tonnes.

CONCLUSIONS

Risk to an individual

The individual risks imposed on the closest resident at Hin Keng Estate, that is Individual A, from the three alternative options as given above are less than the acceptable risk of 1×10^{-5} per year. The conclusion is therefore that the resulting individual off-site risks from various chlorine options are not unacceptable.

Societal risk

As shown in the FN diagrams given in Figs 3, 4 and 5, the societal risks for the three options are not acceptable. It is also evident that 60 tonnes or 5 tonnes are responsible for large proportions of the risks. It infers that even if the existing 60 tonnes storage in bulk tank is reduced to 5 tonnes, the societal risk of the existing works will only be reduced to that for the second option which is still unacceptable. It is therefore considered that the only practical solution to reduce the societal risks is to discontinue use of bulk tanks for liquid chlorine storage.

On-site generation of chlorine is not a generally accepted method of chlorination at large potable water treatment plants for two main reasons. Firstly, it is more complex to operate than using bought-in chlorine, and secondly, it is not economical to generate small quantities of chlorine. Moreover, chlorine produced by on-site generation methods has to be liquefied for removal of impurities (e.g. hydrogen) and storage. At least several tonnes of liquefied chlorine will be required to maintain the balance of flow between the chlorine generation plant and the chlorine dosing plant. This gives rise to unacceptably high societal risk. Hence on-site generation of chlorine is not a practical option.

Though a technically feasible option, on-site generation of sodium hypochlorite is fraught with the uncertainty of a safe limit for chlorate in potable water and the lack of adequate toxicological and epidemiological information on the potential health risk associated with long term exposure to low level chlorate. Meanwhile, the World Health Organization and the United States Environmental Protection Agency are reviewing the chlorate issue alongside other disinfection by-products.

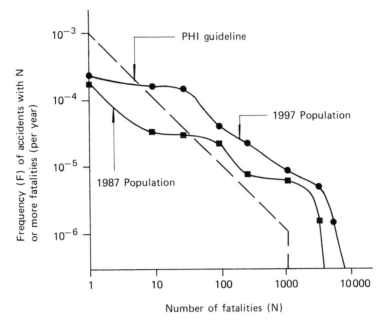

Fig. 3. Societal risks for bulk storage.

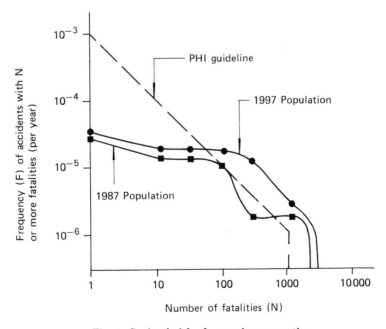

Fig. 4. Societal risks for on-site generation.

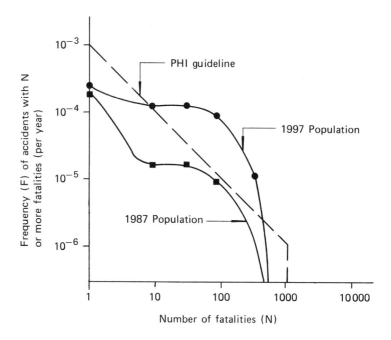

Fig. 5. Societal risks for 1-tonne drums.

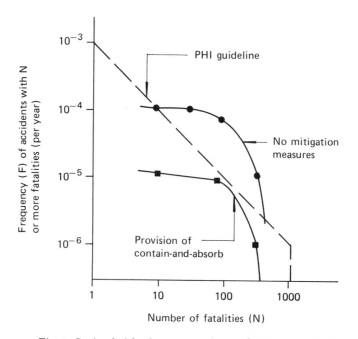

Fig. 6. Societal risks for 1-tonne drums (1997 population).

10

Environmental impact and a new water supply

D. C. Tye, Eur Eng, MICE, MIWEM[†]

INTRODUCTION

Section 11 of the Countryside Act 1968 requires every minister, government department and public body to 'have regard for desirability of conserving the natural beauty and amenity of the countryside'.

Section 48 of the Wildlife & Countryside Act 1981 required water undertakers to exercise their functions so as to 'further the conservation and enhancement of natural beauty and the conservation of flora, fauna and geological or physiographical features of special interest'. Also, they 'shall have regard to the desirability of protecting buildings or other objects of archaeological, architectural or historic interest and shall take into account any effect which the proposals would have on the beauty of, or amenity in, any rural or urban area or any such flora, fauna, features, buildings or objects'.

The Water Act 1989 also contains forms of protection of the countryside. Section 8 defines the general environmental and recreational duties of bodies, including those of the National Rivers Authority (NRA) and Section 9 requires the undertaker to consult with the relevant, notifiable body, such as the Nature Conservancy Council, if works are proposed within a site of special scientific interest (SSSI), or the National Park Authority if works are proposed within a National Park. A Code of Practice has been prepared by the DoE on conservation, access and recreation [1].

[†] Head of Development, Dynamco Limited.

Eastbourne Water has sought to fulfill what it has considered to be its duty as a responsible company for many years in respect of conservation. The implications of this policy can be many and significant and are portrayed in a description of a water supply scheme in a very attractive part of East Sussex, with a long industrial heritage.

THE COUNTRYSIDE AND THE ENVIRONMENT

It is no exaggeration to say that nearly everyone has an opinion on the countryside and the environment: a recent Gallup survey [2] demonstrated that 79% of those contacted considered that 'the countryside is in danger'. Construction work is perceived by 76% as constituting a threat, and industry of all types was seen to be by the respondents to be a source of danger.

Attempts at providing environmental protection in England have included the statutory designation of areas of land as national parks, sites of special scientific interest (SSSI), national and local nature reserves and environmentally sensitive areas. Formal recognition of the material importance of landscape character is given by the designation of areas of outstanding natural beauty (AONB). More recently, environmental issues have been given a monetary value in the Pearce Report [3]. Certainly, the requirements contained within the Town & Country Planning (Assessment and Environmental Affects) Regulations [4], will necessitate far greater awareness by all potential developers, including the water industry, of the need to consider the detailed implications of major developments to their surroundings. Guidance to the Regulations is given in the 1989 DoE document [5].

It is recognized that East Sussex contains many areas of high quality environment, which contributes to the intrinsic quality of life in the area. Two AONBs have been designated, the High Weald [6] and the South Downs. Major new developments, however, planned north of Eastbourne, Bexhill and Hastings, will require the usual infrastructure, including those for water services.

THE COUNTRYSIDE AND INDUSTRY

Like many landscapes, the area of the High Weald, which lies to the north of Eastbourne, Bexhill and Hastings, has been largely fashioned by man. One major influence in this particular section has been the iron industry. The presence of iron ore in the sandstone, large forests and plentiful streams led to the establishment of high technology, firstly in the 14th century and then, in the 15th century, with the introduction of the blast furnace process. Vast areas of woods were coppiced, dams built to form hammer ponds, and furnaces and forges constructed. This industry functioned here until late in the 18th century.

The water industry is now the one that has been bringing the latest technology to the area and some of this work will further change the environment: conservation, however, is now the key word for the remainder of this century.

NEED FOR WATER RESOURCES

Eastbourne Water Company (EW) operates the licence to support potable water for some 215 000 resident customers within the East Sussex County area, which has a present population of some 670 000. The forecast of future population in the Company's area is 277 000 at year 2021.

These increases are by no means spectacular, nor are the actual volumes involved even remotely large. However modest they are, however, they are most significant in that they demand the location and development of new water resources. Moreover, these new resources need to be developed within a part of East Sussex (Fig. 1) that is extensively protected by the establishment of AONB, SSSI and nature reserves, and engineering work, even of a modest nature, can be destructive in this sensitive environment with any adverse results remaining evident for many years.

Poor communication has kept many parts of the area in a virtual preserved state and the character of the countryside remains small-scale, with much interest being very local. In these situations, the utmost care and sensitivity requires to be exercised and an understanding achieved of how best to identify a balanced solution for the provision of the engineering services.

Fig. 1. Location and map of study area.

WATER RESOURCE INVESTIGATIONS AND DEVELOPMENT

In 1976 EW decided, in view of the need for increased resources from 1980 onwards, to carry out comprehensive investigations into the possible availability of new resources, both groundwater and surface water.

The agreed recommendation, jointly taken in 1983, was for EW to develop additional ground water resources and to co-operate with SWA in deriving that supply from an existing reservoir at Darwell, north of Bexhill.

The additional groundwater resources in the Wallers Haven have now been developed and are now operating. The Darwell Reservoir enlargement and transfer proposals are now being actively promoted by Southern Water Services Limited (SWS) and Eastbourne Water Company (EW) and are expected to be affording increased supplies, to both SWS and to EW, from 1996 onwards; a diagrammatic representation of these sources is show in Fig. 2.

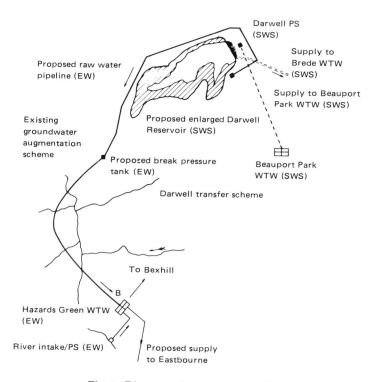

Fig. 2. Diagram of water supply scheme.

Wallers Haven groundwater augmentation

This section of the overall scheme was constructed between 1983 and 1987. However, the investigation into the appropriate form of the development was extensive and considered in detail the appropriate form of design for the particular environment involved [7]. That investigation included further hydrological and hydrogeological work, together with stream quality surveys and water treatment trials [8]. The investigation results were employed to decide whether that water should be transmitted to the water treatment works downstream, at Hazards Green, via the streams or by a pipeline. Of particular concern, was the probably adverse impact on the local area of pipeline construction: these pipelines would have been of relatively small size, with a maximum nominal bore of 400 mm, but they would have been numerous in that they had to connect with seven well sites at different locations.

Historically, three of the sites had been operated in the river augmentation mode since 1954, by the Bexhill Water Corporation, and that use had been accepted over the years by the local community.

The proposal for the river augmentation mode of transfer suffered, however, from two major problems: the groundwater is naturally very rich in iron and manganese and the strata in which the streams flow leak, with the result that the net yield derived can be significantly lower. These problems were answered by the decision to remove some 80–90% of the iron before discharge to the streams and by accepting that the net yield would be no higher than 70%. The iron removal plants selected, following extensive pilot process tests on site, comprise aeration and dissolved air flotation stages (DAF) with sodium hydroxide as the coagulant.

Future Darwell transfer scheme

The installations at the downstream river intake and the treatment plant have been designed for extensions, in stages through to 2021, to handle and treat a maximum flow of 54 000 cubic metres/day. Detailed environmental studies on this transfer also examined the option of river augmentation. It is interesting to note that the Environmental Assessment Directive does not specifically require an environmental assessment (EA) for the development of a water treatment plant.

PLANNING AND ENVIRONMENTAL ASSESSMENT

The Wallers Haven groundwater investigations and developments were planned and implemented before the Environmental Assessment Directive was published. However, consultation was always considered very necessary by Eastbourne Water and was exercised widely. The current Darwell Transfer proposal, however, is being promoted against the background of much-heightened environmentally concerned views and new legislation, and a formal environmental assessment has been completed.

The Directive includes classes of work as requiring specific attention (and which involve the water industry) and for which an environmental statement (ES) may de required to be prepared; it is the planning authority (PA) who actually decides how to interpret the Directive and, if a formal ES is required, what the ES shall include in detail. The National Rivers Authority (NRA) features as a formal consultee in this process.

In the Wallers Haven catchment area, the planning and implementation of the resources investigation and the construction of new installations took cognizance of the rich industrial, architectural and landscape value of the Ashburnham and Penhurst parishes. Of particular significance is the existence of the Ashburnham Estate and Church, near Battle, which were constructed in the 1600s [9]. The lanes here are very narrow and are located along undulating, wooded ridges, and the very limited access has largely helped to conserve the countryside, even though the area lies close to the coastal towns and to the A21 and A22 London roads. One of the ways in which this landscape has been moulded by man, has been the work of Lancelot Brown, who completed commissions for the Earl of Ashburnham in the 1700s [10]. His designs included several dams, including one 17 metres high and which well illustrate his working philosophy for 'properly engineering in the countryside'. It is a matter of conjecture how Brown would have viewed our present environmental concern and values.

Table 1. Consultation and environmental studies for Wallers Haven Scheme

Development	Description	Consultation and studies							
		Land-scape	Water quality	LAs	PCs	Owners	Archeo-logical	Flora/ Wildlife	Fauna
Resources investigation									
●Ground-water	Studies of Chalk Ashdown Sands	–	*	*	*	*	–	–	
●Surface Water	Studies of four possible reservoir sites	*	–	*	*	*	–	–	–
Groundwater augmentation Works development	Installation of four well sitess	–	*	*	*	*	–	*	–
●Phase 1	New river intake and pumping station	–	*	*	–	*	–	–	–
●Phase 2	WTW extension	–	*	*	–	–	–	–	–
●Phase 3 ●Phase 4	Planned WTW extension	*	*	*	*	–	–	*	*
Darwell Transfer	Transmission of bulk supply from Darwell to WTW	*	*	*	*	*	*	*	*
Pipeline supply to Eastbourne	New trunk main from WTW	–	*	*	*	*	*	*	*

An important feature of EAs can be the requirement to include an adequate demonstration that all practical alternatives to the selected development have been examined to a required depth and found less suitable.In the case of the Darwell Transfer proposal, these wider considerations have proved problematical. Detailed consultation with the NRA are, of course, indispensable and are vital, both in respect of the negotiations for licences and consents and in respect of the Authority's obligations under the Act. Final results of the survey work are now available and it is interesting to note that, where extensive water quality and flora/fauna field surveys are required, the planning and timing of this work can prove critical. It is patently clear that without the detailed work now being prepared for the EA, fundamental assumptions could have been made prematurely, and could have resulted in unsatisfactory decisions being made.

Table 1 shows the extent of the consultation programme completed, or underway, for the Wallers Haven Scheme.

ENGINEERING AND THE COUNTRYSIDE

It can be argued that the environmental impact of development by the water industry in the countryside is, very largely, relatively low-key, when compared with other industries' activities. For example, a new clay extraction working or brickworks or a local village bypass will almost certainly create more direct and indirect disturbance and loss of wildlife habitat than most water supply schemes.

However, the environmental impact is usually perceived, not unreasonably, by local people as anything that affects their daily lives and habits or detracts from the enjoyment, whether it is fishing or walking or simply a pleasing, familiar view.

One way that engineers can assist in conserving the countryside is to make an early decision to talk to the local communities and discuss with them alternative ways to achieve the overall aims.

It may be that in the very early, pre-feasibility stage, a new fundamental option may develop or it may be that, in the more detailed approval stage, designs may be identified that could enhance, or further conservation could yield, recreational opportunities.

Examples of ways in which the engineering has been influenced, to date, are given to illustrate this particular approach.

General location and siting

It is suggested that the advice of a landscape architect is obtained, even for modest developments, very early in the life of the project. This advice will help in making a decision on how best to select the location for a new installation, such as whether it is close to existing tree lines or hedges, or whether new planting and landscaping is preferable.

Noise and disturbance

In these very quiet, very natural surroundings, even a small civil construction site creates significant disturbance, although the nearest residents may be several kilometres away. Indeed, it is the very quietness and lack of unnatural activity that constructs the intrinsic value of this particular countryside.

Consideration should therefore be given, in the preparation of construction contracts, to the question of access, type and frequency of vehicles and, most importantly, that of noise.

Well drilling is often completed on a 24-hour shift basis and the provision of floodlighting can cause a great deal of disturbance to the wildlife. Temporary diesel generator sets are another source of disturbance and are probably best provided as packaged units, which already provide for the necessary attenuation to ventilation inlet and outlets.

Building design

Considerable research was undertaken on the local building styles in order to select a suitable structural form and elevational treatment. The vernacular architecture thought to be most appropriate for EW was that of the Sussex outfarm, where historically a small group of buildings provided some accommodation for crops and animals, away from the farmstead. The buildings were usually of brick, timber and slate. For the new buildings required, a timber-frame structure was selected, with part brick and part timber cladding, with roof designs of gable, hip and half-hip ends. Roof tiles were either of handmade clay or of synthetic slate. No windows were included, only intruder-protected ventilators, but a limited amount of roof lighting was included. A typical example of one of these new buildings is shown in Fig. 3.

Fig. 3. New intake building on the Waller's Haven River.

A conscious effort was made to restrain the choice of finishes, with the timber stained black and any paint colours used externally of dark green or olive tone, with a matt finish.

Access roads

Both the local communities and planning authorities were anxious to influence the design of the access roads to new installations within the area. For the road to the new river intake and pumping station, the planning authority stipulated that no road embankment was provided and that the road surface itself was not constructed in concrete. In order to answer the planning authority, EW decided to employ the use of grasscrete plastic formers, which, although still producing a concrete roadslab, provide for a honeycombed structure, through which grass could grow.

Fencing

No attempt was made to protect these rural sites with security fencing. Instead, either post and rail or post and wire fencing was employed throughout, as this was less obtrusive and very much more natural to the environment. Where security fencing was required, as at the treatment plant site, the preference was for the fencing to be of the steel mesh and post variety, with the inclusion of as few structural members as possible, so as to enable the eye, from a distance, to 'look through the fence'.

Pipelines

The most difficult aspect of some water schemes can be the planning and installation of pipelines: certainly, they can create one of the most emotive areas of negotiations.

The landscape of the High Weald presented one of the most difficult tasks to date to achieve satisfactorily. The Hastings Beds strata is of very varied nature, with large sections of rock outcrop; the areas of Tunbridge Wells Sands can either be very well drained, or can give very wet, alluvial type conditions, with sections of rock outcrop; others give mudstones and siltstones, with perched water tables. The terrain can include very steep, forested valleys and natural and planted woodlands; faulting is extensive and slopes have been subject to solifluction.

The recent Code of Practice on Conservation, Access and Recreation requires that pipeline routes should be adjusted to avoid damage to the landscape, conservation sites and features of archaeological, architectural and historic interest. An example is the selection of the route for the EW Darwell Transfer works, where the pipeline route has been adjusted to avoid two SSSIs and the Ashburnham Furnace, where a hammer pond still exists on the site of the Forge and Furnace which operated from the 1600s to the 1800s [11].

Streams and watercourses

Where groundwater abstraction to the streams is carried out, EW has an obligation to the landowners to maintain those sections of watercourse, upstream of the land drainage areas and the NRA-maintained sections. This maintenance work can prove to be most damaging to the immediate environment, where, for example, sections of gabion wall have to be introduced to reinforce the banks and provide for early establishment of vegetation.

CONCLUSIONS

The philosophy adopted in planning and implementing a new water supply development within an area of high scenic and historic value has been discussed. The depth and scope of the environmental studies judged necessary were considered extensive at the time of the resource invesigation programme, but the current national dialogue and the ever-increasing amount of statutory directives will mean that a wider view will need to be adopted in future, even for some modest projects, such as groundwater developments and pipeline construction.

It is recommended that some form of environmental assessment is considered as an integral part of water industry projects, as the cost is relatively minor, but the potential for achieving the best social and engineering solution is high.

It is considered vital that the studies are planned and implemented as an intrinsic part of the overall feasibility and not regarded as some necessary evil, the result of which would add a fashionable gloss to proposals already mentally consolidated. It is equally vital that the appropriate level of expertise required is identified early and that the environmental advisers are valued as essential members of the team.

The local community and their rural landscape is under ever-increasing pressure and environmental assessment can assist in a most positive manner in creating the best balance of needs, often including enhanced consultation and greater understanding with the industry's customers.

ACKNOWLEDGEMENTS

The author wishes to acknowledge his gratitude to Mr G. Hoskins, Executive Chairman, Eastbourne Water Company, for his permission to prepare the chapter.

The views given in this chapter are those of the author and do not necessarily coincide with those of Eastbourne Water.

REFERENCES

[1] *Code of Practice on Conservation, Access & Recreation*, The Water Act 1989, Her Majesty's Stationery Office (1989).

[2] Daily Telegraph, Gallup Poll (1989).

[3] Prof. D. Pearce (1989).

[4] Town & Country Planning (Assessment of Environmental Effects) Regulations (SI 1199), HMSO (1988).

[5] Department of the Environment,(1989) *Environment assessment: a guide to the procedures.*

[6] County Councils of East Sussex, West Sussex, Kent & Surrey, (1988) *High Weald, area of outstanding natural beauty, statement of intent*, First Revision.

[7] Montgomery, H. A. C. (1982) Use of ferruginous groundwaters to augment streamflows, *J. Inst Wat. Engs. Sci.* 1984

[8] Barnhoorn H. T. and Tye D. C. (1984) Treatment of ferruginous groundwaters for river augmentation in the Wallers Haven. *J. Inst Wat. Engs. Sci* 1984

[9] Sowden D. and Starkey D. (1985) *This land of England.* Muller, Bond & White Ltd, London.

[10] Hind T. (1986) *Capability Brown.* Century Hutchinson Ltd, London.

[11] Scholes R. (1985) *Understanding the countryside.* Billings and Sons Ltd, Worcester.

11

Water engineering aspects of tidal power schemes

T. L. Shaw, BSc (Eng), PhD, CEng, FICE, FASCE, FIWEM[†]

INTRODUCTION

The axiom that 'energy and the environment are in conflict' is repeatedly confirmed by experience. Much of the present concern about the side effects of power generation relates to the projected consequences of burning fossil fuels. However, the axiom also applies to renewable energy sources.

Of these sources, the most readily available in the UK on the scale of national energy needs is tidal power. As a form of conventional hydro-power, this source is technically well proven, accurately predictable and, as it happens, geographically well placed (Severn, Mersey, Morecambe, Solway) to meet demands.

This chapter considers those aspects of the design and operation of tidal power schemes that bear on water engineering interests. The list is almost as long as the remit of this discipline. The attention given here to each subject must therefore be brief. Common points between different barrage schemes will be considered rather than issues of specific interest.

HYDRODYNAMICS

The primary requirement that tidal energy barrages are sited in estuaries with high tide ranges means that they are likely to be in ocean-facing locations, though

---

[†] Managing Director, Shawater Limited.

the effects of coastline geometry and offshore bathymetry can produce anomolous conditions.

Although the preferred UK sites are all on the west coast, they exclude the long frontage of much of Scotland where tides are relatively small. The largest tides occur in the Severn Estuary, followed by those along the coast of the Irish Sea. Cornwall is also quite favourable, and the range increases sharply on the south side of the English Channel along the Brittany and Normandy coastline where the pioneering barrage on the Rance Estuary is now in its 25th year of successful operation.

One criterion for tidal magnification is shallowing water. Taken together with the fact that these estuaries are long and have converging coastlines, this usually means that the large volume of water which passes into and out of the estuary to create the tidal rise and fall gives strong currents, these being closely allied in direction to that of the estuary's axis. The rates of exchange of water along the estuary, however, tend to be low, especially in the long and relatively narrow estuaries which typify many high tide sites.

Fig. 1 shows how the Rance Barrage affects the tide range in its basin. The principal change is that the reduced range occupies the upper half of the former range, the latter essentially continuing to occur seaward of the Barrage. Fig. 1 also shows how high water levels in the Rance basin are increased by using the turbines as pumps around the time of high water to add to the volume entering the basin. It is now generally accepted that this is also the preferred method of operating other possible schemes. Its effects on water levels (and consequential flow velocities) within and seaward of the basin relative to existing levels should therefore be noted from Fig. 1.

River flows influence currents in the upper reaches of estuaries but this effect tends to be small for many of the larger tidal power sites in the UK, for which tidal flow rates much exceed river flow rates, even during times of flood.

The tides are therefore the principal cause of advection but even though transport distances per half tide cycle are large, the net seawards transport per full cycle is small.

Wave action in estuaries tends to be determined more by local winds than by ocean waves penetrating in because of the constraint of limited water depths on long-wave propagation without serious dissipation. The shape of estuaries influences the opportunity for wave action to develop in response to winds from any direction, and their limited size ensures that wave characteristics depend more on wind speed than duration. Since limited available fetch means that wavelengths are short, their main influence on sediment movement occurs in shallow water, breaking then occurring close to the water's edge.

A barrage will change the fetch for some wind directions, consequently altering coastal wave climates. The significance of this change is considered below.

In general terms it may therefore be concluded that, in estuaries with high tide ranges, tidal currents form the principal driving mechanism in deeper water, whereas wave action exerts significant control over conditions on the more exposed coastlines. This conclusion is more relevant to conditions in the main body of estuaries rather than in the upper reaches and in any tributary river estuaries, and it is also less valid towards the seaward end of estuaries where a more maritime regime prevails.

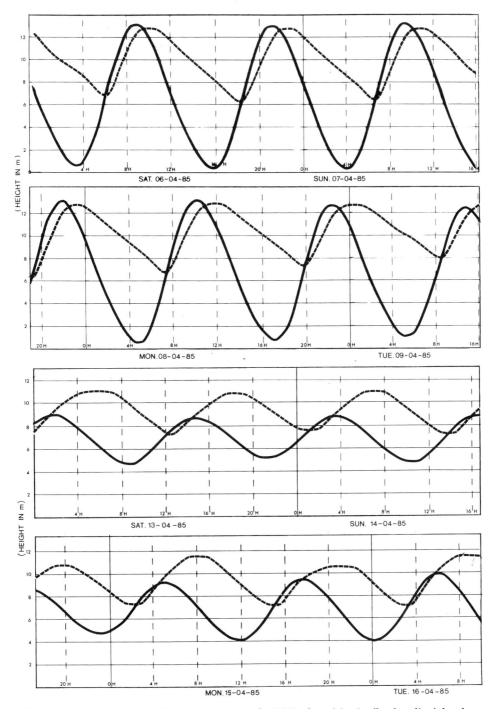

Fig. 1. Rance Barrage: spring and neap sea (solid line) and basin (broken line) levels.

SEDIMENTS

Strong currents make even relatively mud-free estuaries seem turbid, for example the Severn Estuary and Solway Firth. In contrast the Mersey Estuary, for example, more closely resembles many muddy, lower tide range estuaries, for which extensive deposits of intertidal and sub-tidal fine sediments are a prominent feature.

Deposits of sand and exposed bedrock are generally uncommon in estuaries but are not unusual where tide ranges are high. Strong currents give little opportunity for fine material to settle, consolidate and progressively accrete in other than sheltered coastal and protected deep-water locations. Such locations tend not to remain active for long, filling up and then ceasing to provide new storage capacity. The approach channels to some port entrances cut in previously relatively stable deposits show how quickly accretion can occur, making regular maintenance dredging necessary.

High tide range estuaries therefore tend to have mainly sandy or rocky bottoms with peripheral deposits of fine material. Much of these sediments is mobile in response either to the currents in deeper water or to wave action along the coastal margins.

The size and geometry of high tide range estuaries, in particular the fact that they must be relatively shallow to respond actively to tidal forcing, means that their features would have changed appreciably as mean sea level rose since the last Ice Age. Those former alluvial river valleys which, when invaded by the rising sea, allowed the lengths of estuaries to increase substantially provided an opportunity for the formation there of high tides. Superficial fine sediments stripped off former flood plains were in part rolled back to feed salt marshes but were also carried seaward in suspension out of estuaries. The amounts of suspended material per tide cycle, that is suspended concentrations, needed to achieve this seaward loss are suprisingly small and difficult to detect.

The effects of sea wall construction on this erosion process and its consequences are fundamental to the regimes of these estuaries. Containing the tide within a fixed area deprives the estuary of one of its effectively renewable sources of fine sediments. However, since the erosive capacity of waves and currents remains, a fundamental change in coastal zone topography and the general availability of fine sediments to supply various processes will be affected. Beach profiles become more concave as lower deposits of sediment are eroded within the artificial boundary presented by the sea wall. The upper beach profiles will therefore tend to steepen, especially on the more exposed coastlines, leading to the loss of saltmarsh and the sustained seaward movement of fine material.

High tide range estuaries, like the Severn whose perimeters have been formalized by walls, display these characteristic features. Although rates of erosion of historic alluvial silts are small, the process is progressive, especially where the more damaging wave regimes are most prevalent. While continuing inputs of river-borne material lessen the effects of this change, they do not normally play a dominant part in the fine sediment dynamics of high tide estuaries, and will therefore do little more than moderate the changes to that behaviour caused by sea-wall construction. These changes can be extensive, as shown below.

The moderating effects of tidal power barrages on currents and, to a lesser extent, on wave action have been mentioned. The reduced tide range will focus wave action into a narrower intertidal zone though there is some doubt as to how each beach level throughout this zone responds to waves because of the duration of and way in which waves act at each level.

However, the fact that the areal extent of stronger currents would be much reduced will encourage the permanent deposition of fine sediments offshore, reducing seaward transport rates. The ramification of this for turbidity structure and other aspects of ecosystems are considered later.

DREDGING

For economic reasons, the large size of high tide range estuaries usually precludes the removal of dredged material seawards beyond the zone within which the currents are dominated by that estuary's tidal behaviour. In practice, the strong mixing which occurs in estuaries makes it possible to dump dredgings within a short distance to one side of the artificial channel from which they were taken, the tidal currents ensuring that this source is unlikely to add greatly to the subsequent accretion of dredged channels.

However, the fact that accretion still occurs confirms the ability of estuaries to sustain stable geometries, even when the sediment resources needed for this are limited .

It seems reasonable to conclude that the reduced amounts of sediment movement offshore (both fine and coarse) following barrage construction would lessen the need for maintenance dredging. Furthermore, material dumped as now in close proximity to, but away from, the main streams through artificial channels is then less likely to be recycled.

LAND DRAINAGE

It is difficult to discuss the effects of tidal power barrages on land drainage without reference to the use to which the affected area is put and its elevation relative to the limits of neap and spring tide cycles. The differing demands of the urban sector, agriculture and nature conservation on land drainage practice are well known and, within limits, often capable of being reconciled at a price by arranging for the drainage of each area to be suitably managed. Furthermore, in the case of tidal power barrages, the cost of providing and operating the necessary drainage infrastructure is usually small compared with the financial size of the whole project, for the following reasons.

(a) The presence of the barrage and how it is most economically used as an energy source means that higher water levels both landward *and* seaward of it will be reduced following construction. Although this change is more relevant to sea defences than to land drainage, it reduces the cost and consequences of providing sufficient storage to accommodate tide-locked periods, and reduces the maximum lift required for drainage pumps.

(b) The raised low tide levels in the basin (Fig. 1) will mean that a proportion of the land area previously drained by gravity could not be treated in that way. The provision of pumping to deal with that situation is not unusual, and the area so affected is normally only a small part of that area whose free drainage is prevented by tidal movement in the pre-barrage regime. The reason for this relates to the natural shape of drainage basins, the effect of sea-wall construction, mean sea level rise and the dynamic behaviour of a high tide regime on drainage behaviour.

(c) Incorporating a pumping element into the power generating process increases highest water levels in the basin area, but reduces what is sometimes regarded as a high water stand, by causing a sharper peak level in the tidal cycle (Fig. 1). The shape and the peak achieved are both important factors in determining the tidal storage volumes required by any land drainage schemes.

These effects do not apply seaward of the barrage where virtually the full tidal amplitude remains, any small rise in low tide level being least significant on neap tides when gravity drainage is most sensitive to the available duration of free discharge.

COAST PROTECTION/SEA DEFENCES

For the reasons stated above, a tidal power barrage will have little direct effect on the general coastal regime on its seaward side, and some betterment is possible due to the reduction in currents. Both here and landward of the barrage, the most significant effect of the project is likely to relate to how it alters the process of sediment supply to and removal from the coastal zone. Since the most important mechanism for this is usually wave action, it is necessary to establish how the presence of the barrage will affect wave generation by winds from different directions according to wind speed and its frequency of recurrence.

It is unusual for the coastal regime of estuaries to be materially affected by the action of long ocean-generated waves, especially in the relatively shallow conditions often associated with the creation of high tide ranges. The most significant changes will occur on those coastlines where the shortening of wind fetches caused by the barrage alters the incident wave regime. While it may reasonably be assumed that these changes are more likely to occur close to barrages, anomalies cannot be ruled out. Furthermore, as pointed out in the section on sediments, the supply of replacement material to coastal zones following its loss due to wave action is, in the case of fine sediments, more likely to be due to the action of currents than to waves, though both forces affect sand transport.

Another possibly significant effect of the reduction in tide range landward of the barrage is that the water line will persist at each part of the beach for longer than before the barrage was constructed. Wave action will therefore be more concentrated, especially along coastlines exposed to waves whose fetch is unaffected by the barrage. However, since the reduced tide range is still likely to be more than the average natural range around the UK coastline, the effect of range on wave-induced erosion of various sediment types, including the well-consolidated silts which comprise much of the intertidal area of high tide range estuaries can be established by field observations, if similar conditions occur elsewhere.

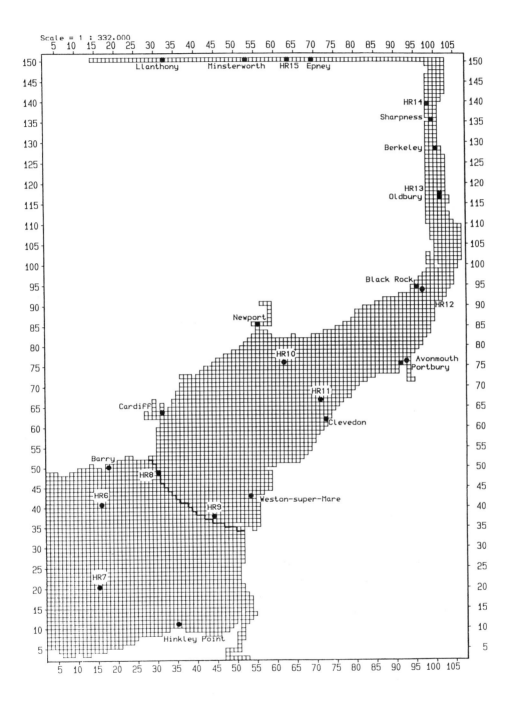

Fig. 2. Location plan of Severn Estuary model.

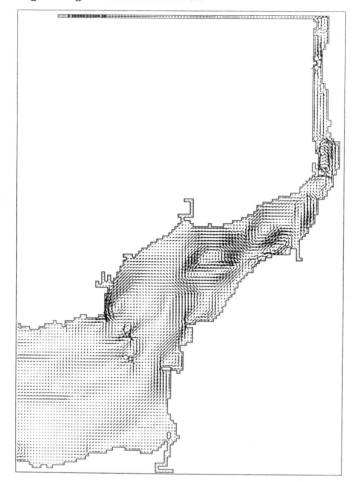

Fig. 3a. Residual velocities, no barrage, mean spring tide.

WATER PARAMETERS

In order to assess the effects of tidal power barrages on the environment of estuaries
it is necessary to know the physical and chemical parameters which the water regime
will change. Earlier sections have considered the physical regime. Principal amongst
the chemical factors are probably salinity, dissolved oxygen and nutrients.

To appreciate how these parameters will change it is first necessary to establish
how water circulations will be affected, and in particular the rates of exchange
of water between different sections of the estuary. It was noted earlier that the
strong tidal flows in high tide range estuaries are relatively closely allied to these
estuaries' axes. This feature of the tide will be substantially changed by the barrage,
especially close to it, due to its design. In particular, much of the incoming flow
passes landward through the sluices, and the outgoing flow is discharged through
the turbines. Barrage operation therefore adds a gyratory component to a basically
linear oscillation.

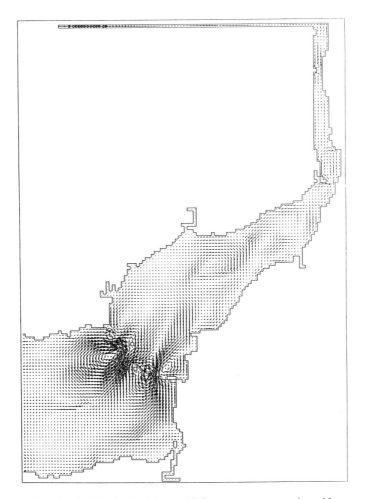

Fig. 3b. Residual velocities, with barrage, mean spring tide.

If all the sluices are together in one section of the barrage and all the turbines are in another, the imposed water rotation will have one axis. However, since it is usually necessary to site the turbines in deeper water than the sluices, two or more rotations may be set up depending on the distribution of water depth along the barrage line.

Fig. 2 shows the extent of the Severn Estuary model, and Fig. 3 shows how the naturally very weak residual flows in the Severn Estuary would be affected by the barrage scheme proposed for the site shown. In that case, the water depth needed for the turbines occurs towards the center of the estuary and there is adequate depth toward each coastline for the sluices. The resultant effect on the currents is self-evident. In particular, the waters for some distance on either side of the barrage will mix much more quickly than in the natural regime, reducing the gradients of salinity and temperature and assisting the movement of larvae and migratory fish. Dissolved oxygen levels relate closely to the demands of suspended organic sediments. It was noted earlier that apart from accretion of fine material along coastlines when wave action is slight, the movement of this material is concentrated into the river estuaries. These act as tributaries to the main estuary and have the highest concentration of suspended solids. The area over which these very high concentrations occur would be reduced following barrage construction, lessening the total demand for oxygen and raising concentrations accordingly.

WATER QUALITY

The steady improvements being made to the quality of UK estuaries against tightening standards makes it difficult to anticipate the conditions which will prevail in any estuary at such time as a tidal power barrage may be constructed. What is certain, however, is that it will be necessary to ensure that the regime created by the barrage shall be no less acceptable than that which would have prevailed in its absence. Maintaining the status quo shall therefore be the minimum target for the project, any improvements being a bonus.

Reference was made earlier to the reduction in peak turbidity levels which a barrage would cause, also to increased rates of exchange of water and the contaminants carried along the estuary, and to reduced oxygen demand in some of the more sensitive locations. Each of these processes will benefit water quality in some way.

For example, contaminants may be divided into those which degrade and those which persist. The process of bacterial decay is one of the most important and demanding in coastal waters but it is made less effective in estuaries where turbidity limits sunlight penetration. Estuaries with high turbidity have very much longer decay times compared with the clear water of open coastlines. The problem is particularly significant in up-estuary locations. These are likely to be well landward of a barrage, hence their turbidity levels will be reduced with resulting benefits for bacterial decay rates.

Conservative (non-degradable) substances, particularly those which go into solution, move with the water mass. When influenced by flows caused by barrage operation, some quickening in dilution rates may be expected depending on the layout of turbines and sluices along the structure and the geometry of the estuary in this reach.

However, it does not follow that near-field pollutant levels will also be reduced. Depending on the location of outfalls relative to the barrage, it is possible for local effluent concentrations to be higher following barrage construction. This is most likely to apply in the upper part of river estuaries, though any adverse trend would be offset should pumping be incorporated into barrage operation, increasing the tide range and hence the currents in these river estuaries.

Although the practice of discharging within the intertidal zone is discouraged, it is likely that after barrage construction most of these outfalls will be subtidal. Early dilution rates will therefore be much increased, with resulting improvement in coastal water quality. The task of containing accidental discharges of oil would be eased, especially if the barrage operation was able to be adjusted accordingly.

These generalized statements of the directions in which important parameters determining the water quality of estuaries are likely to move following barrage construction are not a substitute for the careful investigation of individual outfall conditions. The opportunities which may exist for positive enhancement of the regime also merit serious attention, as shown below.

ECOLOGICAL CONSIDERATIONS

The physical and chemical characteristics of an estuary exert a substantial influence on its ecosystems. High tide range estuaries therefore tend to have different features from the more normal regimes of most estuaries, towards which the ecology of the former is likely to move following barrage construction. It follows that barrages will tend to moderate the more extreme ecological features of high tide range estuaries, probably creating more productive conditions because of the increase in light penetration referred to above.

The biological complexity of estuaries and the relative difficulty of carrying out research in them has meant that they are not as well understood as riverine and coastal conditions. However, much comprehensive work has been done in recent years as part of the national research programme into tidal energy barrages and this work is continuing. Although each estuary has its important and unique features, it has been possible to benefit from generic studies supplemented by estuary-specific work to gain a much clearer understanding of the complexities of these regimes.

It is not possible in the limited space available here to do justice to all this work and its outcome. One important consequence is that it has shown that earlier speculation of the adverse effects of tidal power barrages on the environment seems to be largely unjustified, though the extent to which this lesson applies is not the same for all projects. Moreover, it has become increasingly apparent that there is no case for adopting a passive stance to ecological change. The available evidence firmly shows that there are many ways in which the less welcome effects of barrages may be countered by informed design measures, and the opportunity for enhancement by creating new environments deserves very careful consideration.

Any reduction of the unusual ecological features of high tide range estuaries may therefore be lessened and countered by a creative approach to conservation , for which a comprehensive knowledge of the relevant ecosystems is a prerequisite to constructive planning and design.

CONCLUSIONS – WATER MANAGEMENT

An outline of many of the aspects of water management relevant to tidal power barrages has been given. Much has been learned about these sensitive projects during the past few years and the tempo is being maintained as it becomes progressively more clear that, on the basis of growing knowledge, some possible projects deserve comprehensive investigation and a decision on their acceptability over the next few years.

It seems likely that the broad field of water engineering will continue to play a major part in the development of these projects and in their assessment. The skills of the industry and the leadership which it has shown in estuarine research underline the important role which it has to play.

12

Canals – a focus for regeneration

J. C. Brown, CEng, MICE, MIHT[†]

INTRODUCTION

British Waterways is responsible for 3214 km of canal and river navigation in England, Scotland and Wales, of which 241 km of artificial waterways are in the West Midlands, all used for leisure purposes and built between 1769 and 1858.

GROWTH OF THE SYSTEM

The first canal between Wednesbury and Birmingham was built by James Brindley, and opened in 1769. Although only 16 km long, it resulted in a dramatic fall in the price of coal. New canals, mainly contour, were constructed for craft of 21.34 m length, 2.0 m beam and 1.1 m draft. The capacity of these 'narrow boats' (20 tonnes powered, 30 tonnes unpowered), however, restricted the ability of canals to meet future rail and road competition.

The second generation of canals included the 3.47 km Lapal Tunnel and improvements to the Wolverhampton Canal by John Smeaton.

The third generation were noted for straight lines, huge cuttings and embankments and outstanding structures such as the new Birmingham Canal Main Line, built between 1825 and 1838, which included Galton Bridge and Smethwick Cutting.

During this period railway construction commenced and some canals came under railway control. This resulted in many canal-railway interchange basins being

---

[†] Project Manager, Midlands, British Waterways.

constructed with a consequent upsurge of local traffic. However, by the end of 1858 mining was in decline and railways were gaining long distance traffic.

Many canals were built in this period of 'canal mania' solely to obtain traffic from existing canals, but this was to the financial detriment of the whole network.

DECLINE OF THE SYSTEM

The West Midland canal network peaked at 368 km in 1858, dominated by the Birmingham Canal Navigation's 256 km system, which carryied some nine million tonnes annually. As prosperity declined canals deteriorated and subsequently abandonments took place, eventually totalling 127 km. Some canals which were not officially abandoned become derelict.

As traffic declined, and threatened the existence of canals, early enthusiasts, such as L.T.C. Rolt, cruised the waterways in the late 1930s and campaigned for their restoration.

The canal companies were nationalized in 1948 and administered by the British Transport Commission until 1963, when the British Waterways Board was established to take over canal assets. With an accumulated deficit of nearly £9 million, the future looked very bleak indeed.

THE PRESENT

Today British Waterways administers 3214 km of canal and river navigations, used by 20 000 private and 1600 hire boats, and has an annual grant-in-aid of £45.5 million from Government, amounting to 65% of total income for 1989–1990. The remaining income derives from estates, boat licences and water sales.

British Waterways' duties in respect of canals devolve from the 1968 Transport Act. 'Cruising waterways' are those which are maintained for boating, fishing and recreation and British Waterways has a duty to maintain them for powered boats. The maintenance of these canals is specified as being 'main navigable channels' and so no funds are available for towing path work.

'Remainder waterways' are to be dealt with in the most economic manner possible whilst securing the best financial return, which can include elimination. If a canal is retained, it must not become a public nuisance.

In the West Midlands there are 131 km and 110 km of cruising and remainder waterways respectively (Fig. 1).

PARAMETERS FOR EXTERNALLY-FUNDED SCHEMES

In order to achieve external funding without loss of Government grant-in-aid, the following parameters have to be met: (1) works must be outside British Waterways' statutory duty as defined in the 1968 Transport Act; (2) British Waterways' external funding limit must not be exceeded.

Works eligible include landscaping, structural and lockside enhancement, towing path works and waterway wall repairs where required for towpath support. Some new structures can be included, and for remainder waterways some dredging under certain conditions. Work carried out at Farmers Bridge Top Lock (Fig. 2) is an example.

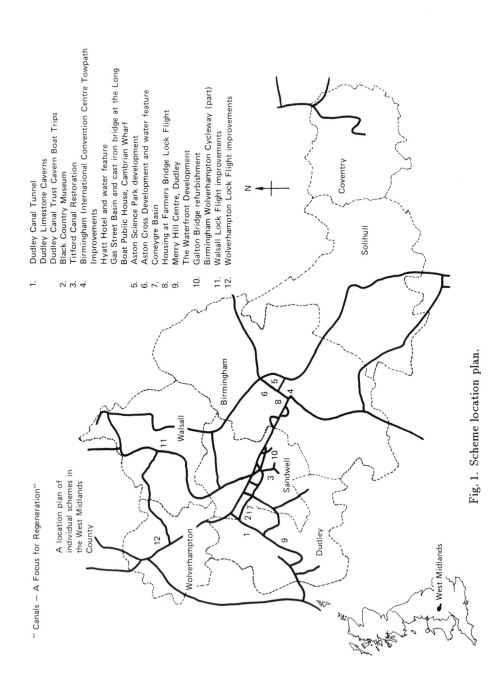

"Canals – A Focus for Regeneration"

A location plan of individual schemes in the West Midlands County

1. Dudley Canal Tunnel
    Dudley Limestone Caverns
    Dudley Canal Trust Cavern Boat Trips
2. Black Country Museum
3. Titford Canal Restoration
4. Birmingham International Convention Centre Towpath Improvements
    Hyatt Hotel and water feature
    Gas Street Basin and cast iron bridge at the Long Boat Public House, Cambrian Wharf
5. Aston Science Park development
6. Aston Cross Development and water feature
7. Coneygre Basin
8. Housing at Farmers Bridge Lock Flight
9. Merry Hill Centre, Dudley
    The Waterfront Development
10. Galton Bridge refurbishment
11. Birmingham Wolverhampton Cycleway (part)
    Walsall Lock Flight improvements
12. Wolverhampton Lock Flight improvements

Fig. 1. Scheme location plan.

Fig. 2. Farmers Bridge Top Lock; (a) circa 1900, and
(b) 1990 showing recent paving works.

Virtually all works of canal enhancement to achieve the following objectives are those which can be achieved within the financial and statutory limitations indicated above.

AIMS OF CANAL ENHANCEMENT

The canal system was of fundamental importance in the expansion of the Black Country and West Midlands, but for many years, since the decline of commercial traffic, local canals were vastly under-utilized, and vandalism and pollution caused virtually all development since 1945 to turn away from the canals. This applied to both industry and housing and all the huge post-war housing estates ignored the potential of the waterways.

With the decline of canal-related industries many of the canals passed through considerable areas of decay, but it is a paradox that these areas offer the finest opportunities if developers can be persuaded to see the benefits of the canals.

Britain as a whole was very slow to exploit canalside developments, but in recent years the attitude has been rapidly changing.

As the West Midlands sits across the watershed of the Rivers Severn and Trent it has no major waterways except the unique 18th century canal network, why then seek to regenerate the waterways for purposes other than those for which they were built?

Significant benefits will result from basic canal improvement schemes and the various partners who actively assist British Waterways to finance these schemes can draw upon a considerably range of opportunities and benefits, as set out below.

THE OPPORTUNITIES

Engineering refurbishment
To enhance, restore, and conserve unique canal structures for future generations.

Landscape improvement
To add to the already significant visual amenity of water.

Interpretation
To provide information to the public by signs, notices and leaflets.

Public use
To increase public use of the towing path and construction of additional accesses.

Public perception
To seek to change the public attitude towards canals by open days, participation and publicity.

THE BENEFITS

Regional image
Canal improvements will benefit the image of the region as a whole by attracting business, tourism and development opportunities in an area formerly in decay and decline.

Tourism
Canal improvements will encourage the development of new and water-based tourism facilities, such as restaurant and passenger trip boats and provide an outlet for business tourism.

The improvements will result in the creation of short and long distance towpath walkways along a peaceful and tranquil route, and connect activities alongside the canal to provide visitors with opportunities for recreation and enjoyment.

Business and development

The presence of a canal will result in an increased environmentally favourably location set against a backdrop of water.

Clients will be favourable influenced by the setting, particularly if related to promotional trips on a canal narrow boat.

Housing development can be related to the canal and made particularly attractive if mooring provisions can be made. Waterside housing, particularly with added moorings, commands a premium of approximately 15%.

Local community needs

Canal improvements will create recreational facilities with water-based activities, including canoeing and towpath activities of jogging and walking. Towpath users will include those who have an interest in nature conservation and industrial archaeology.

Provision of a high quality towing path will integrate the local community by providing local walkways for leisure, shopping and business journeys.

Wherever possible, improved and additional accesses are being provided, capable of being used what is generally a flat and easy walkway after improvement.

EARLY RESTORATION AND ENHANCEMENT SCHEMES

In 1971 the Dudley Canal Trust re-opened the 2.9 km Dudley Tunnel and Branch Canal by major volunteer activity, and now runs boat trips into the adjacent Limestone Caverns from the Black Country Museum. The main tunnel was closed in 1981 due to structural problems but, with grant aid from Dudley Metropolitan Borough Council, was re-opened in 1991. Over the last twelve years the Council and the Trust have contributed over £150 000 toward tunnel maintenance works.

The former Warley Council assisted in the re-opening of the Titford Canal in 1974, subsequently the scene of two national boat rallies.

The former County Council dredged 75 000 tonnes of silt from some remainder canals between 1970 and 1980, and Sandwell, Walsall and Wolverhampton Metropolitan Borough Councils have funded works of improvement in their respective areas.

IMPLEMENTATION OF IMPROVEMENT WORKS

In 1983 British Waterways created, and absorbed the costs of, a project team to maximize the use of external funding, who carried out £3.6 million canal improvements between 1984 and 1987.

The Midlands Regional Project Manager now heads the team, and can obtain assistance from British Waterways Central Engineering Services, consulting engineers, local authorities and developers, co-ordinating their efforts to ensure continuity of design and the retention of traditional details to enhance the unique identity of the canals in a considerably changed environment.

Provided that works undertaken do not have an adverse impact on British Waterways' grant-in-aid, a significant uplift of resources is possible for the West Midlands canal network.

In order to achieve full benefits, superficial environmental improvements were not acceptable, and a sound structural base was required which would reduce British Waterways' future maintenance liability.

Improvements to the waterway structures are followed by environmental improvements, covering the following categories.

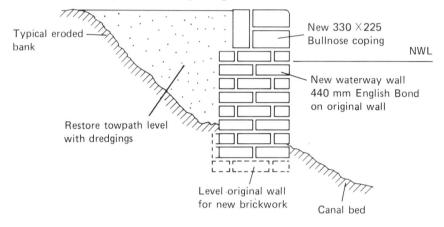

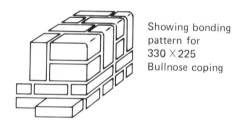

Fig. 3. Typical wall repair and built up bullnose coping.

Waterway walls

Restoration of the waterway wall and copings to 300 mm below water level, gives a solid basis on which to construct the new towing paths (Fig. 3).

Repairs to the waterway wall 300 mm below normal water level are funded by British Waterways from their budget.

Waterway walls are mostly brickwork, with stone or brick copings, and usually the top 600 mm requires complete demolition and rebuilding, normally carried out using blue engineering bricks, re-using existing copings where possible. Otherwise, a built-up coping system is adopted using standard bull-nosed bricks to represent traditional canal construction.

If the waterway wall is in reasonable condition, it is first cleaned with high pressure water jets and then pointed. Where areas have collapsed, brickwork repair is usual. Repairs to waterway walls involve dewatering the canal completely, usually only possible in winter to avoid disruption to boat traffic, and fish rescue operations. If repairs to the existing brick walls are impracticable, or erosion has removed the towing path completely, then light steel galvanized trench sheets are driven to 150 mm above water level, and backfilling takes place to towpath formation, usually dredged from the canal (Fig. 4).

Where a traditional brick coping is required the trench sheets are driven to water level and backfilled with granular fill and concrete to support the coping.

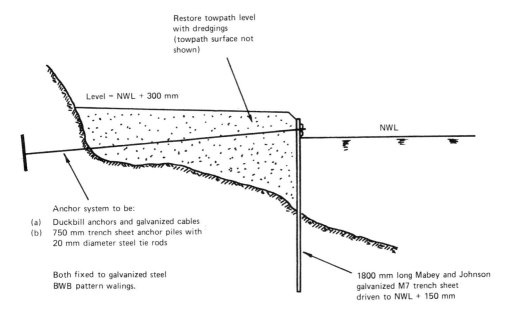

Fig. 4. Sheet piled waterway wall.

Towing paths

The installation of a suitable towing path for areas of intensive pedestrian use requires brick sheeting to provide a high quality surface. Not only does this benefit pedestrians, but offers a high visual impact which will enhance adjacent development, such as the International Convention Centre in Birmingham.

A 50 mm Breedon Gravel surface on 75 mm hardcore is used for other urban areas as an alternative to the traditional black ash. This gives a very hard wearing surface and the dark mustard colour increases the image of the waterways. For the proposed Wolverhampton to Birmingham Cycleway, a 2.0 m width is used, elsewhere a width of 1.5 m–1.8 m is adopted.

On other canals, and on certain historic lock flights, it is intended to use traditional black ash, but mixed with 10% of cement to provide a dust-free surface.

Costs
Some indicative costs (1990) (per linear metre): are:

Waterway wall pointing and light repairs	£34
Waterway wall pointing, heavy repair and copings	£64
Steel-piled waterway wall	£120
Breedon gravel towing path, 2.0 m wide	£20
Brick-sheeted towing path, 2.0 m wide	£80

Information
Suitable signs and direction posts have been installed, and leaflets for towing path users are provided.

Landscaping
Landscaped picnic and other recreational areas are provided.

Towpath access
New and improved accesses to the towing paths are incorporated with provision made for wheelchair users wherever possible, but physical constraints mean that this cannot always be achieved. At many existing accesses ramps are often steep and limitations of space mean that these cannot be reduced, although gaps are left in the brick upstands of ramps to enable wheelchairs to pass. New access points often have to connect towpaths with footpaths at bridges, where considerable vertical differences in levels mean that staircase accesses are often the only possible solution.

Improvements to the towing paths result in their unauthorized use by motorcycles, and barriers have to be installed which will not allow wheelchairs to pass. However, several new access points suitable for wheelchairs have been built.

Structures
New and improved structures, including a new cast iron bridge at Gas Street and a new brick arch bridge at Birmingham University, and the repainting of existing cast iron canal bridges and aqueducts have been carried out.

Individual schemes
These include the Galton Bridge (45 m span cast iron structure at Smethwick, 1988) at a cost of £55 000 and major lockside improvements, paved, at a cost of £25 000.

One of the most difficult tasks faced by contractors is the very restricted access. At Bournville, a 1 km stretch of towpath improvement under construction had only one access point, requiring bridging the canal with a temporary low level bridge, which had to be removed each time a boat required passage. Obviously, this resulted in a high cost for this type of work.

The resulting visual improvements have made the local canal system a much more attractive place and developers are now exploiting the setting in a far more imaginative manner than previously.

Developments which have used the canal setting to considerable advantage include:

Birmingham International Convention Centre
Major pedestrian areas focus upon the canal, and a layby has been constructed to
accommodate trip boats and waterbuses.

Aston Science Park, Birmingham
A modern building, landscaping and absence of a formal boundary integrate an
attractive site with the canal (Fig. 5). A fine setting for a high quality business
park.

Fig. 5. Refurbished towing path at Aston Science Park.

Hyatt Hotel, Birmingham
A water feature, representing the former canal basin, links the hotel with the historic
setting of Worcester Bar.

Aston Cross, Birmingham
An office development with a water feature representing a former canal basin, and
including a long wharf facing the Birmingham and Fazeley Canal. A public house
complete with canalside moorings for customers forms part of this development in
the Birmingham Heartlands area.

Gas Street developments
A public house, offices and restaurant set in the Gas Street Basin area.

Conygree Basin, Tipton
Waterside housing set around a new canal basin, built in 1990.

Housing, Farmers Bridge, Birmingham
Canalside development which recreates the sense of enclosure which was formerly
a feature of the local canals.

The Long Boat, Birmingham
The first development of a modern canalside public house in the county, built in
the late 1960s. A visit to the Longboat is a must for most boating visitors to the
city!

Merry Hill, Dudley

A massive shopping centre includes 'The Waterfront', an office complex built around a new canal basin. A monorail will eventually cross the canal to link the development with car parks, shopping and leisure facilities.

The above examples show how developers have now recognized the value of the local waterways. Many more schemes, some in partnership with British Waterways, are in the planning stage.

The Midland Region of British Waterways moved to a purpose-built office at Fazeley Wharf, just outside the West Midlands County, with associated housing around a new canal basin.

REGENERATION PARTNERSHIPS

The main areas of improvement are listed below. See also Fig. 6.

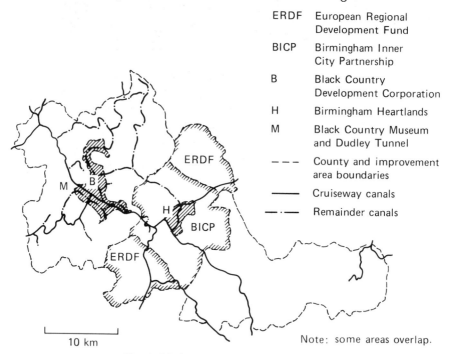

ERDF	European Regional Development Fund
BICP	Birmingham Inner City Partnership
B	Black Country Development Corporation
H	Birmingham Heartlands
M	Black Country Museum and Dudley Tunnel
– – –	County and improvement area boundaries
———	Cruiseway canals
—·—	Remainder canals

Note: some areas overlap.

Fig. 6. Major canal improvement areas.

European Regional Development Fund

The area covers $431 \, km^2$ and has large tracts of urban development where traditional industries have declined.

Canal improvements are undertaken under the tourism budget which aims to enhance the region as a major national and international focus of tourism and sport.

Within the Birmingham Integrated Development Operation, British Waterways has a £1.5 million programme of towpath restoration over 10 km of canal, reclaiming impassable towpaths.

Within the Black Country programme, application has been made for improvement schemes in Dudley.

Works attract grants of 40–50% of the total cost.

Fig. 7. New cast iron footbridge at Gas Street.

Birmingham Inner City Partnership

The programme was established to assist regeneration of the City economy and to overcome problems of obsolescent land, buildings and infrastructure.

Thirty-nine kilometres of canal exist in the Partnership area, improvements being undertaken under the auspices of the Department of the Environment, Birmingham City Council, and British Waterways.

A towpath-based 'city centre walkway' was implemented in 1984, and two major developments were the new cast iron bridge at Gas Street (Fig. 7) and the new brick arch bridge at Birmingham University.

Black Country Development Corporation (BCDC)

Established in 1987, BCDC aims to restore derelict land and buildings for industrial and residential purposes, assist industrial growth, create urban woodlands and improve communications.

Forty kilometres of canal pass through the area and improved waterways provide a major benefit, partly due to enhanced development values.

The improvement programme started with a comprehensive package of measures to enhance some of the canals wich run through areas of industrial decay.

The canals are seen as providing a focal point, providing charm and character, and several major developments are planned to take advantage of the improved waterways.

Black Country Metropolitan Borough Councils

Dudley, Sandwell, Walsall, and Wolverhampton Councils are enthusiastic supporters of canals and examples of their involvement include:

Dudley The Dudley Tunnel Project represents the results of 15 years' active assistance.

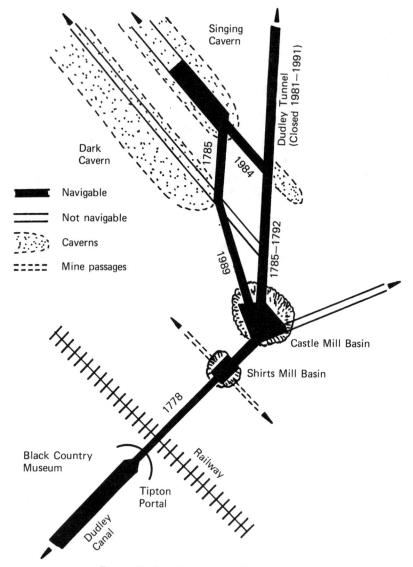

Fig. 8. Dudley Tunnel and Caverns.

Sandwell Enhancement of Brasshouse Lane and Galton Bridges and towpath improvements at Tipton.

Walsall A programme of improvements at Walsall Locks.

Wolverhampton A programme of improvements at Wolverhampton Locks and other towpath schemes.

Cycleway

Improvements to 7 km of the Birmingham-Wolverhampton Canal towpaths for cyclists was funded by the Councils, with assistance from the Sports Council.

Dudley Canal Trust

The Trust re-opened Dudley Tunnel in 1973, and continued commitments with Dudley Council to finance maintenance have secured the future of the 200-year-old, 2.65 km structure, one of the great engineering feats of the Industrial Revolution.

The Trust provided the impetus for the £1.5 million new tunnels constructed in 1984 and 1989 which connect Dudley Tunnel with the Limestone Caverns (Fig. 8).

CONCLUSION

These efforts have been co-ordinated by the Midland Region of British Waterways to enhance considerably the image of the West Midlands region and its canals, to provide a high quality environment for all, and to ensure the waterways heritage is conserved and enhanced for future generations to enjoy.

13

The Birmingham Canal Improvement Programme

T. E. Green, DipTP, DipLA, MRTPI[†]

INTRODUCTION

Many international cities are built on famous rivers, whilst others have become thriving ports as a result of a coastal location. Birmingham, however, stands out amongst other European cities in view of its unique canal network, which in terms of length is greater than that of Venice.

The canals in Birmingham are at the centre of the system of waterways which forms a giant figure of eight and which by 1840 extended to over 6600 km throughout the British Isles. By using waterways which radiate from the city centre it is still possible to travel by boat from Birmingham to the Irish Sea, the North Sea via the Humber, the Thames Estuary via London, and the Bristol Channel at Sharpness. Birmingham is therefore a canal crossroads of the country, a fact which without doubt was the foundation of the growth of the city's importance through the last two centuries.

However, by the beginning of the twentieth century the canal's heyday was already over, and from 1920 onwards they rapidly declined. Birmingham therefore, with its 45 km of canal was left with a problem, albeit a potential asset – considerable lengths of deteriorating canal in the land-locked city.

[†] Department of Planning and Architecture, Birmingham City Council.

THE DEVELOPMENT OF THE PROGRAMME

In the late 1970s and the early 1980s pressure from interested groups and individuals began to exert itself with the demand to know when something was going to be done about the city's canals. But what was it about the canals that generated such interest? The answer could be long and complex, but in simple terms people are drawn to water and the canal environment, with its feeling of enclosure, views along the water, locks, boats, bridges and buildings, provides an attractive contrast to the bustling surrounding urban areas. The canal heritage rekindles pride in the city's past and confidence in its future.

The principal participants in the canal improvement programme were the City Council, British Waterways, who owned the canals, the Department of the Environment (a central government agency), and the West Midlands County Council, which was a regional body, since disbanded. Early discussions quickly agreed on the need for action and five main objectives were decided upon for the Birmingham Canal Improvement Programme. These objectives were:

(a) the attraction of tourists;
(b) the improvement of the quality of residential areas;
(c) the stimulation and regeneration of industrial and commercial areas;
(d) the provision of facilities for recreation both on and off the water, and
(e) the protection and enhancement of habitats for wildlife.

Out of these five objectives many varied and complex issues emerged, including:

(a) the means of funding the project;
(b) the priorities;
(c) safety aspects;
(d) maintenance responsibilities;
(e) balance of new leisure uses with past traditions, and
(f) limitations on uses.

FUNDING AND MAINTENANCE

On the funding issue it was agreed that canal improvements would be part of a partnership that already existed between central and local government to tackle urban deprivation problems in a number of British cities. The joint operation is known as the Inner City Partnership Programme and the basic financial split is 75% central government contribution with the remaining 25% being found by local government, subject to prior approval being gained from the Department of the Environment before projects may be implemented.

The first difficulty encountered on the canal programme concerned the definition of maintenance, as the Department of the Environment would not fund maintenance. The City Council attached priority to the landscaping of the towpath (the pathway alongside the canal), but British Waterways would not permit this until the canal wall retaining the path had been rebuilt. The question of whether this constituted maintenance was eventually resolved with an equal funding split between British Waterways and the Partnership Programme on the re-building costs. The other problems outlined above have been resolved in a similar way, and in the last five years over £5 m has been spent via the Partnership Programme, with additional funding from the City Council on canals that are outside the Partnership area.

TOURISM

The choice of projects has been dictated by the five objectives stated above, with possibly the most popular improvements being carried out primarily for tourism. Tourism is Birmingham's newest industry, and the city has a proud heritage with which to attract tourists, particularly in the field of industrial heritage, of which the canal system is part. The improvement programme was generally based on a centre-out approach, and following this principle Gas Street Basin, which is the hub of a system of radiating canals, was the first project. The Basin lies at the junction of the Worcester and Birmingham Canal and the Birmingham Fazeley Canal, and is enhanced by original canalside housing and the presence of traditional canal narrowboats – a major attraction for tourists. A short distance away along the city centre canal walkway is the new International Convention Centre with its new canalside amenities. In the other direction the Worcester and Birmingham Canal is being enhanced as a leisure route for canal boats to link the city centre with Birmingham University, Cadbury World, and the Patrick Motor Museum at Kings Norton.

RESIDENTIAL AREAS

Canal improvements close to residential areas create safety risks for users of the canal towpath – a problem that had to be resolved in principle at an early date.

Gates leading to towpaths had been kept locked, not only in the interests of safety, but also to prevent danger to or misuse of the lock gates. There was also concern that criminals would use the canal towpaths, raising the problem of policing in these areas. Effective police monitoring of 45 km of canal was considered to be too onerous and did not in practice prove necessary once the public had access to the canal towpaths. Indeed, the presence of people on the towpath acts as a deterrent to crime.

On the safety issue, comparisons were made with other canal cities such as Amsterdam and Venice, and with roadside situations where traffic is a hazard, and it was concluded that no special measures should be taken.

An example of a canal project near to a residential area is Saltley Cottages. These cottages are buildings owned by the city, located next to a canal, and as is so often the case, the design of the estate had 'turned its back' on the canal. In 1987 design work was started on a project to enable residents to derive pleasure from the canalside, and nearly £75 000 of Partnership money has been spent on a new access point and the enhancement of the small woodland area to provide a place for residents to sit and look out onto the canal.

Another example is at Bordesley Village where, as part of the Heartland Initiative, an area around the Birmingham and Warwick junction canal has provided the focus for a new urban village. The canal is used to foster identify, to provide interesting views and activity, and as a pedestrian link to the village centre and the proposed public open space. The village centre will also be linked to the canalside with views over a new quayside.

THE INDUSTRIAL ENVIRONMENT

Studies have shown that an attractive environment was one of the qualities that attracted potential investors when deciding between competing areas in which to invest in industrial regeneration. Whilst the inner city cannot compete with the attractive landscape of green field sites, the canalside can offer investors a different type of site. Industrial units can be planned to overlook the water, whilst the strong design idiom of canal spaces and architecture can inspire more imaginative urban design of new buildings. Historic buildings close to the canalside can be converted to workshops, offices, public houses and restaurants. At lunch time employees can use the towpath for walks and the public houses for refreshment. The canal can upgrade the image of an area and strengthen its identity by building on the very character which makes it different from other locations.

Aston Science Park is an industrial development straddling the Digbeth Branch Canal, close to where it meets the Birmingham and Fazeley Canal at Aston Junction. Here one of the earliest enhancement schemes was completed in 1986. The paving surrounding the locks was restored or replaced; pedestrian access was improved, with landscaping and seating areas provided. The Science Park buildings erected adjacent to the canal area are set in attractive landscaping which slopes down to the banks of the canal, and a new basin is planned to add extra interest to the rather straight waterways of the area.

Birmingham Heartlands was the largest industrial/commercial/residential project being undertaken in the city in 1990. This major initiative to regenerate a large area of East Birmingham was a joint enterprise between the City Council and the private sector. The resultant Urban Development Agency, in adopting the recommendations of a consultants strategy, placed great importance on canals as an environmental asset. Two canals cross the area – the Birmingham and Fazeley and the Birmingham and Warwick Junction Canals – and both are at the heart of two major early action developments: Waterlinks and Bordesley Village.

RECREATION

The use of the canal network to provide recreational facilities, both on and off the water, is another prime objective as adequate provision for recreation is essential for the regeneration of the inner city. Leisure time is increasing but the city finds it difficult to meet the rising demand for water sports, because of the conflicting demands on limited natural resources. Canals provide an ideal opportunity to compensate for some deficiencies in water-based recreation.

As linear walkways, canal towpaths can link other open spaces or provide a relaxing break in the densely built up inner city. The location of the canals in the inner area makes them accessible to many residents whose other opportunities for recreation may be limited. In addition to the established leisure uses of boating, walking and fishing, cycling is another use that has recently been introduced.

However, the towpaths are owned by British Waterways and are not public rights of way. It was therefore necessary for the former West Midlands County Council and the City Council to negotiate a legal agreement to allow public access. Agreement in principle was reached in 1985, and the first specific agreements were signed in 1987.

An area of the City known as 'The Ackers' is managed by a recreational trust, which has taken advantage of the site's canalside location. Following the structural work and towpath restoration funded by the Partnership Programme, the canal is well used for canoeing, narrowboat trips for school children and the disabled, and fishing. The Partnership Programme has also funded the renovation of the canal basin and the construction of a dry dock for boat repairs.

NATURE CONSERVATION

The canal corridors are important for nature conservation, not only for the range of valuable habitats they provide, but also because they are very important links between urban sites and the countryside. This reduces the effects of isolation on plant and animal communities, and in recognition of the canal network's intrinsic value, work was undertaken as part of the Birmingham Habitat Survey to assess the contribution of the canal network to nature conservation. In addition, there are opportunities to enable habitat creation and enhancement schemes to be carried out as part of the Canals Improvement Programme.

CONCLUSION

In conclusion, it is almost certain that without the innovative and collaborative approach of the Birmingham Inner City Partnership Programme, the canals of Birmingham would have remained neglected backwaters. However, their potential as a catalyst for change was grasped and areas not formally known for their environmental quality began regaining their pride of place. The canal system, which was once a classic example of urban dereliction, has now become an instrument to enhance the quality of life.

Part III Water quality

14

The control and monitoring of discharges by biological techniques

D. T. E. Hunt, BSc, MSc, PhD, ARCS, MIWEM[†], I. Johnson, BSc, PhD[‡], and R. Milne, BSc[§]

INTRODUCTION

It is a principle of UK policy for protecting surface water quality that the discharge consent conditions, to which a discharge must conform, are set so that a prescribed level of quality is maintained in the receiving water body. Environmental quality objectives (EQOs) are established to protect legitimate uses of the water, which include general ecosystem protection and the maintenance of quality for such purposes as potable abstraction. An EQO is considered to be met if the water in question complies with a range of environmental quality standards (EQSs) for individual substances, relevant to the use concerned. Discharge consent conditions are set so that the relevant EQSs are met in the waters receiving the discharge, beyond the boundary of the mixing zone.

Accordingly, the EQSs applicable to the most demanding relevant EQO, together with the available degree of dilution and dispersion, normally determine the maximum permitted concentrations of relevant substances in the effluent itself. In the case of specific uses such as potable water abstraction, the EQSs are those of relevant European Community (EC) legislation. For general ecosystem protection the

[†] Senior Principal, Environmental Management, WRc, Medmenham.
[‡] Toxicologist, WRc, Medmenham.
[§] Toxicity Scientist, National Rivers Authority, Welsh Region.

EQSs are either: (a) for List I substances, those of the relevant EC Directives; or (b) for List II substances, those determined by the UK from available toxicological data with appropriate application factors [1].

For cases where an EQS is not available, toxicological data can be used to set a 'likely safe environmental concentration' (LSEC) as a surrogate EQS for consent setting.

The UK is also committed to a precautionary approach to pollution control, through the application of 'best available technology not entailing excessive costs' (BATNEEC) to control the loads of Red List substances to sea. The need to limit the loads, as well as the concentrations, of accumulative substances has also been emphasized by the National Rivers Authority (NRA) [2]. This chapter, however, is concerned with comparing the EQO/EQS and direct toxicity assessment (DTA) approaches. The control of loads will not be considered further, although its importance in discharge control must not be forgotten.

ADVANTAGES AND DISADVANTAGES OF CHEMICAL-SPECIFIC APPROACH

The above approach is satisfactory for simple effluents of well-defined composition, containing only toxicants for which there are adequate toxicological data on which to base an EQS. It leads to simple discharge consent conditions, compliance with which can be assessed by chemical analysis of the effluent. Chemical analysis, particularly 'for traditional' contaminants, is relatively inexpensive and, with appropriate analytical quality control, capable of adequate accuracy . There are, however, a number of disadvantages [3–8] to an approach which is entirely chemical-specific.

(i) Many effluents contain organic chemicals which are not readily identifiable or measurable by analytical techniques.

(ii) Toxicological data are unavailable for many thousands of synthetic chemicals, and when available may not apply to local organisms.

(iii) The complex composition of many effluents can cause problems in applying EQSs, which are derived on the basis of single-substance toxicity, and do not take account of:

(a) chemical interactions between discharge components, or with substances in the receiving waters, which may affect toxicity;

(b) the possible synergistic and antagonistic toxicological effects of substances in complex discharges

(iv) These difficulties may be compounded by the variable composition of many complex effluents, particularly those from batch processes.

ADVANTAGES AND DISADVANTAGES OF DIRECT TOXICITY ASSESSMENT

The above disadvantages of the chemical-specific approach can be partly or wholly overcome by DTA, which considers the effects on organisms of each effluent as a whole [3, 6–8]. Thus, in principle, DTA can:

(a) detect the effect of the combination of all compounds present, even if they cannot be identified or measured by chemical means;

(b) control the toxicity of an effluent which contains substances for which no toxicological data are available;

(c) address complex effluents with due account of chemical and toxicological interactions; and

(d) cope with variations in the composition of complex effluents.

However, in practice, DTA itself has a number of disadvantages.

(i) It is relatively slow and expensive, in comparison with chemical assessment, for small numbers of chemical determinands.

(ii) Because the identity and importance of individual toxicants is not revealed, the approach:

    (a) lends itself less readily than does chemical-specific control to toxicity reduction of the effluent;

    (b) does not provide information on the properties (e.g. bio-accumulation potential) of specific substances;

    (c) may not cover toxicity released downstream by the reaction of effluent components.

(iii) It requires facilities and expertise which are not available in every UK pollution control laboratory; moreover, quality control procedures for toxicity testing are less well-developed than those for chemical analysis.

None of these disadvantages is an overwhelming reason for not pursuing the approach. Thus, point (i) simply indicates that DTA should not be applied when the simpler chemical approach will suffice. Neither is (iia) insurmountable – toxicity reduction can be achieved using a variety of techniques (discussed below). With regard to points (iib) and (iic) above, the chemical-specific approach would not give this information either, if analytical difficulties precluded the identification of toxic species. Finally, point (iii) requires that appropriate techniques should be developed.

EXISTING EXPERIENCE OF DIRECT TOXICITY ASSESSMENT

The US Environmental Protection Agency (USEPA) has a well-developed programme for the biological control of discharges to surface waters, the development of which has been described by Wall and Hanmer [8].

Toxicity-based controls were formally introduced in 1984, by means of a national policy statement recommending their use in combination with chemical controls [9]. In 1985, the USEPA published a manual giving detailed guidance on the regulatory application of biological testing [6]. By 1987, Wall and Hanmer [8] were able to report that toxicity-testing requirements had been written into (a) over 1400 industrial discharge permits (i.e. consent conditions), representing about 38% of the permitted major industrial discharges; and (b) about 400 sewage-treatment works permits, representing about 10% of the major municipal discharges.

The USEPA concluded that toxicity testing, as an adjunct to chemical analysis, improved the assessment and control of potentially polluting discharges, and decided to press for increased use of the approach. A number of other countries have reported their experiences of DTA – including Canada, Eire, Finland, France, Norway, Sweden, West Germany and the UK [10, 11].

There has been wide agreement that DTA is a valuable addition to the chemical control of discharges, and the Organization for Economic Cooperation and Development [12] has produced general guidance on biological testing for water pollution assessment and control.

Although the UK experience of DTA is limited in comparison with that of the US [13], there is growing interest in the approach. The Clyde River Purification Board has been particularly active in this area, applying toxicity testing to discharges from a pharmaceutical plant and an explosives factory [14, 15]. The Microtox bacterial test (see below) has also been used to assess the toxicity of a number of discharges to estuarine and coastal waters in Wales.

TOXICITY REDUCTION

As noted, one of the major perceived disadvantages of DTA is its inability to identify the chemicals causing toxic effects, which makes it more difficult to reduce effluent toxicity. However, Mount [3] has pointed out that BOD and suspended-solids removal has not demanded detailed knowledge of the specific components involved, and a similar point has been made by Wall and Hanmer [8]. Three basic approaches to toxicity reduction can be adopted.

(a) *Causative agent* – in which toxicity testing is followed by chemical analysis to identify causal agents, allowing treatment and substitution options to be assessed.

(b) *Fractionation* – in which toxicity testing is applied to effluent fractions separated by physico-chemical means, to trace toxicity to a specific fraction without attribution to specific compounds.

(c) *Toxicity treatability* – in which treatment options are applied at bench scale, and their efficacy is assessed by toxicity testing.

Wall and Hanmer [8] reported that the USEPA experience favours the causative agent approach, because it is more effective to keep a substance out of a waste-stream after contamination. They also point out that, even if a unique attribution of toxicity is not achieved, the particular process in a complex plant which contributes the causative toxicant(s) may be identified; therefore only its waste-stream, rather than the entire effluent, need be treated.

ROLE OF DIRECT TOXICITY ASSESSMENT

As a result of an all assessment of all the above issues, it was concluded [16] that:

(i) DTA has a potentially important role in controlling discharges to UK surface waters, both fresh and saline;

(ii) it is complementary to, rather than a substitute for, conventional chemical-specific controls. For discharges containing well-known substances for which suitable toxicological data are available, chemical-specific control will usually be more cost-effective;

(iii) DTA is particularly advantageous for complex discharges containing substances for which suitable toxicological data are not available;

(iv) wider use of DTA requires a general protocol for its sound application and the establishment of appropriate quality control; and

(v) the adoption of DTA should not draw effort away from the planned acquisition of toxicological data for specific contaminants, upon which sound EQSs may be based; otherwise, the effectiveness of chemical-specific control will be impaired.

PROPOSED UK APPROACH

Introduction

The proposed approach [16] is based particularly upon that of the USEPA, but adapted to UK circumstances. It would be wasteful of resources for DTA to be introduced without a clear assessment of its likely value in controlling individual discharges, therefore a three-stage strategy involving preliminary discharge screening has been proposed; this is shown in Fig. 1 and is outlined below [16].

Stage 1

This involves the selection and prioritization of all discharges with the potential to benefit from DTA, on the basis of an assessment of existing data and background information, including: (a) existing knowledge of environmental impacts; (b) presence of potentially toxic substances not subject to EQSs, or for which toxicological data are lacking; (c) complexity and variability of composition; (d) flow relative to the receiving water; and (e) toxicity of whole effluent or waste-stream constituents.

Those discharges unsuitable for DTA continue to be dealt with by the chemical-specific approach alone; the others are subjected to:

Stage 2

This involves the application to be the selected discharges (in priority order determined by the Stage 1) of a simple, rapid acute screening test using (for the present, at least) the commercially-available Microtox system.

This is based on the bioluminescent response of the marine bacterium *Photobacterium phosphoreum*, which emits light as a natural by-product of respiration [17]. This light output can easily be quantified by a sensitive photometer. When exposed to a toxicant, the change in light emitted by the bacteria is proportional to the ability of the toxicant to inhibit metabolism, which in turn gives an indication of its toxicity. Microtox test results are reported as median effective concentrations (EC_{50}) which are defined as the concentrations of toxicant which result in a 50% reduction in bacterial light output after a 5–30 minute exposure period, relative to the control. The lower the EC_{50} for a toxicant the greater the toxicity.

The sensitivity of this organism has been shown to be similar to that of other commonly-used aquatic tests requiring 24–96 hours to complete [18, 19], and (for example) it has been advocated as a screening tool for pulp and paper-mill effluents in North America [20, 21]. However, other tests may also be appropriate, depending on the particular circumstances; for example, an algal test (see below) might be an appropriate addition at the screening stage for a herbicide factory effluent.

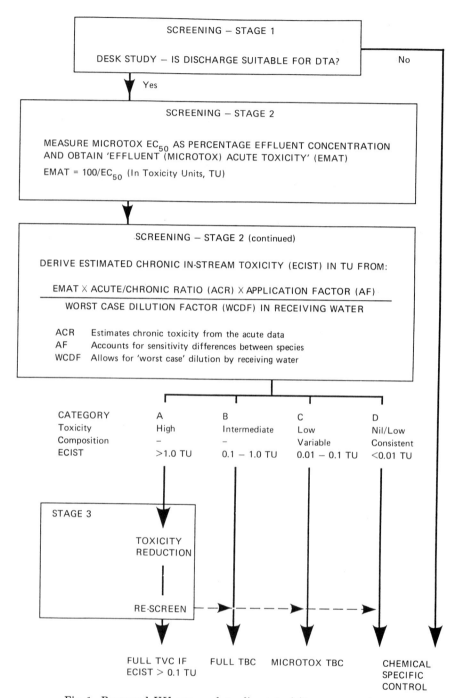

Fig. 1. Proposed UK approach to direct toxicity assessment.

The screening test is followed by the derivation, from the acute data and other factors (Fig. 1) of the 'estimated chronic in-stream toxicity' (ECIST). Tested effluents are then allocated to one of four categories (categories A, B, C and D of Fig. 1) on the basis of their ECIST values.

Stage 3
Further action, depending on category:
  (a) Further investigation of those (category A) discharges estimated to cause chronic toxicity in the receiving waters, with a view to toxicity and/or discharge reduction, probably followed by the establishment of a full toxicity-based consent (TBC).
  (b) Establishment of full TBCs for those (category B) discharges showing intermediate toxicity.
  (c) Establishment of Microtox-based TBCs for those (category C) discharges showing low toxicity, but whose composition is likely to be variable.
  (d) Continued reliance upon chemical-specific consent conditions for those (category D) discharges which : (i) show little or no toxicity; or (ii) show low toxicity and limited variability of composition.

For category C discharges, it is considered that a TBC can be set on the basis of the Microtox test alone. For categories A and B, however, it is proposed that 'full' TBCs should be set, based on acute tests using at least three representative species. However, the Microtox should then be 'calibrated' against such species, to provide a simple and relatively inexpensive test for routine monitoring purposes.

It is proposed that the three tests should normally be well-established standard protocols, for an alga, an invertebrate, and a fish. Thus, for freshwaters, it is currently recommended that acute tests with the alga *Chlorella vulgaris*, the waterlea *Daphnia magna*, and the rainbow trout *Oncorhyncus mykiss*, should be used. Similarly, for the saline waters, tests using the alga Phaeodactylum tricornutum, the brown shrimp *Crangon crangon* or the larvae of the Pacific oyster *Crassostrea gigas*, and juvenile plaice *Pleuronectes platessa* or turbot *Rhombus maximus*, are recommended.

In setting full TBCs, it is recommended that the combination of the three main tests and the Microtox test should be applied on no fewer than four occasions over a minimum period of three months. The Microtox response is then to be calibrated against the most sensitive of the other three tests, and a consented 'effluent (Microtox) acute toxicity' (EMAT) (Fig. 1) set, equivalent to an ECIST value of 1 toxic unit for the most sensitive test species.

The Microtox system can be used by both discharger and regulator for routine effluent monitoring; however, periodically there will be a need to retest the effluent with the most sensitive species and adjust the consented 'effluent (Microtox) acute toxicity' by recalibration. The frequency of recalibration will depend on such factors as: (i) the closeness of routine results to the consented EMAT; (ii) the expected variability of effluent composition; (iii) changes in the nature or operation of the plant; and (iv) the importance of the discharge to the receiving water quality.

Note also that there may be circumstances in which the Microtox test lacks the necessary sensitivity, and another test will be needed instead; similarly, for large and important discharges, sub-lethal toxicity testing (e.g. *Daphnia* reproduction studies) may be an appropriate addition to the suite of tests.

Table 1. Derivation of Microtox consent limits for category A and B discharges

Effluent	Most sensitive species in the in-depth study and mean ES/LC$_{50}$ value (% effluent)	Estimated (Microtox) acute toxicity (EMAT) for most sensitive species EMAT = 100/(EC/LC$_{50}$)	Acute to chronic ratio (ACR) used	Application factor (AF) used
AI	Oyster embryos 0.55	181.8	10	1.0
AII	Oyster embryos 0.46	217.4	10	1.0
BI	Oyster embryos 10.66	9.38	10	1.0
BII	Oyster embryos 8.65	11.57	10	1.0

Table 1. (continued)

Effluent	Worst case dilution factor (WCDF) used	Estimated chronic 'in-stream' toxicity (ECIST)	For ECIST = 1.0 for most sensitive species		Calibrated consent limit (% effluent)
			Required EMAT	Required EC/LC$_{50}$	
AI	1250	1.45	125	0.8	18.6
AII	1000	2.17	100	1.0	4.22
BI	250	0.38	25	4.0	20.41
BII	250	0.46	25	4.0	23.26

CASE STUDIES

This interim protocol has been assessed in a series of case studies involving both WRc, NRA and Clyde River Purification Board staff, with the objectives of:
  (a) evaluating the value and effectiveness of Microtox-based screening;
  (b) testing the procedures for establishing full TBCs; and
  (c) assessing the efficacy of TBCs, with particular respect to:
      (i) their ability to protect and improve the receiving environment;
      (ii) their ability to help detect failure to comply with chemical consents; and
      (iii) their ability generally to assist pollution officers in carrying out their responsibilities.

A study of 15 industrial and domestic sewage discharges to the marine waters of Irvine Bay, Scotland, has been carried out in conjunction with the Clyde River Purification Board. This showed that, at the screening stage, the Microtox test was effective in discriminating between highly toxic and low/non-toxic effluents. However, Microtox was less effective at classifying the discharges of intermediate toxicity, and the oyster (*Classostrea gigas*) embryo-larval test was also used at the screening stage to effectively prioritize these effluents. On this basis, a battery of complementary simple and rapid acute tests should be used at the screening stage to ensure that effluents are appropriately classified.

After classifying the discharges, two effluents from each of categories A–D were selected, and the toxicity was assessed with Microtox and the higher organism algal (*Phaeodactylum tricornutum*) growth inhibition, oyster embryo-larval and juvenile turbot (*Rhombus maximus*) lethality tests. Ranking the effluents on the basis of the toxicity values generated, generally confirmed that the effluents were assigned to the appropriate category at the screening stage (Fig. 2).

In-depth toxicity testing of the category A (high toxicity) and B (intermediate toxicity) discharges was continued on three additional occasions. The results were used to calibrate the Microtox test and derive a consent limit, against which compliance can be judged by routine monitoring. The derivation of Microtox limits for the category A and B effluents are detailed in Fig. 3 and Table 1. The most sensitive test for all the effluents was the oyster embryo larval development test. Mean values from these tests with the appropriate 'acute to chronic ratios', 'application factors', and 'worst case dilution factors' were used to derive the proposed Microtox limits. The derived Microtox consent limits are currently being monitored monthly to assess compliance.

CONCLUSIONS

  1. Direct toxicity assessment is effective for controlling complex effluents.
  2. Toxicity tests have to be defensible in a court of law, and must be (i) simple, robust and unambiguous, and (ii) based on standard methods.
  3. A uniform, cost-effective system is needed throughout the UK.
  4. Case studies are important in developing the UK approach.

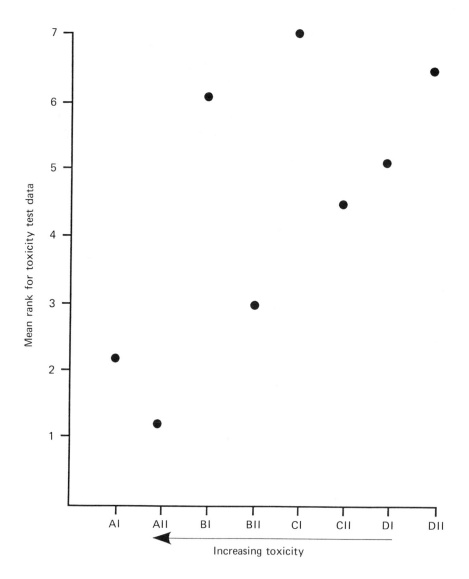

Fig. 2. Mean ranking of category A–D effluents based on toxicity tests carried out
on 22/1/91 and 6/2/91.

ACKNOWLEDGEMENTS

The authors wish to acknowledge the funding of the work described by the National
Rivers Authority and equivalent authorities in Scotland and Northern Ireland.

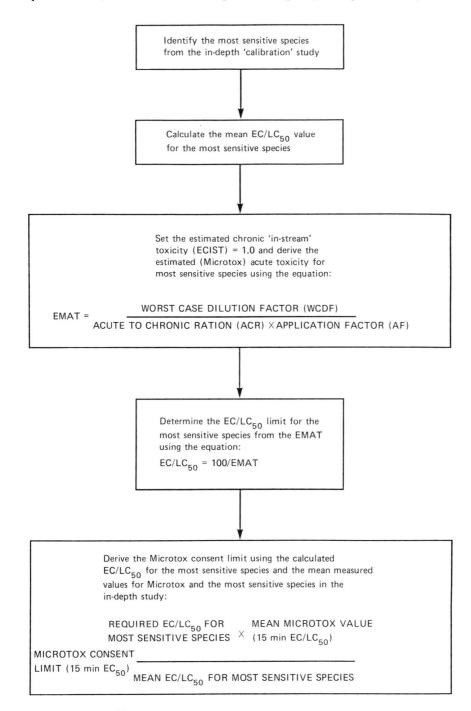

Fig. 3. Derivation of calibrated Microtox consent limits for effluents to be controlled by toxicity-based consents.

REFERENCES

[1] Gardiner, J. and Mance, G. (1984) *United Kingdom water quality standards arising from European community Directives.* WRc Report TR 204. WRc.

[2] National Rivers Authority (1990) *Discharge consent and compliance policy: A blueprint for the future.* Water Quality Series No. 1, NRA, London.

[3] Mount, D. I. (1984) The role of biological assessment in effluent control. In: *Proc. of Int. Workshop on Biological Testing of Effluents,* 10–14 September 1984, Duluth, Minnesota, USA. Organisation for Economic Cooperation and Development, United States Environmental Projection Agency and Environment Canada, 15–30.

[4] Sloof, W., Van Leeuwen, C. J. and Van Donk, E. Use of biomonitoring in water quality regulatory decisions. *Ibid.*, p. 61.

[5] Hanmer, R. W. and Newton, B. J. (1984) Utility of the effluent toxicity tests from the US regulatory perspective. *Ibid.* p. 353.

[6] US Environmental Protection Agency. (1985) *Technical Support Document for Water Quality-Based Toxics Control.* Report Number EPA-440/4-85-032, US EPA, Office of Water, Washington DC.

[7] US Environmental Protection Agency. (1986) Biological testing to control toxic water pollutants: The United States experience. In: *Proc. Int. Seminar on the Use of Biological Tests for Water Pollution Assessment and Control* 25-26 June 1986, ISPRA Research Centre, Varese, Italy.

[8] Wall, T. M. and Hanmer, R. W. (1987) Biological testing to control toxic water pollutants. *J. Wat. Pollut. Control Fed.,* 59 (1), 7.

[9] US Environmental Protection Agency. (1984) Development of Water Quality-based Permit limitations for Toxic Pollutants: National Policy. *Federal Register,* 49 (48), 9016.

[10] Organisation for Economic Cooperation and Development and the Commission of the European Communities. (1986) In: *Proc. Int. Seminar on the Use of Biological Tests for Water Pollution Assessment and Control,* 25-26 June 1986, ISPRA Research Centre, Varese, Italy.

[11] Bengtsson, B-E., Norberg-King, T. J. and Mount, D. I. (1987) Effluent and ambient toxicity testing in the Gota Alv and Viskan Rivers, Sweden. A US-Swedish OECD cooperative project. *Naturvardsverket Rapport 3275.*

[12] Organization for Economic Cooperation and Development. (1987) The Use of Biological Tests for Water Pollution Assessment and Control. *OECD Environment Monograph No. 11,* Paris.

[13] Pearce, A. S. (1984) Biological testing of effluents and associated receiving waters. In: *Proc. Int. Workshop on Biological Testing of Effluents,* 10–14 September 1984, Duluth, Minnesota, USA. Organisation for the Economic Co-operation and Development, United States Environmental Protection Agency and Environment Canada, 297.

[14] Haig, A. J. N., Curran, J.C., Redshaw, C. J. and Kerr, R. (1986) Use of mixing zone to derive a toxicity test condition. *J. Instn. Wat. & Envir. Mangt.,* 3. (4), 356.

[15] Mackay, D. W., Holmes, P. J. and Redshaw, C. J. (1989) The application of bioassay techniques to water pollution problems – the United Kingdom experience. *Hydrobiologica*, 188/189, 77.

[16] Hunt, D. T. E. (1989)Discharge control by direct toxicity assessment (DTA) – Interim protocol. *WRc Report PRS 2160-M/1.*

[17] Bulich, A. A. (1979) Use of luminescent bacteria for determining toxicity in aquatic environments. *Aquatic Toxicology* (Second Symposium), ASTM STP 667, edited by L. L. Marking and R. A. Kimerle. American Society for Testing and Materials, Philadelphia, 98.

[18] McFeters, G. A., Bond, P. J., Olson, S. B. and Tchan, Y. T. (1983) A comparison of microbial bioassays for the detection of aquatic pollution. *Wat. Res.*, 17 (12), 1757.

[19] Tarkpea, M., Hanson, M. and Samuelsson, B. (1986) A comparison of the Microtox test with the 96-hr LC 50 test for the Harpacticoid *Nicotra spinipes*. *Ecotoxicol. Env. Safety*, 11, 127.

[20] Blaise, C., Van Coillie, R., Bermingham, N. and Coulombe, G. Comparison of the toxic responses of three bioindicators. *Rev. Int. des Sci. de l'Eau*, 3, (1), 9.

[21] Firth, B. K. and Backman, C. J. (1990) A comparison of Microtox testing with rainbow trout acute and Ceriodaphnia chronic bioassays using pulp and paper mill wastewaters. *Proc. Techn. Assoc. of Pulp and Paper Industry*, p. 621.

15

Discharge consents –
are percentile standards still relevant?

J. P. Lumbers, BSc, MSc, DIC, PhD, CEng, MICE, MIWEM[†] and
S. J. Wishart, MA, MSc, DIC, PhD, CEng, MICE, MIWEM[‡]

INTRODUCTION

The formation of the National Rivers Authority (NRA) and the privatization of the water authorities to form private water companies (PWCs) has given new impetus to the search for consistently derived and applied effluent discharge standards. Consistency is required not only among the different PWCs but also between industry in general and the PWCs. Consistency is also required regarding the methodologies used by the different regions of the NRA in the translation of environmental quality objectives (EQOs) into river water quality standards (ROSs) and then the subsequent calculation of discharge consent conditions.

In the past there have been clear differences in the approaches to the establishment of ROSs, the derivation of consent conditions and the subsequent assessment of compliance. [1–3]. Personal observation leads the authors to conclude that there still remain more variations in methodology and interpretation than is desirable, especially given the new structure of the UK water industry and the apparent public desire to see the prosecution of polluters.

[†] Chairman, Tynemarch Systems Engineering Ltd.
[‡] Principal Engineer, Watson Hawksley.

DISCHARGE STANDARDS

Where new or revised discharge standards are to be more stringent than previously required, it is essential that such standards take into account the cost-benefit with respect to the local beneficiaries, for example riparian owners, and the PWC customers in general. Furthermore the discharge standards formulated should be achieveable using known technology, appropriate for the particular location. In this respect the 5th and 8th Reports of the Royal Commission on Sewage Disposal 1878 and 1912, [4] provide some guidance. The fifth report stated that;

paragraph 311: under the Rivers Pollution Prevention Act, 1876 in the case of sewage entering non-tidal waters, the duty is imposed on the local authority of adopting the best practical and available means to render the sewage harmless before it enters the rivers.

paragraph 312: our terms of reference require us, however, to have regard to the 'economic and efficient discharge' of the duties of the local authority, and in view of the importance of not requiring a local authority to incur any further expenditure on sewage disposal than the circumstances of its area require, we feel strongly that local circumstances be taken into account.

Prior to the Water Act 1989 [5] the approach to establishing discharge standards and assessing compliance differed considerably between the effluents from water authority wastewater treatment works (WWTW) and industrial discharges to watercourses. Industrial effluents were subject to maximum or absolute limits with respect to concentration and flow, whereas the water authority WWTW effluents were assessed using statistical criteria. The intention of applying statistical criteria can be regarded as having two objectives: firstly in some way to reflect the inherent variability in WWTW effluents and secondly to reduce the risk of unnecessary prosecution for occasional infringements of a limit. Clearly these two objectives are not necessarily compatible and the use of 95 percentile based standards satisfied neither fully. Again the Report of the Royal Commission 1912 has some continuing relevance in that it stated that it would not be expected that a single violation of an effluent standard would lead to prosecution (recognizing variability), but went on to recommed that, despite this, the power to prosecute for any violation should remain.

A further complication prior to the 1989 Act was that industrial effluents were subject to the requirements of tripartite sampling as a basis for prosecution, whereas the water authority effluents were not. The logic for this difference may seem obvious with the benefit of hindsight in that the number of samples required to obtain a reasonably accurate estimate of a 95 percentile is far greater than could be obtained if tripartite sampling were required, because of the limited resources available for sampling and analysis. That the water authorities were charged with the task of setting standards for their own works, subject to DoE approval, and of assessing their own compliance is a further factor which may have influenced the formulation of the different approaches to establishing discharge standards and assessing compliance vis-à-vis industrialists and the then water authorities. The desirability for 'even-handedness' was emphasized by the National Water Council [6], although the formulation and achievement of such a policy remains a challenge.

Apart from the various difficulties of framing effluent discharge standards with statistical criteria, that is 95 percentiles, and ensuring their consistent application, it is worth noting that there is no firm basis for supposing that compliance with a 95 percentile standard is related to the achievement of a desired environmental quality (EQO). This point is particularly relevant as it is understood that the Statutory River Quality Objectives required by the 1989 Water Act will incorporate a classification system for river water quality which will include reference to EIFAC standards [7], which are also couched in terms of 95 percentiles.

An example of the apparent poor association between 95 percentile values of unionized ammonia and fishery status is given in WRc Report TR 260 [8]. The results for the Severn Trent data are shown in Fig. 1, from which no clear relationship can be deduced. Statistical analysis of data from 285 sites from 1981 to 1985 by Anglian Water revealed no significant correlation between fish biomass and ammonia in either the ionized or unionized form. In fact the data identified several very good and good to moderate fisheries in rivers where the 95 percentile of unionied ammonia exceeded 0.21 mg/l.

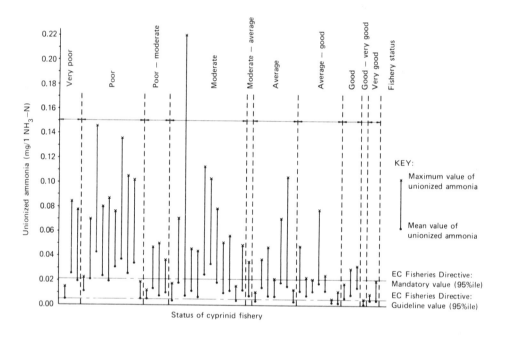

Fig. 1. Mismatch between fishery status and 95 percentile water quality concentrations of unionized ammonia.

In the context of fish mortality, 95 percentiles can be interpreted in a third way, that is if for only 5 per cent of the time a particular concentration (the 95 percentile) is exceeded, then there will be no fish kills. Or alternatively there is only a 5 percent chance of a fish kill. In this sense the true meaning of a 95 percentile value is that it represents some measure of acceptable risk of damage to the environment. Of course this concept is far from that, previously mentioned, of incorporating a measure of variability as a means of reducing the likelihood of prosecution.

The interpretation of 95 percentile standards comprising several determinands can vary according to whether a sample is deemed to fail if any one determinand fails or whether compliance is assessed for each determinand, separately, independent of sample. It is interesting that the latter approach has been suggested by the recent NRA report [9]. Some statistical considerations of alternative forms of effluent consents and compliance have been discussed in detail by Ellis [10,11] generally assuming that the determinands are independent. However, there can be no doubt that many effluent and river quality determinands are inter-related and may act in combination with respect to fish toxicity. It is also apparent that toxicity to fish does not necessarily depend upon compliance with a particular 95 percentile, but is influenced more by: (a) the magnitude of the exceedance; (b) the duration of the event; (c) the frequency of occurrence; (d) the concentrations of other compounding substances or factors.

For example the toxicity of variable concentrations of unionized ammonia to fish depends not only on the magnitude, duration and frequency of concentration peaks, but is also a function of the pH and temperature. Thus the objective of any RQS and subsequently derived discharge standard is to accommodate the state of knowledge with respect to the above factors in a suitable manner for implementation. In order to investigate the response of fish to, for example, variable ammonia concentrations, it is necessary to construct a dynamic model of the toxic response, that is, to be able to predict the lethal concentration at the point of toxic action within the fish as a result of various pollution event scenarios. Such a model would allow the evaluation of the efficacy of different methods of defining discharge standards in relation to the estimated water quality immediately downstream. Where a river quality model is used to predict downstream conditions the fish toxicity effects model would be used to evaluate the impact of the predicted conditions.

DYNAMIC MODELLING OF TOXIC EFFECTS OF AMMONIA ON FISH

The first part of this chapter has argued that a discharge consent needs to take account of the effect on the environmental quality objectives for the receiving water. This part of the chapter illustrates how the interactions between determinands and temporal variations in water quality both influence the effect. The example chosen is the effect of ammonia on freshwater fisheries. Ammonia limits are already included in many discharge consents and are likely to become increasingly widespread if the recommendation of the NRA report [9] is adopted.

Water quality standards for ammonia have been proposed by EIFAC [7], the United States Environmental Protection Agency [12] and the WRc [8]. The standards for the protection of fisheries are based on the concentration of unionized ammonia, which is the form most toxic to fish. The reviews included in the standards show that the toxicity of unionized ammonia to fish is influenced by several factors. The most important of these are: (a) the duration of exposure; (b) the time for recovery between exposures; (c) pH; (d) temperature; (e) dissolved oxygen concentration.

The interaction of these factors has been explored in a mathematical model [13,14]. This model simulated the uptake and depuration of ammonia by fish by modelling the exchange of ammonia between 'compartments' (hypothetical regions within the fish such as blood and tissues). This type of compartmental modelling approach has been widely used in physiological and pharmacological studies. In the field of aquatic toxicology, the most complex model of this type is probably the FGETS model developed by the USEPA to simulate the bioaccumulation of organic pollutants by fish [15].

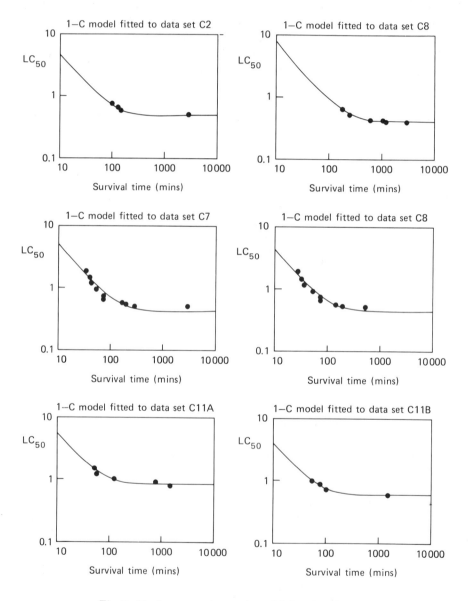

Fig. 2. Single compartmental model fitted to bioassay data.

In the model of the effects of ammonia on fish, one 'compartment' represents the water column, and the other 'compartments' represent the blood and tissues of the fish. The transfer of ammonia from one compartment to another is described by first order kinetics, using an appropriate rate constant. The death of fish was taken to occur when the concentration of ammonia in the compartment representing the site of lethal action reached a critical level. Values for the model parameters (rate constants) were estimated from data obtained from laboratory determinations of the median lethal concentration (LC_{50}) of ammonia for different durations of exposure. Published data were used for this purpose and parameters were estimated by fitting the survival time/concetnration curve predictced by the model to the observed values. Examples of the fit obtained are shown on Fig. 2.

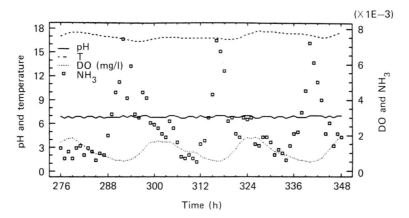

Fig. 3a. Typical variations in river quality for the River Mole, August 1988.

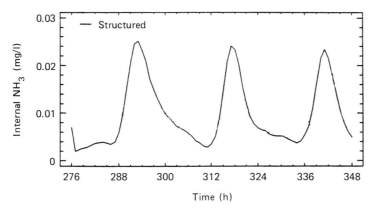

Fig. 3b. Predicted internal concentrations of unionized ammonia in fish
in the River Mole, August 1989.

The effects of pH, temperature and dissolved oxygen (DO) on ammonia toxicity were incorporated in the model by making the model parameters functions of these determinands. The functions were selected so that the behaviour of the model matched the variations in the symptotic LC_{50} concentration of ammonia observed in laboratory tests. The functions chosen were based on the pH and temperature dependence of ammonia toxicity estimated by the USEPA (1985), and Erickson [16], and DO dependence [17].

The model was used to investigate the effect of different patterns of water quality using data for the River Mole at Kinnersley Manor, collected by Thames Water and subsequently the NRA. An automatic water quality monitor provided hourly data for nine determinands including ammonia, pH, temperature and DO. Water quality at the site is influenced by upstream discharges from several wastewater treatment works. Fig. 3a is an example of the type of variations recorded in river water quality. The data exhibit a strong diurnal variation in the concentrations of unionized ammonia and DO, the lowest DO concentrations occuring at around the same time as the highest ammonia concentrations.

Fig. 3b shows the internal concentration of ammonia predicted by the model from the input data given in Fig. 3a. The internal concentration responds relatively quickly to the external ammonia concentration with the uptake rate of ammonia being increased at low levels of DO.

The dynamic behaviour of the model is illustrated in Fig. 4 showing combinations of concentration, duration of exposure and frequency of exposure which are predicted to give a 50% mortality. These curves are based on the response of the model to rectangular pulses of concentration and demonstrate the rapid uptake and depuration of ammonia. The maximum internal concentration of ammonia is reached after about four hours of exposure. It can be seen that for a given interval between pulses a reduction in the duration of the pulse leads to increased concentrations within the fish, that is, there is insufficient time for the fish to recover from the previous pulse. Where the interval between pulses is greater than four hours the response of the fish is predicted to be independent of the duration of the pulse. Thus the model is able to give some indication as to the relative importance of the pattern of exposure, which can then be translated back into criteria for setting the discharge standard. In practice this would require an estimate of the diurnal variations expected in the effluent and the typical variations in the upstream water quality so that a time-series of downstream water quality can be generated through mass balance calculations as an input to the model. It is realized that such information will only be an estimate in most circumstances, but the approach would be able to accomodate such knowledge as exists.

Regarding the calibration of the fish toxicity model it should be noted that only a limited amount of data were available, particularly with respect to the dependence upon pH, temperature and DO. More extensive observations of the effects of fluctuating ammonia concentrations are needed to verify fully the model. In particular the model does not take account of the possible effects of acclimation. However, it should noted that the objective of presenting the results of the model is not to claim a complete mathematical representation of the effect of ammonia on fish, but rather to illustrate a means of identifying those aspects of the discharge which would most influence the environmental quality.

CONCLUSIONS

It has been argued that standards defined in terms of percentiles have been subject to different interpretations and have led to difficulties in assessing compliance. Retrospective analysis to determine performance is no longer appropriate under the new operating conditions of the UK water industry. Even-handedness between the PWCs and industry also requires that the same approach be employed for the setting and assessement of discharge consents.

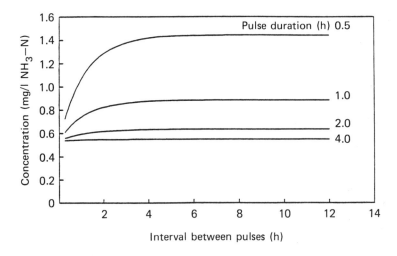

Fig. 4. Predicted combinations of magnitude, duration and interval between pulses of unionized ammonia concentrations geving a 50% fish mortality.

In an ideal world where all the necessary continuous monitoring data are available it would be possible to use a fish toxicity model, such as that described above, to develop discharge standards that are directly related to the desired EQO and vary according to the predicted conditions in the receiving waters. However, for the present such a model can only be used to provide some insight into the relative importance of deviation from the usual effluent quality achieved. The most frequent value arising in a distribution is the mode. It is suggested that rather than any particular percentile, the most sensible measure of central tendency in the context of discharge quality determinands (which may have a variety of parametric and non-parametric distributions) is the mode, which can then be adopted as a realistic operating target. In relation to process evaluation and design the most frequently arising value also seems the most appropriate to adopt.

It is considered therefore that discharge consents be defined in terms of a maximum limit plus an operating target based on a range about the mode of the expected distribution. Allowable deviation from the mode would take into account analytical error. Compliance with the modal range could be defined simply in terms of an allowable number of consecutive determinand values falling outside the specified range. Where reasonable knowledge exists regarding the typical time series of water quality data in the receiving waters, application of the fish toxicity model would help establish the relative values required for the modal range and the maximum limit.

REFERENCES

[1] Price, D. R. H. and Pearson, M. J. (1979) *Wat. Pollut. Cont* pp. 118–138.

[2] Warn, A. E. and Matthews, P. J. (1984) *Wat. Sci. Tech.* 16, 183–196.

[3] Casapieri, P. (1977) *Publ. Hlth. Eng.* 5 (3), 76–79.

[4] Royal Commission on Sewage Disposal (1912) *Standards and tests for sewage and sewage effluents discharging into rivers and streams.* Her Majesty's Stationery Office.

[5] The Water Act (1989) Her Majesty's Stationery Office.

[6] National Water Council (1978) *River water quality – the next stage: review of discharge consent conditions.*

[7] Alabaster, J. S. and Lloyd, R. (1980) *Water quality criteria for freshwater fish.* Butterworths.

[8] Seager, , Wolff, E. and Cooper, V. A. (1988) *Proposed environmental quality standards for List II substances in water – Ammonia.* WRc Technical Report TR 260.

[9] National Rivers Authority (1990) *Discharge consent and compliance policy: a blueprint for the future.* Water Quality Series No. 1.

[10] Ellis, J. C. (1985) *Determination of pollutants in effluents. Part A – Assessment of alternative definitions of effluent compliance.* WRc Report TR 230.

[11] Ellis, J. C. (1986) *Determination of pollutants in effluents. Part B – Alternative forms of effluent consents: some statistical considerations.* WRc Report TR 235

[12] United States Environmental Protection Agency (1985) *Ambient water quality criteria for ammonia.* Environment Protection Agency, Washington DC, Criteria and Standards Division EPA/440/5-85/001.

[13] Lumbers, J. P. and Wishart, S. J. (1989) *The use of a compartmental model to develop rules for the interpretation of water quality data.* No. 2.

[14] Wishart, S. J. (1990) *Environmental quality assessment methods for river pollution control.* PhD thesis. Imperial College of Science Technology and Medicine, London.

[15] Barber, M. C., Suarex, L. A., and Lassiter, R. R. (1988) FGETS (Food and Gill exchange of toxic substances): *A simulation model for predicting bioaccumulation of nonpolar organic pollutants by fish.* US EPA, Environment Research Laboratory, Athens GA 30613.

[16] Erickson, R. J. (1985) *Water Res.* 19, (18), 1047–1058.

[17] Thurston, R. V., Phillips, G. R., Russo, R. C. and Hinkins, S. M. (1981) *Can. J. Fish. Aquat. Sci.* 38, 983–988.

16

Long-term benthic monitoring of the effects of a dense effluent discharged to North Sea coastal waters

N. C. D. Craig, BSc, N. Shillabeer, MSc, and J. C. W. Parr[†]

INTRODUCTION

Cleveland Potash Ltd operate a mine situated in a National Park on the Yorkshire coast north-west of Whitby (NE England). The tailings from the mining and extraction processes are discharged via a 1.25 kilometre long sea outfall which terminates in a depth of 23 metres (Fig. 1). The waste consists mostly of clays and salt and is of a density greater than seawater. To ensure its maximum possible dispersion it is discharged at high velocity from two closely sited outfalls, each of which consists of a pair of 10 cm diameter nozzles. The outfalls are situated in an area of soft sediment that is ringed by outcropping rocks (Fig. 1).

Although both these substrates have been monitored since 1970, this chapter concentrates on the benthic fauna of the soft sediments. This chapter has two objectives. The first is to establish whether the discharge at Boulby has had an effect on the benthic fauna of the sand patch, whilst the second relates to the techniques used.

The aim of the monitoring has been to detemine whether the mine wastes have had any effect on the benthic fauna, and, if so, to define the affected area and to quantify the extent of changes.

---

† ICI Group Environmental Laboratory, Brixham.

Traditional marine benthic monitoring surveys involve the enumeration and identification to species of all the animals sampled. However, recently it has been suggested that the same, or even improved, information may be obtained when the
animals are only identified to the family level [1,2].

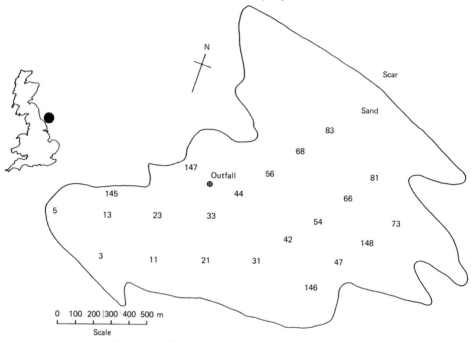

Fig. 1. Boulby sand patch sampling positions.

Much of the work studying the effects of limiting identification to the family level
has involved areas of organic enrichment. The Cleveland Potash monitoring programme provides the opportunity to compare results of data analysed to the species
level, and data analysed to the family level in a physically disturbed environment
that is not subject to organic enrichment.

METHODS

During 1969 a detailed biological survey was made of the whole soft area together
with two potential comparative areas. This survey consisted of more than 200
stations, all identified by Decca Navigator co-ordinates and individually numbered.
The most appropriate stations were selected from the initial survey to form the
monitoring programme; these stations retained their original numbers.

Monitoring commenced in 1970, but discharge did not start until late 1974,
therefore five years pre-discharge data are available. During the 1974 to 1985 period
there was a steady increase in the amount of insoluble material discharged (Fig. 2).

The benthic microfauna of the Boulby sand patch were sampled in the spring
and autumn of each of the 20 years of the monitoring programme. With very few
exceptions, generally related to bad weather or equipment failure, single samples
were taken from each of twenty-two stations within the sand patch (Fig. 1).

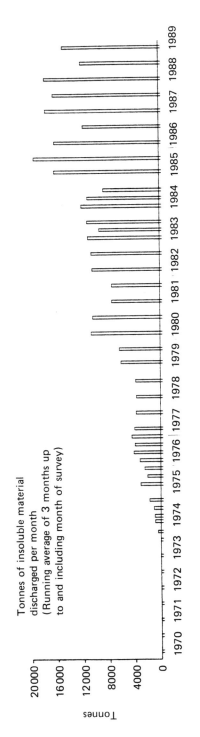

Fig. 2. Insoluble material discharged.

Initially the fauna were sampled using a $0.1\,\mathrm{m}^2$ Smith–McIntyre grab [3], however in 1984 this grab was superseded by a $0.1\,\mathrm{m}^2$ Day grab [4]. The results obtained from the use of these grabs have been compared [5,6]. Tyler and Shackley, working in an area with coarser sediments than those at Boulby, did find significant differences between the sampling effeciencies of the two types of grab. However, work in an area with a very similar sediment type to that found at Boulby [6] has indicated that, provided the Day grab is weighted in the correct way, it samples as efficiently as the Smith–McIntyre.

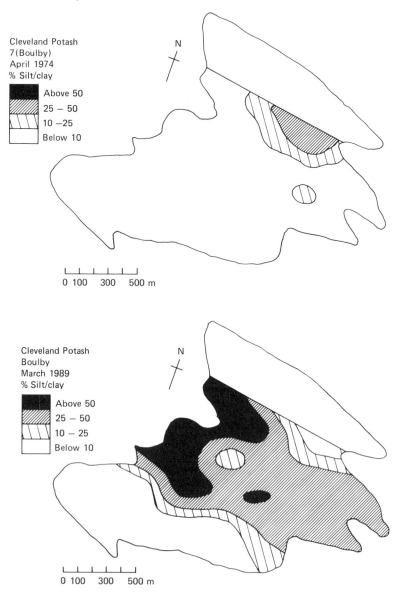

Fig. 3. Percentage of silt and clay.

The sediment in each sample was sub-sampled with a 21-mm diameter corer for particle size analysis and the remainder was washed through a 1-mm mesh sieve. The retained fauna were preserved in a mixture of 4% formalin in seawater together with the dye Rose Bengal. The fauna were counted and, wherever possible, identified to the species level.

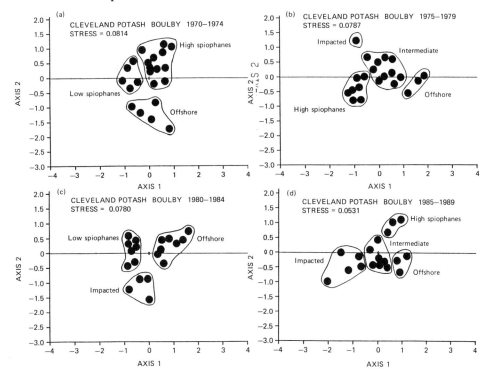

Fig. 4. MDS plots (species level analysis).

From 1970 to 1986 sediment particle size analysis was completed by dry-sieving and weighing [7]. After that date the analysis was completed by use of a laser particle sizer. The results from these two forms of analysis are comparable after the application of a conversion model [8].

The statistical analysis of the data has been completed using the technique of multi-dimensional scaling [9]. Multi-dimensional scaling (MDS) is a technique that positions stations on a graph according to their similarity. Stations with a similar fauna in terms of species and abundance are placed close together, those with a dissimilar fauna are placed widely apart. The graph can then be interpreted as representing zones within an area that have a similar fauna. The data were assembled into four, five-yearly blocks so that each point in the analysis represents ten grab samples. The data were neither truncated to remove rare species nor statistically transformed. Although these latter techniques are often used during the statistical analysis of biological data, the data from Boulby present a clear pattern without their use.

The same analyses were also repeated with species grouped together as families, following the nomenclature of the Marine Species Directory [10].

RESULTS

Before the mine waste discharges began in 1974 there was a limited area (< 10 hectares) of fine sediment at the outer margin of the survey area. This fine area has remained in place, but since 1974 the proportion of the survey area covered by fine sediments has increased. Between 1985 and 1989 fine sediments have frequently covered more than 80 hectares, equivalent to 40% of the sand patch. Fig. 3 illustrates the area of sediment high in silt/clay (< 63 microns) in 1974 and 1989. The fine sediments centre around the outfalls.

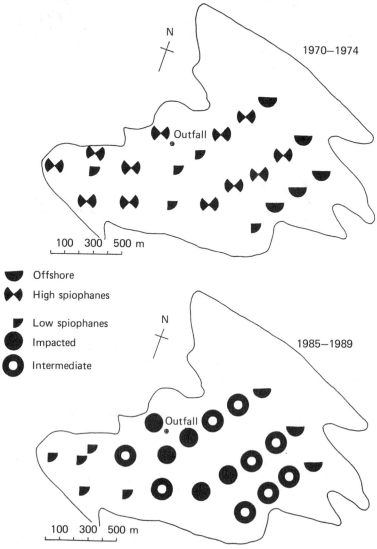

Fig. 5. Boulby sand patch fauna zones.

The results of the MDS analysis of the four datasets 1970–1974, 1975–1979, 1980–1984 and 1985–1989, are set out in Figure 4. This analysis used data in which the fauna have been identified to the species level.

Prior to discharge (Fig. 4a), three zones of fauna type were evident at Boulby sand patch. They consisted of an offshore fauna and two other fauna types, one with high numbers of the polychaete worm *Spiophanes bombyx* and one without.

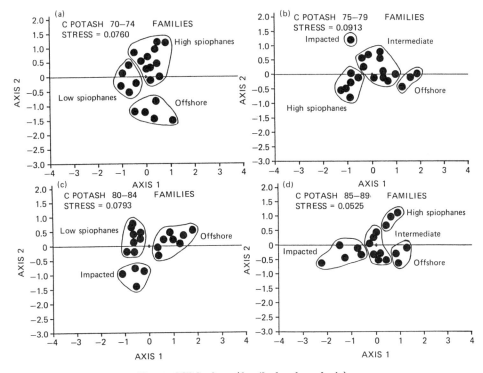

Fig. 6. MDS plots (family level analysis).

During the 1975–1979 period (Fig. 4b) the first evidence of the impact of the discharge was clear. The analysis of these data identified four fauna zones, two of which were similar to those observed during 1970–1974, namely the offshore fauna and the fauna with high numbers of *Spiophanes*. The remaining two fauna groups comprised a group consisting of a single station, station 33, with a sparse fauna, and a grouping consisting of a stations intermediate between the offshore stations and the stations with high numbers of *Spiophanes*. Station 33, the sparse fauna station, is the station closest to the outfall.

Four impacted stations were evident in the 1980–1984 data (Fig. 4c). It is also interesting to note that the number of stations classified as 'offshore' stations increased to nine in this analysis. During the 1970–1974 period there were five, and in 1975–1979, three stations in this group. This increase may have also been related to the effluent discharge, inshore stations whose sediments became only slightly finer supporting a fauna more typical of the finer offshore sediments.

Analysis of the most recent dataset (Fig. 4d) has identified five impacted stations. The remainder of the stations divide into an offshore zone and stations with and without high numbers of *Spiophanes*.

The distribution of the fauna zones pre-discharge (1970–1974), and in the most recent period (1985–1989), are presented in Fig. 5.

When the MDS analysis was repeated using the same data identified to the family level (Fig. 6), the MDS outputs presented the stations in identical groupings to those in the original, species level, data analysis. As in the species level analysis, station 33 was identified as impacted during the 1975 to 1979 period, and the succession of impacted stations in 1980–1984 and 1985–1989 was the same.

DISCUSSION AND CONCLUSIONS

The original objectives of the monitoring programme were to identify whether the discharge had an effect on the benthic communities and, if it did, to describe the limits of that effect. The work was funded by Cleveland Potash Ltd, not as an academic exercise, but as a means of understanding the effects of their effluent on the local environment and thereby maintaining their right to discharge.

The effects of the disposal of fine, inorganic, sediments on marine benthos in British coastal waters have been studied before. Probert [11] and other earlier workers studied the effects of china clay waste on the benthos of the predominately gravel substrates of Mevagissey Bay, Cornwall. This discharge did have an effect on the local benthos but it did not produce the severe effects that have been observed off Boulby. Bamber's work [12] on the disposal of pulverized fly ash (pfa) off the north-east coast of England describes effects that were similar to those observed at Boulby. In this case the disposal of fine ash into an area already containing a fine substrate led to the establishment of a clear gradient of effect running from the dumpsite. At the station with the highest level of pfa a $0.2\,\mathrm{m}^2$ sample yielded only eight individuals.

The changes caused in the sediment characteristics of the Boulby sand patch by the effluent discharge are comparable to those caused elsewhere by pfa dumping, and one would therefore expect a similar response to be evident in the benthic fauna. A marked gradient of effect is obvious around the outfalls. The area in which the benthic fauna is affected covers approximately 40 hectares of a sandy patch which has a total area of 200 hectares.

Therefore the primary objective of this study has been achieved in that the effect of the discharge on the local benthos has been described, as has the size of the affected area.

The large data base generated by this study has also provided an opportunity to compare the results from analysis of data generated at the family level with that from species level identification. The application of a multivariate analysis to these data produced the same results from both the family and species level data. Multivariate analysis is most commonly used to assess the information obtained from benthic monitoring, and the results obtained suggest that, in future, the identification of the fauna at Boulby could be limited to the family level without any loss of accuracy.

This conclusion can be easily reached for the Boulby monitoring programme where there are pre-discharge data, twenty years of results and a market gradient of effect. Such datasets are rare and the application of family level monitoring to all benthic monitoring studies must be approached with great caution.

The advantages of only analysing samples from monitoring surveys to the family level are obvious. There will be a major saving in the time spent processing the samples. This should not be regarded as simply a means of saving costs but as an opportunity to expand the other areas of monitoring programmes which are often deficient, such as an adequate sampling regime and sample replication.

The question of the relative accuracy and value of species and family level identification causes considerable discussion amongst benthic biologists. It may be that the design of future monitoring programmes may incorporate a compromise whereby baseline surveys involve identification to species level but that established programmes only involve family level identification. The results obtained add to the body of literature supporting family level studies. The authors are not aware of any published work on marine benthic monitoring that has highlighted a case where species level identification has been a proven necessity.

ACKNOWLEDGEMENT

This chapter was produced with both the permission and support of Mr G. Holyfield of Cleveland Potash Ltd.

REFERENCES

[1] Warwick, R. M. (1988) Analysis of the community attributes of the marcrobenthos of Frierfjord and Langesundfjord at taxonomic levels higher than species. *Mar. Ecol. Prog. Ser.* **46**, No. 1–3, 167.

[2] Warwick, R. M., (1988) The level of taxonomic discrimination required to detect pollution effects on marine benthic communities. *Mar. Poll. Bull.* **19**, No. 6, 259.

[3] Smith, W. and McIntyre, A. D. (1954) A spring-loaded bottom sampler. *J. Mar. Biol. Ass. (UK)* **33**, 257–264.

[4] Warwick, R. M. and Davies, J. R. (1977) The distribution of sublittoral macrofaunal communities in the Bristol Channel in relation to substrate. *Est. Coast. Mar. Sci.* **5**, 267–288.

[5] Tyler, P. and Shackley, S. E. (1978) Comparative efficiency of the Day and Smith-McIntyre grabs. *Est. Coast. Mar. Sci.* **6**, 439–445.

[6] Shillabeer, N. (1985) *A comparison of the sampling efficiency of Smith-McIntyre and Day grabs when used on varying sediment types.* ICI Brixham Laboratory report BL/B/2640.

[7] Holme, N. A. and McIntyre, A. D. (1971) *Methods for the study of marine benthos.* IBP Handbook 16. Blackwell Scientific Publications.

[8] Shillabeer, N. *et al.* (in press) *A comparison of the results achieved by dry-sieve and laser sizer analysis of marine sediments.*

[9] Field, J. G., Clarke, K. R. and Warwick, R. M. (1982) A practical strategy for analysing multispecies distribution patterns. *Mar. Ecol. Prog. Ser.* **8**, 37–52.

[10] Howson, C. M. (1987) *Species directory to British marine fauna and flora.* Marine Conservation Society.

[11] Probert, P. K. (1981) Changes in the benthic community of china clay waste deposits in Mevagissey Bay following a reduction of discharges. *J. Mar. Biol. Ass. (UK)* **61**, 789–804.

[12] Bamber, R. N. (1984) The benthos of a marine fly-ash dumping ground. *J. Mar. Biol. Ass. (UK)* **64**, 211–226.

17

Monitoring of pollutant fluxes in point source inputs to rivers and coastal waters

R. F. Rayner, BSc, MSc, PhD, CBIOL, MIWEM[†], M. Tidman, BSc, MIWEM[‡], and A. Fitzgerald BSc[§]

INTRODUCTION

Measurements of pollutant fluxes from domestic and industrial discharges are usually undertaken for two purposes. Firstly, as a means of monitoring compliance with individual discharge consents and secondly, to determine inputs to river catchments and coastal regions when examining their pollutant loading, their receiving capacity and their general water quality status.

The formation of the National Rivers Authority and the the general adoption of the principle of 'polluter pays' has increased the need for reliable monitoring of pollutant fluxes by both regulator and discharger.

The development of management plans for rivers and coastal regions which meet the requirements of ever more stringent environmental legislation necessitates access to reliable data on all major inputs both in terms of mean fluxes and those under extreme conditions such as during the operation of storm overflows.

For all of these applications it is desirable to be able to monitor flows and the concentration of particular water quality determinands on a continuous basis. In practice this is technologically difficult for most of the parameters of interest.

[†] Director, Wimpey Environmental Limited.
[‡] Senior Environmental Scientist, Wimpey Environmental Limited.
[§] Senior Environmental Scientist, Wimpey Environmental Limited.

PRESENT TECHNIQUES

To quantify volume and rate of pollutant input from a point source some form of flow measurement is required. There are three broad options available for achieving this: (a) use of data from permanent flow measurement installations; (b) carry out spot measurements; (c) temporary installation of continuous recording flow measurement equipment.

In the absence of permanent installations the choice is between taking spot measurements and installing continuous recording flow monitors for a specified period; to obtain a detailed and comprehensive set of data the latter option is usually chosen.

Flow monitors come in many shapes and sizes. For reasons of versatility and reliability in use those based on pressure depth measurement and doppler velocity measurement are best suited to the task. Pressure and velocity sensors are usually incorporated into a single sensing head with typical operating ranges of 25-4000 mm for depth of flow and 0.15-4.0 m/s for velocity. The streamlined sensors are fitted to the wall or invert of the sewer and interference to the effluent flow is minimal.

Data is usually logged onto solid state memories at selected time intervals (typically every two minutes). Manual flow measurements are taken at each site to calibrate the equipment and to determine the velocity profile. Once installed and calibrated these instruments may be left in place for considerable periods of time although to ensure continuity of data it is advisable to visit each measurement site on a weekly basis to retrieve data and check that the instruments are working correctly.

Comprehensive guidelines on the measurement and analysis of sewer flow data are provided in the WAA/WRc Flow Survey Guide [1].

Having established a means of reliably monitoring flow velocity and depth, and hence flow volumes, it is necessary to determine flow composition in order to compute the fluxes of chemical and microbiological pollutants from a discharge. Fluxes need to be determined for a range of operating conditions since they may vary over both short and long timescales. For example, pollutant fluxes from an operating storm overflow will vary considerably over timescales of less than one hour, whereas fluxes from a typical wastewater outfall will vary more gradually through a day but may exhibit marked seasonal variability.

Ideally, the measurement of chemical and microbiological composition of inputs should be conducted on a continuous basis so that events occurring on differing timescales are adequately detected. In practice, this is only possible for a very limited range of determinands. For many determinands collection of discrete samples on a time interval basis with subsequent laboratory analysis remains the only viable technique for deriving time histories of pollutant flux. Intensity of measurement is therefore often a balance between desirable sampling rates and the cost of laboratory determinations.

Where concentrations are relatively constant and there is no necessity to detect periods where flux is greater than some predetermined value, then composite sampling techniques may be employed. These rely on drawing a continuous sample at a known rate and accumulating it in a suitable container. Analysis of this sample provides a time integrated concentration. However, the period over which samples may be collected remains constrained for many determinands by considerations of sample viability over time.

Automatic samplers and composite samplers have the obvious advantage over manual sampling in the reduction in time spent on sample collection. However, samples once collected within the sampler can degrade until collection, sampling flexibility is reduced since sampling rate must be predetermined and the use of these techniques is limited where separate samples are needed for a number of different determinands. When used in a synoptic study of a number of point source inputs, they also have the disadvantage that failure of the system will not be detected until the time of retrieval thus creating the potential for loss of a vital input to an intergrated study.

For the purposes of a relatively short term but intensive study of a system the most suitable method remains manual collection of a large number of samples, both on a time basis and a determinand by determinand basis. Manual collection gives considerable flexibility in the timing and nature of sampling. When properly conducted it ensures that scientific integrity is maintained through correct use of different types of sample containers, suitable pretreatment and preservation of samples, and sufficiently rapid transport to the laboratory for analysis.

The main disadvantage is the high cost per sample both in terms of labour and sophisticated analytical techniques. For major studies adequate handling and pretreatment of samples can only be successfully accomplished through the use of on-site laboratory facilities which further adds to the cost and logistic complexity of such work.

In situ techniques give the obvious advantages of immediate availability of data, potential for rapid or even continuous measurement, and the potential for acquisition of data from remote sites by data telemetry.

Although in principle a number of determinands may be measured by *in situ* techniques, in practice long-term measurement, particularly at temporary locations, has many limitations. Problems such as sensor survivability, sensor drift, interference between chemical species and the difficulties associated with keeping sensors clean and free from fouling all serve to limit the effective use of in situ techniques. For this reason only a very limited number of determinands are routinely monitored in this way. Examples are *in situ* determination of parameters such as temperature, conductivity (salinity), suspended solids, dissolved oxygen and pH. Even for these, at first sight quite simple, parameters great care must be taken to use these methods only under suitable field conditions. For example reliable *in situ* measurement of pH under estuarine conditions is particularly difficult because of the effects of varying salinity on the performance of pH electrodes.

It is necessary to ensure that the results of *in situ* measurements are carefully and frequently calibrated by reference to laboratory analysis of samples. Sensors and associated electronics will often require frequent calibration and cleaning if they are to continue to function correctly.

More sophisticated monitoring systems do exist. Examples are those for measurement of ammonia and other nutrients using ion-selective and colorimetric techniques, TOC monitors employing chemical digestion and infra-red measurement, hydrocarbon by fluorescence and complex oxygen demand systems. However, these systems are not yet at a stage in their development where they can readily be used in temporary installations.

Determination of the flux of microbiological constituents of wastewater inputs is a common requirement particularly at coastal sites. This can only be achieved by the collection of discrete samples with subsequent laboratory analysis. There is also the added problem that the organisms within a collected sample only remain viable for a limited period of time. There remains considerable debate about the effects of sample storage on the count of particular micro-organisms, and even the currently accepted maximum storage period of six hours for determinations of coliforms is open to question. It is therefore vital that samples are analysed as quickly as possible. In many cases this will necessitate the establishment of temporary laboratory facilities at the site of interest.

CASE STUDIES

Some of the practical difficulties associated with monitoring fluxes are best illustrated by reference to actual studies undertaken at two UK sites.

Site 1

A major environmental investigation was conducted in the Taw and Torridge region of North Devon over a two-year period between October 1987 and the autumn of 1989. The study area comprised the riverine and estuarine regions of the Rivers Taw and Torridge and the coastal region of Barnstaple Bay. The study focused on an assessment of seasonal variability in physical, chemical, biological and microbiological conditions within the region. The programme of work also included the determination of point source fluxes of a variety of determinands from all major discharges. Fluxes were determined under conditions of dry and storm flows during summer (peak holiday period) and winter. Measurements were conducted over two one-month periods during 1988.

Simultaneous monitoring was undertaken at each of the 32 major discharges into the study area. Monitoring sites included: (a) Gravity outfalls of crude sewage effluent: (b) Gravity discharges of treated sewage effluent: (c) Gravity storm overflows: (d) Pumped discharges of crude sewage effluent and industrial effluent.

Measurement of the gravity flows was achieved using continuous recording velocity and depth monitors. The two pumped discharges were measured by calibrating the pumps and monitoring periods of pump operation. At one site measurements had to be taken directly from a sea outfall where the only access was well below the high tide mark on an exposed coastal site. For this site it was necessary to construct a custom designed flow monitor capable of operating whilst submerged and housed in such a way that it would not be damaged by storms.

A major problem encountered during the fieldwork was the interference to flows caused by tidal water intrusion at coastal and estuarine sites. Surcharging occurred at many sites at high tide and analysis of the free catchment discharge was therefore complex. Wherever possible, the monitors were located far enough upstream above the level of high water. However, such was the extent of tidal influences that in many cases this would have resulted in the flow measurement point being too far from the outfall and therefore not representative of the total catchment discharge. In these cases the data analysis techniques used to compute fluxes had to be modified so that the effects of tidally induced flows could be removed.

Flow composition was determined during a number of intensive sampling periods during which samples were taken and analysed for the following: nutrients (phosphate, ammonia and TON); metals; BOD; COD; chloride; total and faecal coliforms, and suspended solids.

Typical summer and winter dry weather flows were monitored by three-hourly sampling of all foul sewers over two sixty-hour periods. This proved to be a complex logistic exercise, particularly during the peak summer period when traffic conditions made sample collection and transport particularly difficult.

Sampling programmes had to be designed so that all volatile and degradable determinands could be fixed and analysed as soon as possible. In the case of nutrients this entailed filtering of samples on site followed by immediate freezing to -20 °C for transportation. Sample shipment logistics had to be established so that all BOD and bacterial enumeration samples could be delivered to the laboratory within six hours of collection. Metal analysis samples were collected in pre-acidified containers to that the sample remained stable whilst in transit to ensure the laboratory.

Monitoring of fluxes during major storms was also undertaken on a simultaneous basis for all discharges. This was a particularly difficult exercise since it was necessary to mobilize all sampling equipment, personnel and laboratory facilities at very short notice based on weather radar forecasts of storm events. Storm overflows were monitored both manually and with automatically flow-triggered composite samplers. After the onset of heavy rainfall major flows were sampled manually every ten minutes for the first hour, and at fifteen minute intervals for the following two hours. This proved to be a satisfactory time interval for detecting 'first flush' effects. Minor flows were sampled by flow-depth triggered samplers which provided only a single measurement of mean concentration over the period of flow. An unusual limitation of such samplers was illustrated by this study in those storm sewers that became tidally surcharged so initiating sampling on every surcharge event. It proved necessary to incorporate a manual override so that sample pumping would only begin when an additional trigger was activated.

Site 2

The second case study related to work conducted in the Plymouth area. This work comprised a baseline assessment of the physical, chemical and microbiological conditions of the waters of the Tamar, Tavy and Plym estuaries and Plymouth Sound, conducted during August and September of 1990 and a detailed study of the discharge from the Camel's Head treatment works conducted during May 1990.

A total of 13 point source inputs were monitored to determine mean and maximum dry weather loading during the peak of the tourist season. As with the Taw and Torridge study manual sampling was chosen as the most effective means of obtaining the necessary information. Flow monitoring was undertaken over a single 30-day period with constituent sampling at three-hourly intervals over two separate 24-hour periods and two ebb tide periods. There were a number of problems for the flow survey engineers to overcome including selection of suitable sites for monitors, problems of access to sewers and once again the problems associated with tidal surcharging.

A further aspect of the work in the Plymouth region was an intensive study of the nature and fate of the discharge from the Camel's Head treatment works. The

Camel's Head works handles primary treatment from a population equivalent of 63 000. Treated effluent is discharged into Weston Mill Lake, a tidal inlet off the Tamar Estuary. Work at this site comprised the study of the hydrodynamics of the local waters and the quantification of fluxes of ammonia from the discharge over a 12-day period.

The study was used as an opportunity to develop and evaluate continuous monitoring techniques for ammonia which were employed alongside conventional sampling and laboratory analysis. This allowed inter comparison of results and well illustrates the problems associated with producing reliable measurements by in situ techniques.

Direct monitoring of ammonia was conducted using a system based on a potentiometric ion-selective probe. Such ion-selective probes offer the potential of obtaining continuous measurements free from operator error and sample transit degradation problems. The ion-selective probe was used to monitor a filtered extract from the treated effluent flow using a system of pumps, and filters. A Minivac GP50 diesel pump was used to provide the required 3-m suction head and supply the filter with effluent at the required flow rate of 70 litres per minute. The continuous effluent stream was filtered by a Kent Ultrafilter, a standard self-scouring cross-flow filter.

As in the earlier case studies effluent volume flow was measured using pressure depth and doppler velocity sensors. Ammonia concentrations were determined by manual sampling with a higher sampling rate during peak flows (samples taken every 30 minutes) than at other times (samples taken every hour).

Ammonia concentrations were monitored using a Kent ammonia ion-selective electrode. This detects internal pH changes in response to ammonia movement through a semi-permeable membrane, which changes the potential difference between a gas-sensing electrode and a reference electrode. Potentiometric probes measure total ammonia only; for this reason, a system of mixing tubes is needed to mix an alkaline reagent with the sample and provide sufficient time for all the ammonium to be converted to gaseous ammonia. The whole system was circulated by a ten-channel peristaltic pump. Probe responses were passed to a Corning Delta 250 which provided a digital readout and an analogue output.

A separate section of the system also provided continuous monitoring of pH and temperature. Results of continuous monitoring and discrete sampling of ammonia showed similar trends, but concentrations obtained by the probe system were consistently lower than those obtained by conventional laboratory analysis.

A major contributor to the difference is thought to be the presence of high levels of amines within the effluent. These cause positive interference to the chemical analysis of ammonia, and negative interference to the ion-selective probe. Algal growth within the transmission lines to the probe may also have caused reduction in ammonia prior to the sample reaching the probe.

The most notable discrepancy between the discrete sample results and continuous monitoring was the occurrence of unexpectedly high short-term fluctuations in concentration in discrete samples, which was not mirrored in the continuous monitoring results. Results from the continuous monitor tended to vary smoothly throughout the measurement period.

The occurrence of rapid fluctuations in ammonia concentrations has been traced back to works operations such as weir cleaning. These result in transient spikes of high concentration which are detected if sample collection occurs at the right time. These rapid changes were not detected by the continuous monitor because the mixing characteristics of the cross-flow filter were thought to have effectively buffered any sharp changes in concentration.

CONCLUSIONS

The above case studies help to illustrate that there are no simple solutions to overcoming the scientific, technical and logistic complexities of accurately and reliably monitoring pollutant fluxes from point source inputs.

The problems of physical measurement of flow have largely been overcome in existing sensors, but great care must be taken to ensure that flow measurements are conducted at appropriate locations, particularly in discharges to tidal waters.

For chemical and microbiological measurement the scientific and technical problems remain complex. There is currently no one system that will provide high quality, high density data for a number of determinands in all types of use. At present the advantages and disadvantages of different techniques are such that the choice must be based on an assessment of each individual application. The range of parameters required, the time period over which monitoring is to take place, the frequency of sampling and the nature and location of the discharge must all be taken into account in making the choice [2]. It is also necessary, in most cases, to strike a balance between desirable objectives of monitoring and cost.

For long-term monitoring commitments, such as assessment of consent compliance, continuous monitoring is the ideal sought by many development programmes. For shorter-term more generalized surveys of inputs to catchments and coastal regions, an approach which combines *in situ* techniques with autosampling and sample pretreatment is desirable. However, development in this area remains slow.

At first sight the use of *in situ* probes has many attractions. However, as the case study illustrates there is still some way to go before measurement of systems are available where a probe can be simply immersed in the flow and will then provide reliable long-term measurements. Nevertheless the increasing variety and sensitivity of ion-selective probes is such that they have considerable potential, particularly if the problems of long-term stability and use in saline waters can be overcome. A useful review of available ion-selective electrodes is provided by Whitfield [3].

A number of other techniques are also advancing to the stage where they may be of use in portable systems, including voltametry for the detection of trace metals [4].

A recently produced report by the Department of Trade and Industry [5] highlights the considerable advances that are taking place in solid state and optical sensors. Biosensors are also under development for specific analytes and broad band measurements.

ACKNOWLEDGEMENTS

Case studies described in this chapter draw on work conducted by Wimpey Environmental on behalf of South West Water Services Limited. The authors would like to thank South West Water for granting permission to make use of this material.

REFERENCES

[1] Water Authorities' Association/WRc (1987) *A guide to short-term flow surveys of sewer systems.*

[2] Rayner, R. F. and Pagett, R. M. (1989) Monitoring water quality in the marine environment – present methods and some likely future improvements. In: *Advances in water modelling and measurement* (M. H. Palmer, Ed). BHRA Information Services pp. 1–10.

[3] Whitfield, M. (1985) Ion-selective electrodes. In: estuarine analysis. In: *Practical estuarine chemistry.* (P. C. Head, Ed). Cambridge University Press pp. 201–272.

[4] Anon (1987) One step ahead for water analysis. *Water and Waste Treatment,* September 1987, p. 16.

[5] Department of Trade and Industry (1990) *Advanced sensors in the environment.*

18

Water quality in London's Docklands and management strategies

**J. B. Dawson, BSc, C.Eng, MICE (Dip WEM), FIWEM[†],
K. Guganesharajah, BSc, MSc, C.Eng, MICE[‡], and B. Thake, BSc[§]**

INTRODUCTION

The London Docklands extend for 17 km on both sides of the River Thames, downstream of Tower Bridge, as shown on Fig. 1. They include some of the largest man-made docks in the world, but changes in shipping practice resulted in the docks becoming increasingly uneconomic, and by 1980 they had all closed.

In 1981 London Docklands Development Corporation (LDDC) was established with the objective of regeneration an area of 21 km², which encompassed the docks, and a massive redevelopment programe is now in hand. Many of the projects being undertaken make a feature of the water environment. In addition, access to the docks is being increased and leisure use is being encouraged. Water sports such as canoeing, sailing, windsurfing, water skiing and water bikes are already permitted. At the same time, some of the docks are used by community groups for angling and nature conservation. These activities have to complete with the use of the docks for the occasional berthing of large vessels, and their possible use a heat source/sink for heat pumps.

[†] Associate, Mott MacDonald Water Supply and Wastewater.
[‡] Divisional Director, Mott MacDonald Water and Land Development.
[§] Lecturer in Marine and Freshwater Biology, Queen Mary and Westfield College, London University.

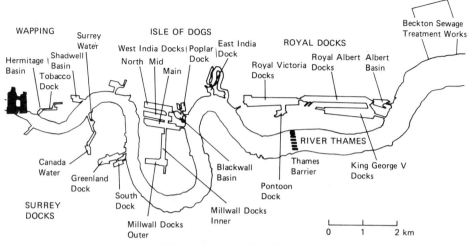

Fig. 1. London's Docklands.

Table 1. Average depths, areas and volumes of the dock system

Dock system	Depth (m) area	Water surface area (ha)	Volume (hm³)	Runoff area (ha)
Royal Dock	9.6	95.8	9.1	5.7
Isle of Dogs	8.7	54.0	4.7	6.8
East India	6.8	2.4	0.16	0.0
Shadwell Basin	7.2	3.2	0.23	0.7
Hermitage/				
Tobacco Dock	1.0	1.4	0.01	0.18
Surrey Water	4.0	1.2	0.05	1.6
Canada Water	9.0	1.5	0.14	0.3
Greenland/South Dock	7.8	10.9	0.85	10.5

There have been complaints from users of unsightly litter, oil films and surface scum, and in 1988 a water sports event had to be cancelled owing to the presence of algal blooms.

The Corporation therefore commissioned a study of water quality in all docks. This chapter describes the results of the study to investigate the factors influencing water quality; the aim is to provide water which is: (a) aesthetically pleasing, and (b) acceptable for water contact activities. In addition, a mathematical model was to be developed to predict water quality changes and assist in the evaluation of development options.

DESCRIPTION OF THE DOCK SYSTEM

The area for which the LDDC is responsible contains 20 different docks, which are shown in Fig. 1. These docks are in four main groups: (1) Royal Docks; (2) Isle of Dogs; (3) Wapping; (4) Surrey Docks.

Fig. 2. On-site sampling and monitoring in West India Dock.

There are large variations in size as shown in Table 1 from which it can be seen that the Royal Dock system is much larger than the others.

Inflows and outflows

Principal flows into and out of the docks are from the river except for Hermitage Basin/Tobacco Dock and Canada Water, where flows are from Thames Water potable supply and groundwater, respectively. Surface water runoff only enters from the paved area immediately adjacent to the docks, and foul sewers have been diverted so that they do not discharge into the docks. Average retention time in each of the dock systems varies between two days for East India Dock, and about 2900 days for Canada Water. Except for East India Dock, the residence time is extremely long and short-term variations in input flows and quality are unlikely to have much influence on the overall quality of the dock waters.

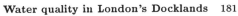

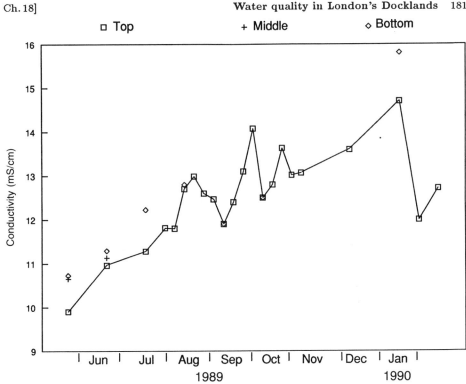

Fig. 3. Royal Albert Dock: plot of conductivity v. time.

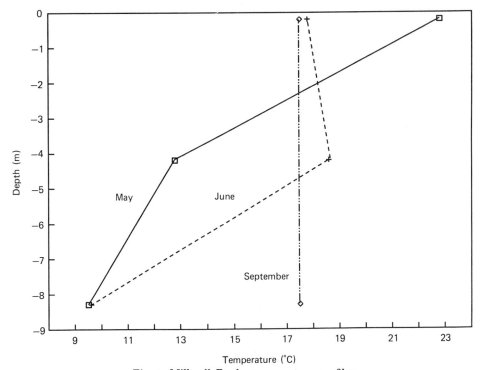

Fig. 4. Millwall Dock: temperature profiles.

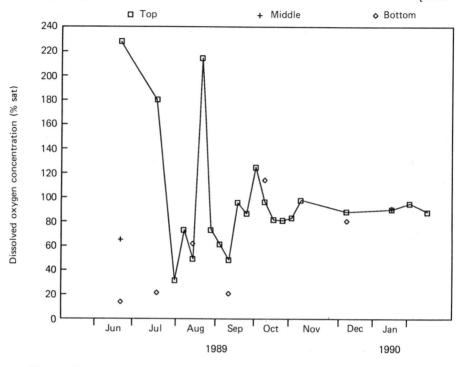

Fig. 5a. King George V Dock: concentration of dissolved oxygen v. time.

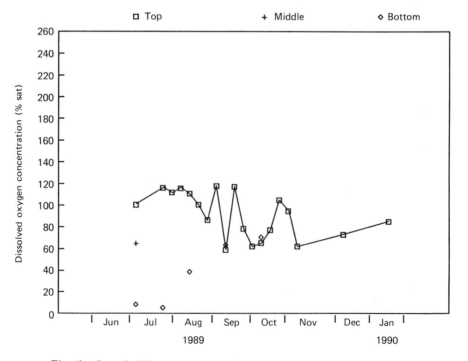

Fig. 5b. Canada Water: concentration of dissolved oxygen v. time.

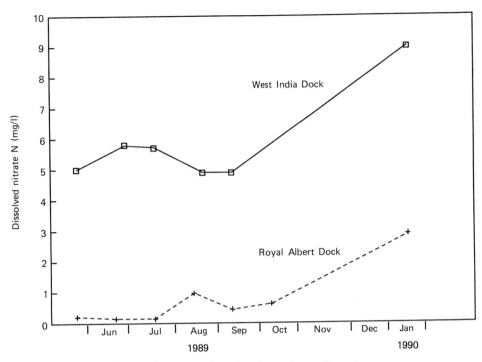

Fig. 6. Concentration of surface nitrate-N v. time.

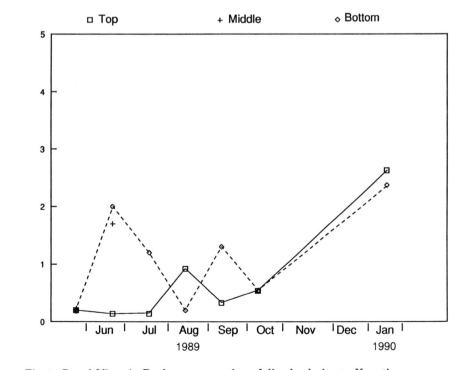

Fig. 7. Royal Victoria Dock: concentration of dissolved nitrate-N v. time.

SURVEY OF THE DOCK WATERS AND INPUTS

To determine the characteristics of the waters, monitoring sites were established within the docks and at location where significant flows could enter the docks (Fig. 2). Samples were then collected, nominally at monthly intervals. On-site measurements were taken and analyses of determinands which could not be preserved were made shortly after sampling. Analyses for other determinands were carried out off site. Determinands can broadly be classified as physical, chemical, algological or microbiological.

Surveys to monitor the phytoplankton crop were conducted weekly, and monitoring in respect of the parameters specified in the EC Bathing Water Directive [1] was generally carried out fortnightly.

RESULTS AND DISCUSSION

Conductivity

The river is tidal, and since most of the docks are connected to the river, changes in the salt concentration of the river are reflected in the dock water. The Royal Docks have the highest salt concentration, and there is a reduction in the salinity the further upstream a dock system is located. Variations in the conductivity with time in the Royal Albert Dock are shown in Fig. 3. There was a general increase in the conductivity during 1989 because of the low rainfall and an increasing proportion of saline water in the river.

Temperature

Maximum surface water temperature of about 24 °C was reached in July after which it declined gradually to a minimum winter temperature of 8 °C. There was a fairly uniform decline in temperature with depth, and in May, temperatures at the bottom were much colder than on the surface. By August most docks were mixed and similar temperatures were recorded throughout the system.

The temperature profile for Millwall Docks in May was very steep, as shown in Fig. 4, with very cold water at the bottom caused by the presence of a barrier restricting water movement. Uniformity of temperature was not obtained until mid-September.

Frequent boat movements in West India Docks resulted in the water being well mixed, and the contents were at a similar temperature throughout the dock for the duration of the study.

Oxygen

Variations in the concentration of dissolved oxygen (DO) with time for King George V Dock are shown in Fig. 5a. At the surface peak concentrations of over 220% saturation were measured, and when DO concentrations were high at the surface they were very low at the bottom. In July oxygen concentrations at the bottom fell to a minimum in Millwall Docks, Shadwell Basin, Surrey Water, Canada Water, Greenland Dock and South Dock. In Millwall Dock 2% saturation was measured and 6% in South Dock. The low DO at the bottom of South Dock was associated with 230% saturation on the surface and a pH of 10.48. From Fig. 5b in Canada Water where the principal of inflow is from groundwater, the surface concentration of DO was much less than for the other docks.

Table 2. Average surface concentrations of combined
inorganic nitrogen (N) and phosphorus (P) (mg/l)

Dock (sites)		May	June	July	1989 Aug	Sept	Oct	1990 Jan
Royals	N	0.68	0.19	0.23	1.42	1.15	1.01	2.98
(1-5,7)	P	1.73	0.97	0.77	2.04	1.82	2.42	2.2
West India	N	5.56	6.05	5.74	3.44	3.45	–	6.12
(9-14)	P	1.22	1.35	1.37	1.02	1.27	–	1.27
Millwall	N	3.36	5.75	5.09	5.14	4.95	–	8.89
(17-19)	P	0.67	1.67	0.63	1.6	1.45	–	1.95
Shadwell	N	0.49	0.46	0.21	–	–	–	1.13
	P	0.10	0.61	0.19	–	–	–	2.20
Hermitage	N	–	0.18	–	–	–	–	1.79
	P	–	0.50	–	–	–	–	1.00
Surrey Water	N	1.18	0.47	0.19	0.08	0.70	–	1.15
	P	0.30	1.10	0.36	0.03	1.20	–	–
Canada	N	0.25	0.18	0.20	0.12	0.20	0.50	0.74
	P	0.10	0.13	0.02	<0.02	<0.02	0.27	0.02
Greenland	N	0.73	1.28	0.70	–	–	–	5.22
	P	1.50	0.94	0.40	–	–	–	2.70
South Dock	N	0.76	0.97	0.21	–	–	–	5.73
	P	1.20	1.00	0.13	–	–	–	2.60

Nutrients

In the Isle of Dogs, the surface concentration of nitrate nitrogen was in the range 5 to 6 mg/l during the peak algal growing season, increasing to more than 8 mg/l in January, as shown in Fig. 6, the same order of magnitude as in the river. Because waters in the West India Docks are mixed by boat movements there is little difference between measurements made at the surface and at depth.

In Table 2 the concentration of inorganic nitrogen during the peak algal growing season was much less in the dock systems where river water enters less frequently.

Fig. 7 shows that in the Royal Docks reduction in the surface concentration of nitrate was accompanied by an increase in its concentration in the bottom waters. Phytoplankton growth resulted in the depletion of all forms of inorganic nitrogen and decomposition of phytoplankton cells at depth with the liberation of nitrogen.

It is considered that phosphorus concentrations should be below 10 μg/l to limit algal growth [2]. Inorganic phosphorus concentrations at the surface and bottom were approaching this limit in Canada Water between July and October, and in Surrey Water in August, although, as shown in Fig. 8, this did not occur in the Royals or the Isle of Dogs.

Nitrogen may have limited phytoplankton production, particularly since phosphate concentrations were relatively high (and N:P ratios were low) and there were low epilimnetic concentrations [3]. However, a low concentration in the water is not necessarily an indication that a nutrient is in short supply for algal growth [4]. The concentration at which a given nutrient becomes limiting varies according to the level of other factors, and many algae, particularly blue-greens, can store nutrients within the cell.

Silicate concentrations were high in all docks and at all depths with values between 0.6 and 15 mg/l SiO_2 being observed. The concentrations are greater than the 0.5 mg/l SiO_2 figure considered to be limiting for the growth of diatoms [5].

At all times the observed alkalinity was in the range of 50 to 270 mg/l CaO_3, typical for lowland revers. The pH values were generally between 8.5 and 9.5, although values up to 10.5 were observed in surface waters of docks with large phytoplankton populations. These values are greater than those observed in the Thames of 7.2 to 9.1 (NRA 1989), and result from the growth of algae in the docks. Since the dock water is 'naturally' alkaline and highly buffered, the rise in pH is more notable. At the higher pH values it is doubtful if there was any free CO_2. However, since many phytoplankton taxa are able to utilize bicarbonate as a source of CO_2 for photosynthesis and some can use carbonate ions, algal growth may not be severely limited by carbon supply [6].

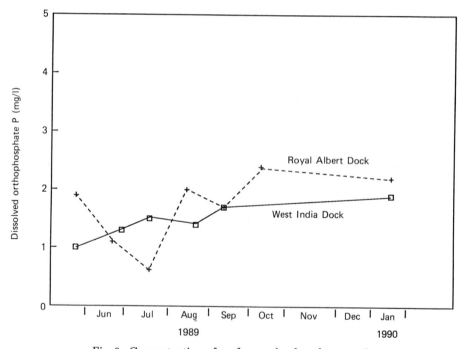

Fig. 8. Concentration of surface orthophosphate v. time.

Algae

All the docks with the exception of Canada Water contained very large standing crops of phytoplankton during spring, summer and autumn. Even in mid-winter, chlorophyll concentration were in the order of 20 to 30 μg/l. In the growing season, very large concentrations (200 to 300 μg/l) were common, and on this basis the docks would be categorized as hypereutrophic [8].

Table 3. Light/depth relationships

Weather	Site†	Date	Secchi (cm)	Depth (m)	Surface	0.5	1.0	1.5 (m)	2.0	2.5	3.0	3.5
Overcast	RV1	06.06.89	50	7.5	180	127	30	11	<5	–	–	–
Sunny	RV2	06.06.89	50	7.5	438	192	89	<5	–	–	–	–
Sunny	RA3	06.06.89	50	9	95	55	12	<5	–	–	–	–
Overcast & raining	RA4	06.06.89	40	9	67	18	5	<5	–	–	–	–
Sunny	RA5	05.06.89	40	9	468	159	10	<5	–	–	–	–
Sunny	AB6	05.06.89	50	7	392	75	7	<5	–	–	–	–
Overcast	KG7	05.06.89	40	10	224	17	5	<5	–	–	–	–
Overcast	KG8	05.06.89	40	10	247	100	10	<5	–	–	–	–
Overcast	W19	08.06.89	120	8	344	190	87	43	28	<5	–	–
Overcast	WI10	08.06.89	110	8	571	244	95	45	22	<5	–	–
Overcast	WI11	08.06.89	80	5	371	194	63	23	12	<5	–	–
Overcast	WI12	08.06.89	80	7	314	170	72	32	13	–	–	–
Overcast	WI13	09.06.89	70	7	192	89	35	16	<5	–	–	–
Sunny	WI14	09.06.89	70	8	688	400	207	80	15	<5	–	–
Sunny	P15	09.06.89	70	3	900	421	200	100	40	15	8	<5
Sunny	B16	09.06.89	50	3	1249	312	195	60	35	<5	–	–
Sunny	M17	09.06.89	60	9	1165	529	187	66	15	<5	–	–
Overcast & raining	E120	06.06.89	70	3.5	128	37	10	<5	–	–	–	–
Sunny	M19	09.06.89	50	7	267	137	32	<5	–	–	–	–
Overcast & raining	E120	06.06.89	50	3.5	172	107	23	<5	–	–	–	–
Overcast	SB21	08.06.89	90	7	279	124	58	32	13	<5	–	–
Very sunny	H22	07.06.89	70	1	66	32	–	–	–	–	–	–
Very sunny	H23	07.06.89	>30	0.3	233	–	–	–	–	–	–	–
Dull	S24	07.06.89	90	4	107	66	43	18	7	<5	–	–
Very sunny	C25	07.06.89	>600	6	379	397	312	268	203	168	100	<5
Overcast	G26	07.06.89	50	7	469	78	20	<5	–	–	–	–
Very dark	G27	07.06.89	50	7	153	43	7	<5	–	–	–	–
Overcast	SD28	07.06.89	20	7	232	92	8	<5	–	–	–	–

†RV=Royal Victoria; RA=Royal Albert; AB=Albert Basin; KG=King George V; W=West India; P=Poplar; B=Blackwall; M=Millwall; EI=East India; SB=Shadwell Basin; H=Hermitage Basin; S=Surrey Water; C=Canada Water; G=Greenland; SD=South Dock.

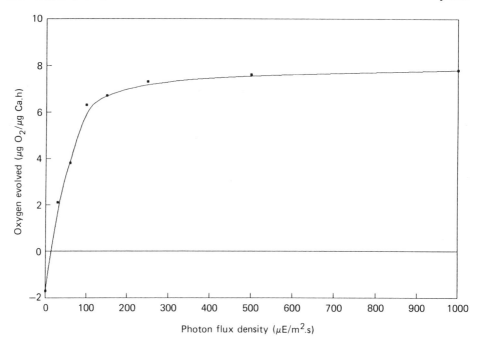

Temp. 15 °C   —— Net photosynthesis

Fig. 9. Royal Victoria Dock: photosynthesis v. irradiance.

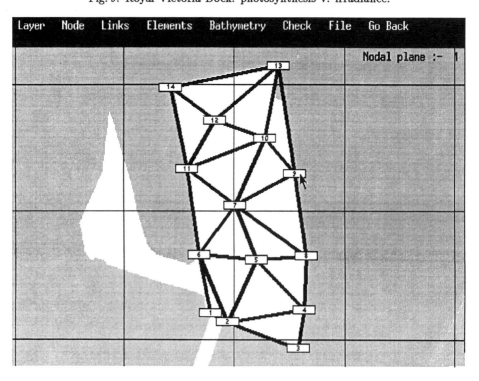

Fig. 10. Poplar Dock: typical network of triangular elements.

The dominant taxa varied somewhat from dock to dock. In general, the Royal Docks were notable for the presence of *Oscillatoria* sp throughout the study period. This is a non-nitrogen fixing, filamentous blue-green alga (or cyanobacterium). In large concentrations, such as those present in the Royal Docks in the summer of 1989, this organism, like many blue-greens, has a characteristic, strong 'earthy' smell which is not particularly unpleasant. When examined microscopically, the *Oscillatoria* invariably contained gas-vesicles, so that the algae were potentially buoyant. However, large surface scums were rarely observed, probably due to the well-mixed surface water in the Royal Docks, which are oriented along the path of the prevailing wind.

In contrast, algal populations in the West India Docks were dominated by diatoms in the spring and autumn, and by green algae in summer, although bluegreen algae were frequent components of the more mixed algal flora of the West India Dock system.

The most likely reason for the difference between these two major dock systems is the frequency of inputs of water from the Thames. The Canary Wharf development in the Isle of Dogs has made great use of river transport and the barge traffic has not only caused nutrients and bacteria to enter the docks, but has also increased the amount of surface mixing.

The Surrey Docks (excluding Canada Water) were in general dominated by blue-green algae. In contrast to the Royal Docks, the taxonomic composition of the dominant algae changed considerably through the growing season, with diatoms and yellow flagellates predominating in spring and early summer.

At an early stage in the study, since the river was the main input source, it became clear from the water chemistry that in many docks algal growth was unlikely to be limited by nutrient supply. Light limitation was therefore likely to be a major factor. Thus, measurements of light penetration (as photon flux density versus depth) were made in situ as well as the Secchi disc measurements of water transparency. In addition, field sampling and analysis was supplemented with laboratory experiments, so that the dynamics of the relationship between light and algal photosynthesis could be monitored. The key parameters of the photosynthesis: irradiance curves are used in the mathematical model referred to later, together with the light penetration values from field measurement to predict algal growth rates under different degrees of light limitation.

Photosynthesis/irradiance (P/I) relationships were established for early summer phytoplankton populations from 24 sites and a representative example is shown in Fig. 9. Although the species comparison of algae in different docks was quite different at the time of sampling, there was little difference between the P/I curves, when expressed as $\Delta[O_2]$ per unit chlorophyll.

The compensation point (where net photosynthesis exactly balances respiration) in the graph occurs at a photon flux density of 10–20μmol/m²s. Table 3 shows the light penetration for one sampling week at all sites. The depth to which 10–20μmol/m²s penetrates varies for each dock, but for most sites this compensation depth is only 1.5–2.5 m.

The P/I curves also provide evidence that the respiratory rates of the phytoplankton community constitute a significant sink for oxygen beneath the photic zone. Within the photic zone, the P/I relationship indicates the net amount of oxygen generated by the algae at different light supply rates. Unfortunately not all of the oxygen remains dissolved in the water and, when surface waters become supersaturated, increasing amounts are released into the atmosphere.

As the 1989 summer progressed the chemistry data showed that the algae were affecting a number of water quality variables. Phosphorus concentrations remained relatively high in all docks (except Canada Water) at all times, but nitrate values in some such as the Royals, (Fig. 7) approached limiting values in mid-summer. These low concentrations were not, however, accompanied by a change of species to nitrogen-fixing blue-green algae, as has often been predicted for low N and high P waters. The other water quality variable altered by algae, and of importance as it is a constituent variable in the EC Bathing Water Directive, is pH as described earlier.

Trace determinands

Samples were collected from all docks for the analysis of trace inorganic and organic substances. Concentrations were generally all below permitted levels.

Microbiology

The presence of the pathogenic organisms salmonella, *Pseudomona aeruginosa*, enterovirus and rotavirus was not detected in any dock system. Whereas the docks are not designated as Bathing Waters, mandatory total and faecal coliforms standards were complied with for 56% of the dock systems between May 1989 and May 1990. During this period there was only one dock which failed to meet the bacterial standards where water contact activities regularly took place. The results were also assessed to cover the period May to September when more water sports took place, during which 61% of docks were in compliance. As expected, failure to comply was generally in those docks receiving flow directly from the river.

Inflows

The River Thames is the main source of inflow to the docks. Average concentration of oxidized nitrogen and orthophosphate of 8.7 mg/l N and 2.9 mg/l P, and 9.7 mg/l N and 4.2 mg/l P, for the Royals and Isle of Dogs respectively were recorded. There is not much variation with the position of a dock system along the river, but the concentration of silicate increases with the distance from the sea. Nitrite N and ammoniacal N concentration in the river was small compared with nitrate concentration.

Bacterial quality of the river is much worse than that of the docks and total and faecal coliform counts of over 47 000 and 5000 MPN per 100 ml, respectively, were recorded.

Canada Water is no longer connected to the river and the main inflow is groundwater which is pumped in from the London Underground. Nutrient concentrations were much less than in the river and ranged between < 0.14 and 0.29 mg/l N as nitrate and 0.34 and 0.21 mg/l P as orthophosphate.

Other nutrient sources

Inputs of nutrients can be received from rainfall, surface water runoff, bird droppings, leaf litter, deposited rubbish, and exchange with the sediment. For dock systems connected to the river, all these sources were small in comparison with the river.

Sediment sampling

None of the dock basins appears to have significant quantities of silt, except for localized areas of the Albert Basin and the West India Docks. Average values for the major constituents are shown in Table 4.

MANAGEMENT IMPLICATIONS

Water quality objectives

Although the docks are not designated Bathing Waters, in situations where water contact sports are encouraged, the standards specified in the EC Directive [1], are probably the most appropriate against which to assess dock water quality. At present in the UK reporting on compliance with the Directive only refers to the mandatory values and emphasizes the bacterial parameters.

In general there are three main parameters in the Directive which are of particular note for the docks, low transparency, high pH and bacterial concentrations.

Table 4. Average composition of silt

Determinand		Mean
Specific gravity		1.14
Moisture content	% w/w	80
Total N	% dw	0.78
Loss on ignition at 500°C	% dw	15
Total P	% dw	0.38
Sulphide	% dw	0.44

Algal growth

The high pH and low transparency were largely the result of algal growth. A pH of 10 or above is likely to cause skin irritation problems for water users and dense algal crops are unwelcome for a variety of reasons. These include the risk of deoxygenation through algal bloom collapse, aesthetic problems (visual and olfactory) and the possibility that some strains are toxic.

The objectives therefore are to increase water transparency to that stated in the EC Bathing Water Directive and to reduce the frequency of occurrence of high pH in the surface waters. In order to do this it is necessary to have a clear understanding of the factors for algal growth.

Oxygen concentration

The largest oxygen demand during the period covered by this survey was exerted via the respiration of phytoplankton, and the breakdown of dead algal material. This

entails the oxidation of both carbonaceous material and ammonia. The balance
in oxygen supply is made up by aeration at the surface and by oxygen production
from photosynthesis. Since both these processes provide oxygen near the surface of
the docks and the oxidation processes tend to occur near the bed, the imbalance
results in anoxic conditions at depth unless vertical mixing provides sufficient exchange. The prevention of anoxic conditions therefore depends on the control of
algal growth and the maintenance of adequate mixing between surface and bottom
waters. Should this prove inadequate under extreme conditions, direct aeration of
anoxic bottom waters by direct diffusion of air or oxygen may be necessary. Maintaining minimal levels of dissolved oxygen in the water column will also control
sulphide formation.

Bacterial concentration

The highest bacterical concentrations were generally found in the basins with the
greatest exchange with the river. This included West India (Main), West India
(Middle), Blackwall, Millwall Inner, East India and South Dock. Higher concentrations tended to occur in the winter and spring, because of the combined effects
on the die-off rate of lower temperatures, light intensities, and increased turbidities
in the river.

The most obvious way to reduce bacterial populations is to limit the inflow of
river water, however the ability to do this is limited. Nevertheless, it may be
possible to limit surface populations by restricting river inflows to occur only at
depth.

MANAGEMENT STRATEGIES

The two practical means of controlling the growth of algae are via the management
of the nutrient input and/or the light input.

Control of nutrient inputs

Since the survey showed that nutrients, particularly phosphorus were present in
excess, nutrient control to limiting concentrations for algal growth, particularly
in the large dock basins, would be difficult. Nevertheless, since the river is the
major source of nutrients there is merit in reducing the amount of river water
entering the docks. Residence times would be increased, which is probably why the
concentration of inorganic nitrogen in the Royal Docks is lower than in the West
India Docks.

In the smaller docks (East India, Shadwell Basin, Hermitage Basin, Tobacco
Dock, Surrey Water, Canada Water) which no longer have useable lock gates, make-
up water is required to replace evaporation and seepage losses. These rates have
been estimated for each dock to vary between $30 \, m^3/d$ and $50 \, m^3/d$, and could be
provided from the main supply or groundwater, instead of the river Thames. These
flows could also be provided in association with heat pumps utilizing groundwater.
Utilization of groundwater with its lower concentration of nutrients has been beneficial to the water quality in Canada Water and has the added advantage of helping
to control the rising groundwater levels [9].

Most of the smaller docks are still linked to the Thames by sluices or weirs which permit the inflow of river water at high spring tides; excess water is discharged at low tide. Elimination of river inflows would reduce the inputs of nutrients and increase the residence times to values similar to that for Canada Water.

For the docks with usable locks future use and therefore incursions of Thames Water, is likely to be less frequent than when the docks were used commercially. Most locking will be for smaller vessels and the volume of water lost could be considerably reduced by limiting locking to periods around high water, and by redesigning the locks to accommodate smaller craft. In these docks the system of exchange with the river could also be reviewed to prevent unnecessary inflows at high spring tides.

Even without these measures, much of the make-up water for the Greenland/South Dock system could be provided by groundwater, which at present provides the input to Canada Water. In the West India Docks and Royal Docks it will be more difficult to find suitable alternative water sources. However, the proposal to construct an underground railway station at Canary Wharf in the Isle of Dogs may provide an opportunity to consider groundwater for replenishment. In addition, the nutrient-rich water could be liminated from some of the basins by installation of permanent or temporary barriers across the connecting passages.

Artificial mixing and aeration

Induced mixing can slow the rate of algal growth by reducing the time individual cells spend within the photic zone [10], and has been used successfully to reduce algal growth in small impoundments [11]. Paradoxically, there is also evidence that algal growth can be increased by such a strategy. This apparent contradiction can be resolves by considering the total depth of a water body in relation to its photic zone and mixed depth. It is now agreed [2] that the total depth of a water body must be at least twice the depth of the photosynthetic zone for induced mixing, that is, increasing the mixed depth to be successful as a means of algal biomass control. When artificial mixing merely recirculates algae within, or just below the photic zone, growth increases, since light interception is increased.

Results from the P/I experiments show that for the majority of docks the photic zone in the growing season is less than 2.5-m deep, and for many sites, less than 1.5 m. Thus, for the docks with a total depth of 5 m or more, artificial mixing will be a strategy worth pursuing.

Mixing can also reaerate anoxic bottom water by bringing it to the surface where it can dissolve oxygen from the air, and where the oxygen production from rapid photosynthesis at the surface can be entrained. Mixing may be carried out by proprietary mechanical mixing systems [12], or by creating 'bubble curtains' by discharging compressed air through submerged diffusers [13]. Care would have to be taken in siting these systems to avoid disturbance of the bottom sediment, and near-bottom velocities of less than 10 cm/s may be necessary. To achieve such low near-bottom velocities, and yet provide sufficient mixing, a number of small units would probably be necessary. The computer model described in the next section will be used to assist in determining the best configuration and the most economical solution.

MODELLING OF WATER QUALITY

The main objective of the hydraulic and water quality models is to investigate the water quality parameters under a wide range of possible environmental conditions and engineering scenarios. The model suite HYDRO-3D developed for the study is capable of simulating the spatial and temporal distribution of water elevation, current speed, temperature, nitrate, orthophosphate, dissolved oxygen, biochemical oxygen demand, total and faecal coliform bacteria and concentration of chlorophyll. In addition to being able to simulate the inflows to the dock system, the model is also capable of simulating the effects of water mixing equipment, which is incorporated as source and sink terms in the system and all of the above parameters under winds of different speeds and directions.

Theoretical basis of the model

The model suite consists of routines to simulate the hydrodynamic and water quality parameters which depend on the hydraulic conditions in the system. The hydro-dynamic component simulates current velocities and water levels, while the water quality components simulate the concentrations of water quality parameters.

Hydrodynamic model

The model is based on finite element methodology making use of Galerkins' weighting procedure. The prototype is subdivided into tetrahedral elements, and the water levels and current velocities in three orthogonal directions are simulated at the vertices of each element. The momentum equation is represented by Reynold's theory which accounts for stresses due to turbulence. Both wind and bed stresses are accommodated in the model. The equations used in the model are as follows:

x-direction

$$\frac{\partial u}{\partial t} + \overline{U}\nabla u = 2\omega\nu\sin\phi - \frac{1}{\rho}\frac{\partial p}{\partial x} + \frac{1}{\rho}\frac{\partial \tau_{xx}}{\partial x} + \frac{\partial \tau_{xy}}{\partial y} + \frac{\partial \tau_{xz}}{\partial z}$$

y-direction

$$\frac{\partial \nu}{\partial t} + \overline{U}\nabla \nu = 2\omega u\sin\phi - \frac{1}{\rho}\frac{\partial p}{\partial y} + \frac{1}{\rho}\frac{\partial \tau_{yx}}{\partial x} + \frac{\partial \tau_{yy}}{\partial y} + \frac{\partial \tau_{yz}}{\partial z}$$

z-direction

$$\frac{\partial p}{\partial z} + \rho g = 0$$

Continuity equation

$$\nabla\overline{U} = 0$$

where:

$$\nabla = i\frac{\partial}{\partial x} + j\frac{\partial}{\partial y} + k\frac{\partial}{\partial z}; \quad \overline{U} = iu + jv + kw$$

where: x, y, z = Cartesian co-ordinates aligned along north, east and vertical direction; t = time; u, v, w = velocity component along x, y, z directions; ω = angular velocity of earth; ϕ = latitude; p = pressure; ρ = density, and τ_{xx}, τ_{xy}, τ_{yx}, τ_{yy}, τ_{xz}, τ_{yz} = shear tensor.

The interface between the layers is fixed and the water elevation is adjusted at each time step. The eddy viscosity terms for each nodal point are calculated by assuming a linear stress variation at the top surface (wind stress) and at the bottom surface (bed stress). The viscosity terms are defined at the beginning of every time step using the local velocity field.

Water quality model

The model is based on the advection dispersion equation (ADE) which includes the source or sink terms due to: influx or efflux of pollutants; interactive first order decay process.

The equation represented in the model is as follows:

$$\frac{\partial c}{\partial t} + \nabla \overline{U}_c = \frac{\partial}{\partial x}\left(D_x \frac{\partial s}{\partial x}\right) + \frac{\partial}{\partial y}\left(D_y \frac{\partial s}{\partial y}\right) + \frac{\partial}{\partial z}\left(D_z \frac{\partial s}{\partial z}\right) + S$$

where: c = concentration of water quality parameters; D_x, D_y, D_z = dispersion coefficients in x, y, and z directions, and S = source and sink terms.

Modelling dissolved oxygen (DO)

The oxygen cycle in the model is represented by the interaction of various determinands in the system. This includes the effects of the nitrogen and phosphorus cycles.

In order to model the DO concentrations the source term in the ADE is represented by the following equation:

$$S = -K_{BOD}(BOD) + K_S(DO_S - DO) + \alpha_A(G_p - D_p)P$$
$$- \alpha_B K_{NH_3-N}(NH_3 - N) - \alpha_C K_{NO_2-N}(NO_2 - N)$$

where: K_{BOD} = BOD reaction rate; K_S – re-aeration rate; $\alpha_A \alpha_B \alpha_C$ = stochiometric constants; K_{NH_3-N} = rate of oxidation of ammoniacal nitrogen; K_{NO_2-N} = rate of oxidation of nitrite nitrogen; DO_S dissolved oxygen saturation concentration; G_p = specific growth rate of phytoplankton, and D_p = specific death rate of phytoplankton; p = concentration of phytoplankton.

Modelling algae

In order to model the algae, the concentration of phytoplankton is represented in terms of chlorophyll concentration. The differential equation representing the growth and production of algae is:

$$\frac{\partial P}{\partial t} = (G_p - D_p)P - K_p P,$$

where: G_p = local specific growth rate of phytoplankton; D_p = local specific death rate of phytoplankton, K_p = setting rate for algae.

In HYDRO-3D the coefficient D_p includes the conversion of algae phosphorus to organic phosphorus and algae nitrogen to organic nitrogen. The local specific growth factor G_p is dependent on both the availability of nutrients and light. The factor G_p is represented as a function of growth limitations due to light and nutrients. The expression representing G_p is as follows (EPA):

$$G_p = G_{p_{MAX}} K_l K_n K_p$$

where: K_l = algae growth limitation factor for light; K_n = algae growth limitation factor for nitrogen, and K_p = algae growth limitation factor for phosphorus.

A variety of mathematical relationships between photosynthesis and light are available. HYDRO-3D incorporates Steel's function which assumes an exponential function to model the effect of photoinhibition in the algal growth rate. The relationship as follows:

$$K_l = \frac{I_Z}{K_m} \exp(1 - \frac{I_Z}{K_m})$$

$$I_Z = I \exp(-\lambda_2)$$

where: l_Z = light intensity at depth z; I = intensity of light at the surface; λ_2 = light extinction coefficient, and K_m = light intensity corresponding to maximum algal growth rate.

The algal nutrient relationship in HYDRO-3D is defined by the Monod expression as described below:

$$K_n = \frac{(NH_3 - N) + (NO_3 - N)}{(NH_3 - N) + (NO_3 - N) + M_n}$$

$$K_p = \frac{P}{P + M_p}$$

where: M_n, M_p = the Michaelis–Menton half saturation constant for nitrogen and phosphorus; $NH_3 - N$ = concentration of ammonia nitrogen, and NO_3-N = concentration of nitrate nitrogen. The factors G_p and D_p are also temperature dependent and these factors are adjusted within the model for temperature correction.

A finite element technique similar to the hydrodynamic model is employed. The same geometric configuration defined in the hydrodynamic model is adopted and the velocity field for the water quality model is obtained from the hydrodynamic model. The water quality model then combines these advecture fields with both the dispersive component and the relevant rate expressions to produce an overall advective dispersion equation which is applied to the base data of the water quality model. The base data is not only both the initial concentrations of water quality parameters and their incoming loads at nodal locations, but also the rate coefficients and stochiometric constraints of the kinetic expressions which describe their interaction. The rate expressions and dispersion coefficients are fundamental to the water quality model. These coefficients are estimated during the model calibration process.

Solutions method

In the above models, the Newton–Raphson method is used to derive the linear system of equations which comprise the correction required for the variables at each interation. A sparse matrix routine is used to solve the system of equations.

For steady-state runs the solutions are derived by omitting the time related-terms in the governing equations.

Data file

The model requires a network file which defines the configuration of the system. This file defines the linkage of the nodes with the element and the co-ordinate of the nodes. In order to run the model the fcllowing additional data files are required: inflow data files; pollutant concentration details; climatic data files; chemico-biological coefficents data file.

The preprocessor will allow the model user to select, edit or create these model files prior to a model run. The network created by the preprocessor for the Poplar Dock is shown in Fig. 10.

Data management

A computer-based data management system is used for model development. In this context, the data management covers several activities such as digitizing and presenting the prototype system, entering of data files, calibration of model and presentation of results. All these activities are controlled interactively by template facilities on the screen and functional keys to select different tasks which include graphic dislpay facilities to animate the results and to produce hard copies.

Calibration

In any situation where a mathematical model is employed, the final assessment of its value is based on how well it calibrates with reality. In this case some of the mechanisms influencing algal growth are not fully understood and in consequence it is unrealistic to expect that the model will calibrate all parameters perfectly. Nevertheless that does not mean that the model will not be an effective tool in the overall management of water quality. At the time of preparing this chapter, model calibration is taking place.

CONCLUSIONS

1. The docks are being used for activities which require a much improved quality than for their previous use.
2. In comparison with the EC Bathing Water Directive the main parameters of particular concern are: low transparency, high pH, and bacteria concentrations. These are a direct consequence of inflow from the river, and the resulting large algal biomass. Dock systems with long residence times, and which are not connected directly to the river have the best water quality.
3. There are only two practicable means of controlling algal growth and hence, biomass: limiting light and limiting nutrients. The strategy for Docklands currently being developed is to use these factors separately and in combination. The understanding which has been gained from the monitoring sections of the study describe is an essential prerequisite for management. In such eutrophic systems as the London Docklands, the principal strategy will be to limit light via artificial mixing systems. This strategy will be supplemented wherever

possible by limiting the input of nutrients from the Thames. This can be achieved by: reducing the exchange of water, and replacing Thames water with ground water for 'top-up' purposes wherever appropriate.

4. A mathematical model of the water quality is being developed which will permit qualitative evaluation of the effects of various management options. It will also be used to evaluate development options, and make short-term predictions of changes in water quality.

REFERENCES

[1] Council of European Communities (1976) Directive concerning the quality of Bathing Water 76/160/EEC. *Official Journal*, L 31.

[2] *National Rivers Authority (1990) Toxic blue-green algae*, pp. 62–69.

[3] Reynolds, C.S. (1984) *The ecology of freshwater phytoplankton.* Cambridge University Press, p. 384.

[4] Fogg, G.E. and Thake, Brenda. (1987) *Algal cultures and phytoplankton ecology*, 3rd Edn. Univ. Wisconsin Press, p. 269.

[5] Lund, J.W.G. (1950) Studies on Asterionella formosa Hass II Nutrient depletion and the spring maximum. *J. Ecol*, **38**, 1–14, 15–35.

[6] Lucas, W.J. and Berry J.A. (1985) Inorganic carbon uptake by aquatic photosynthetic organisms. *Am. Soc. Plant. Physiol.* Maryland.

[7] Dong, L.F., Thake, B.A. and Heathcote, P. (1991) Carbon acquisition by three species of seaweed. *Oebalia* (in press).

[8] OECD (1982) *Eutrophication of waters, monitoring, assessment and control.* OECD, Paris.

[9] McAvoy, A. (1990) That sinking feeling. *Group Focus* (Institution of Civil Engineers Newsletter).

[10] Lorenzen, M. and Mitchell, R. (1973) Theoretical effects of artificial destratification on algal production in impoundments. *Env. Sci. and Technol*, **7**, 939–944.

[11] Raman, R.K. and Arbuckle, B.R. (1989) Long-term destratification in on Illinois lake. *J. Am. Wat. Wks Assoc.* **81**, (6), 66–71.

[12] Radway, A., Walker, L.S. and Carradice, I. (1988) Water quality improvement at Salford Quays. *J. Instn. Wat.& Envir. Mangt.*, **2**, (5), 523–531.

[13] Tolland, H.G. (1977) Destratification and aeration in reservoirs. *Technical Report, No. 50*, Medmenham, England.

19

The application of QUASAR (quality simulation along rivers) for environmental impact assessment

P. G. Whitehead, PhD, BSc, MSc. MIWEM[†]; R. J. Williams, BSc[†]; J. Clark, BA[†]; N. Murdock, PhD, BSc[‡]; H. Sambrook, BSc, MSc[‡]; and B. Mann, PhD, BSc[‡]

INTRODUCTION

The water quality model QUASAR has been designed at the Institute of Hydrology to assess the impact of pollutants on river systems. The model was originally developed as part of the Bedford Ouse Study, a Department of the Environment (DoE) and Anglian Water Authority funded project initiated in 1972. The primary objective was to simulate the dynamic behaviour of flow and water quality along the river system [1, 2]. Initial applications involved the use of the model within a real-time forecasting scheme collating telemetered data and providing forecasts at key abstraction sites along the river [3].

The model was also used within a stochastic or Monte-Carlo framework to provide information on the distribution of water quality within river systems, particularly in rivers subjected to major effluent discharges [4]. This technique was also used [5,6] to assess mass balance problems within river systems. There has also been a range of model applications to other UK rivers such as the Tawe in South Wales to assess heavy metal pollution, and the Thames to assess the movement and distribution of nitrates and algae along this river system [7,8].

[†] Institute of Hydrology.
[‡] National Rivers Authority, South West Region.

A total of eight water quality variables are simulated in addition to flow, including nitrate, ionized and unionized ammonia, dissolved oxygen (DO), biochemical oxygen demand (BOD), nitrate, ammonia, pH, temperature, *E. coli* and any conservative pollutant or inert material in solution. A wide range of inputs can be investigated including tributaries, groundwater inflows, direct runoff, effluents, storm water and the model can allow for abstractions for public water supply or irrigation. A multi-reach approach is utilized so that the user specifies reach boundaries at locations of primary interest.

The model can be used deterministically or stochastically and is essentially physically based with some empirical formulations such as velocity-flow relationships. In the deterministic or dynamic mode pulses of pollutant can be traced downstream, while in the stochastic or planning mode, effluent consent conditions can be established given a river quality objective.

MODEL DESCRIPTION

A detailed description of the underlying model structure is given by Whitehead *et al.* [1,2]. Essentially the model performs a mass balance on flow and water quality sequentially down a river system. The model takes into account inputs from tributaries, groundwaters and effluent discharges, and allows for abstractions, chemical decay processes and biological behaviour along the river system. In addition it accounts for the varying travel times operating at different flow conditions so that pollutants are transported along the river at realistic velocities. The model operates in two modes: planning mode (stochastic simulation), and operational mode (dynamic simulation).

Planning (or stochastic) mode

In the planning or stochastic mode a cumulative frequency curve and distribution histogram of a water quality parameter are generated by repeatedly running the model using different input data selected according to probability distributions defined for each input variable. This technique, known as Monte-Carlo simulation, has been used [4] to provide information which aids in long-term planning of water quality management. In this mode statistical data of the water quality and flow in the first reach at the top of the river and in tributaries, discharges and abstractions at key locations along the river, are required. These data include, for each variable input to the model, the mean, standard deviation and distribution type, that is lognormal, rectangular or gaussian. Random numbers are generated and water quality and flow values are then chosen from these characterized distributions. A mass balance is performed at the top of each reach to include tributaries, discharges and any other inputs to the river at that point on the river for each run of the model. The values generated by the model equations represent the water quality or flow at the end of the reach. The model equations are run using the random numbers as the input values either until steady state has been reached or for a maximum of 30 time periods. Steady state is said to have been achieved when the result of successive runs is less than 1%. Five hundred and twelve random simulations are generated. The output is stored and used to produce cumulative frequency distributions and distribution histograms (Fig. 1).

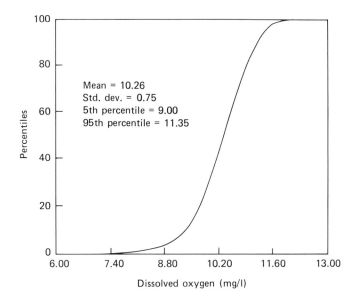

Fig. 1. Output from QUASAR in planning mode showing distribution of water quality.

Operational (or dynamic) mode

In the operational or dynamic mode, the water quality and flow are simulated over selected periods. This allows the possible effects of a pollution event on a river to be investigated. In this mode time series data are required for water quality and flow parameters for the first reach of the river and for tributaries, effluent discharges and abstractions along the section of the river of interest. A mass balance is performed over each reach of the river and a time series of downstream flow and quality computed. A typical model output is given in Fig. 2. The model has been set up in a real time operational model for the whole of the Great Ouse catchment and used to provide forecasts to operational managers during pollution events [3]

Data requirements

Three sets of data are required by QUASAR; a catchment structure consisting of a river map, boundary conditions which define the water quality and flow of the tributaries and of the water at the top of the river, and reach parameters consisting of data specific to each reach. Detailed information on data requirements, chemical process equations and the velocity flow relationships is given by Whitehead *et al.* [9].

APPLICATION OF QUASAR TO ROADFORD AND THE TAMAR

Roadford is a new reservoir which impounds and regulates the flows of the River Wolf, a tributary of the River Tamar. Quasar has been adapted to the River Tamar and has been used to assess the impact of the reservoir and its operation on the receiving waters. The model will also aid in the assessment and determination of operational procedures for the reservoir, regulation releases to and abstractions from the River Tamar. Fig. 3 details the study area to which Quasar has been applied and traces its source from Roadford Lake to the head of tide at Gunnislake. The pathway of the model starts at Roadford Lake which discharges in the rivers Wolf, Thrushel and Lyd before joining the main steam Tamar at Lydfoot (see Fig. 4). This system is referred to as the Wolf-Thrushel-Lyd-Tamar river system or more concisely the Wolf-Tamar systems.

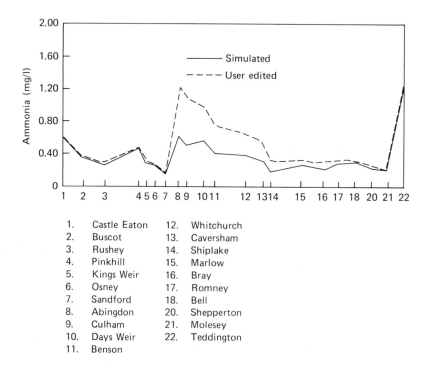

1.	Castle Eaton	12.	Whitchurch
2.	Buscot	13.	Caversham
3.	Rushey	14.	Shiplake
4.	Pinkhill	15.	Marlow
5.	Kings Weir	16.	Bray
6.	Osney	17.	Romney
7.	Sandford	18.	Bell
8.	Abingdon	20.	Shepperton
9.	Culham	21.	Molesey
10.	Days Weir	22.	Teddington
11.	Benson		

Fig. 2. Output from QUASAR in operation mode.

The primary impacts of impoundment and the operation of this strategic river regulation reservoir will be changes to the hydrological regime and water quality of the receiving waters. Any impacts will be greatest in these rivers, local and downstream of the reservoir. The NRA as the regulatory body will be required to minimize the impact of the scheme on the environment, while promoting the proper use and conservation of water resources. More specifically in terms of water quality, the study must ensure that all EC Directives and other statutory requirements are achieved, existing quality objectives are met and quality maintained so that waters are capable of supporting their identified uses.

Quasar is considered a useful model for this type of assessment. To date, the key variables modelled in this application include flow, temperature, dissolved oxygen, biochemical oxygen demand (BOD), nitrate, ammonia and pH.

The approach taken in the use of Quasar has progressed from model validation to scenario modelling.

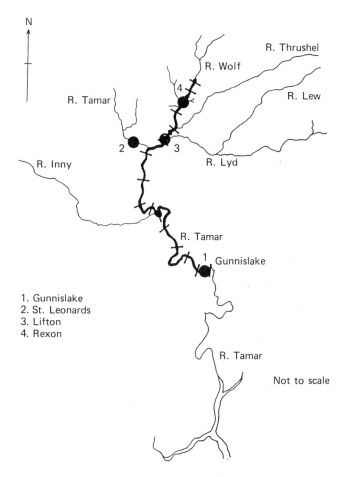

Fig. 3. Map of the Wolf-Tamar system

Model validation

Environmental models are calibrated against observed data. In the case of Quasar this involves adjusting the determinand process parameters so that reasonable agreement results between the simulated and observed values.

With a spatial data set, where the data are for specific locations at a fixed time, this is relatively straightforward. Table 1 shows the simulated and observed values of the principal water quality determinands for such a data set. The data are for January 20 1986 at various locations along the Wolf-Tamar river system. Mean errors are relatively low and well within typical measurement errors [2]. Results from this validation process justified the use of Quasar and its application to the Wolf-Tamar system.

For a data set with both spatial and temporal variations, the calibration is more involved. This is because the 'process' parameters represent not just the natural riverine chemical processes but also the effects of diffuse inputs and other time varying mechanisms. It should be possible to allow the process parameters to vary in space and time and hence fit the model to the data set. However, if the equations describing the processes represent the dominant effects, then the error of using fixed process values should be acceptable. This is the approach used here. It is recognized as a compromise between the 'classical' approach, where processes are described in ever greater detail in order to obtain a fit, and the 'statistical' approach in which full empirical fitting occurs. Retention of the 'classical' approach ensures that the model is physically based, a required attribute in this study where dynamic rather than stochastic simulations are performed. The 'statistical' approach is exemplified by Simcat [10] where an autocalibration algorithm is used in a stochastic simulation.

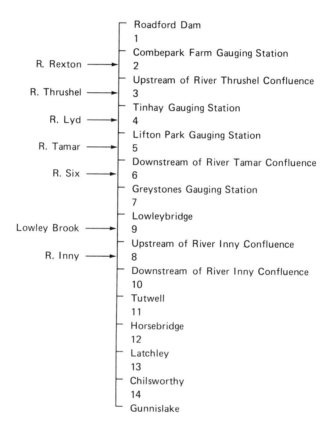

Fig. 4. Reach structure using QUASAR.

Annual data sets for the years 1982 and 1984 were used to calibrate Quasar for the Roadford project. These two years were chosen on the basis that, hydrologically, 1982 was considered a typical year whilst 1984 was a drought year. Data were readily available from the hydrological and water quality archives. Daily mean flows (cumecs) and monthly summary statistics from the routine sampling (spot) programmes were used. The high degree of variability inherent in such data is known and recognized. Even so, Quasar can be readily adapted to utilize this type of archived data. This has an added benefit in this type of overview study as it negates the need for specialized and expensive surveys.

Output for the model allows comparison between the archived and simulated data (Fig. 5). This example shows an annual profile for dissolved oxygen at Greystones on the main Tamar. The simulated data are observed to follow the general trend of the actual data. Once the model has been set up and validated Quasar can perform annual simulations relatively quickly, approximately 10 minutes (running on a DEC VAX 3100 computer). This facility is particularly useful when the model is used to assess seasonal effects.

Having calibrated the model with 1982 and 1984 data, numerous simulations of Roadford operational scenario were performed.

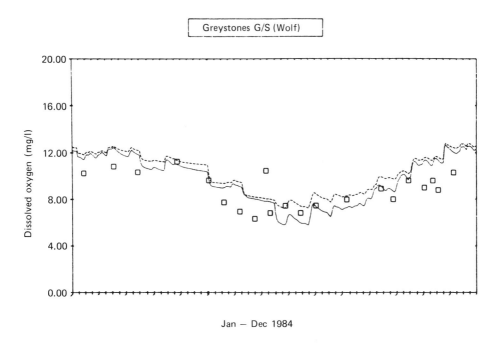

Fig. 5. Simulated and observed oxygen profiles at Greystones.

Scenario modelling

Within the model there is the facility for altering flows, effluent discharges, abstractions etc. to simulate changes in the environment. This facility has been used to simulate possible Roadford releases into the top of the Wolf-Tamar river system. The model allows the flows rate and water temperature to be varied, together with DO, BOD, nitrate, ammonia and pH values.

Various operational scenarios have been modelled. In these the flow and water quality profiles were compared with the simulated profiles for a system without a reservoir at Roadford. This method of comparing the simulations has two benefits.

Firstly, in any model, calibration will have defined limits of acceptability and hence the accuracy is not precise but an approximation. However, it is considered that by performing comparative simulations errors common to both the 'with' and 'without' reservoir models, will tend to cancel out.

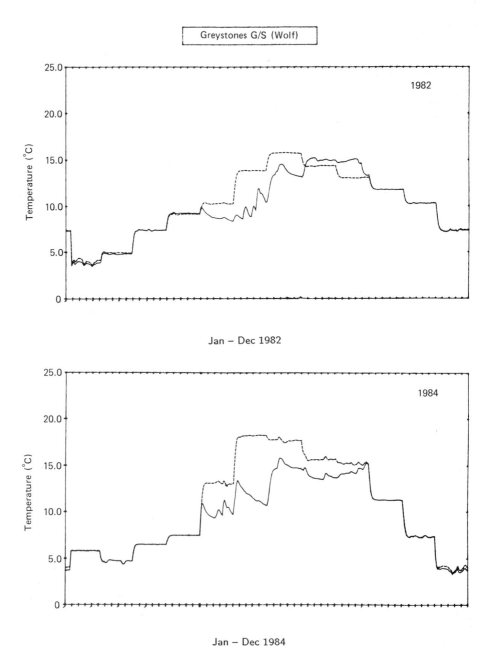

Fig. 6. Temperature at Greystones using maximum demand flow.

Table 1. Comparison of observed and simulated values of principal water quality parameters

Site	BOD			DO			Nitrate			Ammonia		
	Observed	Simulated	Error	Observed	Simulated	Error	Observed	Simulated	Error	Observed	Simulated	Error
Gunnislake	2.3	2.2	-0.1	11.0	10.8	-0.2	3.2	2.8	-0.4	0.20	0.22	0.02
Horsebridge	2.7	2.3	-0.4	10.1	10.8	0.7	3.2	2.9	-0.3	0.20	0.22	0.02
Greystones	2.9	2.4	-0.5	10.5	10.7	0.2	3.1	2.8	-0.3	0.29	0.27	-0.02
Lifton	2.8	2.9	0.1	11.0	11.0	0.0	2.3	2.3	0.0	0.29	0.29	0.00
Tinhay	3.6	3.4	-0.2	10.5	10.1	-0.4	1.9	1.8	-0.1	0.42	0.38	-0.04
R.Wolf	3.1	2.4	-0.7	10.8	10.7	-0.1	1.7	1.4	-0.03	0.33	0.22	-0.11
MEAN ERROR			-0.36			0.04			-0.28			-0.026

Date: 20.01.86

The second benefit relates to interpretation. Presentation of individual profiles for each Roadford scenario is not very informative. In cases where poor water quality exists a proposed development may improve the water quality but still leaves it poor. Thus it is relevant to know what the effect of the development is relative to the prior situation than the absolute values themselves.

Specimen outputs from a modelled scenario are shown in Figs 6–8. The basis of this example scenario is a large and continuous release from Roadford of approximately 2 cumecs. This release is made during the summer. The reservoir water quality has been estimated in order to simulate the scenario. Certain values for BOD, ammonia and pH which were derived for the River Wolf were used for the reservoir discharge. The discharge temperature was modified to include the thermal inertia effects likely to occur in a large reservoir. Drawing on observations for the nearby Wimbleball reservoir [11] amplitude of the annual temperature variation was damped by 20% and time-lagged by one month. The DO values are modified so as to be consistent with the temperature change. An 80% saturation value was assumed. These values and assumptions are not definitive and are used here for illustration.

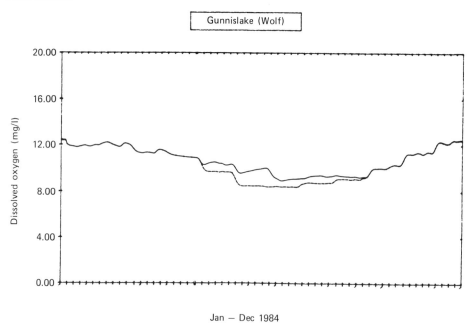

Fig. 7. Dissolved oxygen profiles at Gunnislake for 1984 using maximum demand flow.

Example output

1. *Temperature.* Fig. 6 shows the temperature response at Greystones. The thermal inertia of the reservoir has resulted in a cooling in the early summer for 1982. In the 1982 scenario a reversal occurs in August as the non-reservoir flow contribution cools at a faster rate than the reservoir discharge. For 1984 this effect is not present because of the higher ambient temperatures.

2. *Dissolved oxygen.* Fig. 7 shows the DO profiles at Gunnislake. Under the assumptions of the modelled scenario the effect has been to raise DO values during the summer months.
3. *BOD.* Fig. 8 shows the BOD profile at Greystones. The figure shows that the BOD values during the summer months are substantially reduced. In this scenario there is a greater flow of better quality water from the reservoir into the River Wolf. When mixed with the Tamar flow this has the effect of reducing the BOD concentration in the Tamar.

The above scenario was modelled prior to Roadford becoming operational. As the water quality data base is established for the new reservoir various scenarios will be remodelled in order to check on the original assumptions and assessments.

The use of the model in the Roadford project has been to illustrate the consequences of various operating conditions for Roadford. This allows the interested parties to debate more objectively the benefits/disbenefits of a particular operating strategy. The above scenario is one example from a wide range of possible scenarios. By simulating different scenarios, those operating strategies which are protective and beneficial to the environment may be identified.

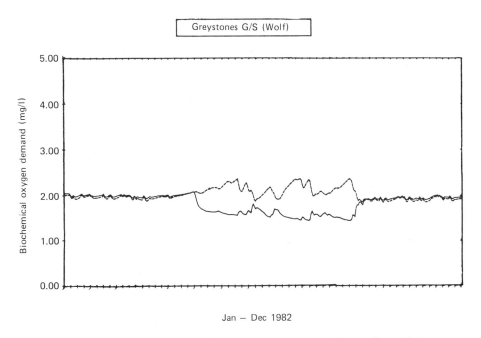

Fig. 8. BOD profiles at Gunnislake for 1982 using maximum demand flow.

CONCLUSIONS

The basis of the Quasar model has been described. Useful features of Quasar are its ability to model several determinands, flexibility on river inputs and fast execution time which permits annual simulations to be performed. These give a better overview of environmental response than single event modelling would give.

Quasar was calibrated for the Wolf-Tamar system using archived data for the years 1982 and 1984.

The methodology of applying Quasar in the Roadford project has been described. This involved running comparative simulations to find the environmental response with and without Roadford.

An example scenario showed the temperature DO and BOD response at two locations on the Wolf-Tamar system. The calibration and scenario modelling methodology is now being applied on other river systems.

ACKNOWLEDGEMENTS

The authors N. Murdoch, H. Sambrook and B. Mann, acknowledge the facilities provided by the NRA and South West Water Services Ltd (SWWSL) in preparing this chapter. The views expressed in this chapter are not necessarily the views of the NRA or South West Water.

REFERENCES

[1] Whitehead, P.G., Young and Hornberger (1979) A systems model of stream flow and water quality in the Bedford Ouse river – I Stream flow modelling. *Water Research*, **13**, 1155–1169.

[2] Whitehead, P.G., Beck and O'Connel (1981) A systems model of flow and water quality in the Bedford Ouse river system – II Water quality modelling. *Water Research*, **15**, 1157–1171.

[3] Whitehead, P.G., Caddy and Templeman (1984) An on-line monitoring, data management and forecasting system for the Bedford Ouse river basin. *Wat. Sci. Tech.*, **16**, 295–314.

[4] Whitehead, P.G. and Young (1979) Water quality in river systems, Monte-Carlo analysis. *Water Resources Research*, 451–459.

[5] Warn, A. and Brew (1980) Mass balance. *Water Research*, **14**, 1427–1434.

[6] Warn, A. and Matthews (1984) Calculation of the compliance of discharges with emission standards. *Wat. Sci. Tech.*, **16**, 183–196.

[7] Whitehead, P.G. and Williams, R.J. (1982) A dynamic nitrate balance model for river basins. In: *IAHS Exeter conference proceedings*, IAHS publication no 139.

[8] Whitehead, P.G. and Hornberger (1984) Modelling algal behaviour in the River Thames. *Water Research*, **18**, No. 8, 945–953.

[9] Whitehead, P.G., Clark, J. and Williams, R. (In preparation) A dynamic stochastic model for *Water Quality Simulation Along River Systems* (QUASAR).

[10] Warn, A. (1990) *SIMCAT Manual*, Anglian NRA.

[11] Webb, B. (1980) The Influence of Wimbleball Lake on river water temperature. *Rep. Trans. Devon Ass. Advist. Sci.*, **120**, 45–65.

20

The development of a control policy for fish farming

H. Smith, DipWEM, MIWEM and A. J. N. Haig, BSc, PhD, CBiol, MIBiol, MIWEM[†]

INTRODUCTION

It is doubtful if any industry has attracted as much attention as fish farming in the last ten years. Environmental pressure groups state that the industry is responsible for pollution of the aqueous environment. Fish farmers are accused of killing birds and seals, while escaped farmed fish are claimed to transmit disease to the wild fish population.

There is no doubt that the industry has experienced a rapid growth which for a time outstripped exisiting legislation. This chapter looks at the growth of the industry against existing control legislation and attempts to describe the development of the Clyde River Purification Board's control policy.

TYPES OF FISH FARMING

The intensive cultivation of fish may take place in land-based tanks or in large cages or nets suspended from floating platforms in freshwater lochs or marine waters. Each type of farm causes the release to the receiving waters of large quantities of organic waste in the form of surplus food and faecal matter. In addition, soluble

---

[†] Chief Inspector and Chief Scientist respectively, Clyde River Purification Board, East Kibride, Glasgow.

components of the waste, for example nitrogen and phosphorus, may cause algal blooms. Chemicals are also used to control disease but each type of farm can cause particular problems for the controlling authority as follows.

Land-based farms

A land-based fish farm is shown in Fig.1 and a typical layout in Fig.2. These farms operate through water being abstracted from a river or coastal water and passed through the rearing tanks before being discharged back to the source. River purification boards, unlike the National Rivers Authority, have no power to control abstractions and there have been instances when rivers have dried up through over-abstraction. Boards have generally tried to overcome this problem within their conditions of consent by limiting the volume of the discharge and thereby limiting the abstraction. There are at present 28 land-based farms in the Clyde Board's area, with 18 discharging to freshwater streams and 10 discharging to coastal waters.

Although there is no problem in imposing consent conditions on the discharge, it is unfair to the farmer if no consideration is given to the quality of the supply. The Clyde Board favours the imposition of incremental consents whereby the conditions to be achieved at the outlet are based on the quality of the inlet water.

Fig. 1. Land-based fish farm at Ardtaraig, Loch Strivin.

Cage farms

An illustration of a cage farm and the waste produced is depicted in Fig.3. This type of installation caused severe problems for the boards in that the discharge was not from a point source, and legal debate still continues as to whether caged farms were subject to control under the Control of Pollution Act, 1974 (COPA). This Act still applies in Scotland, but it has been amended by Schedule 23 to the Water Act 1989, which makes it clear that, for the purposes of defining industrial effluent, floating cages shall be considered trade premises.

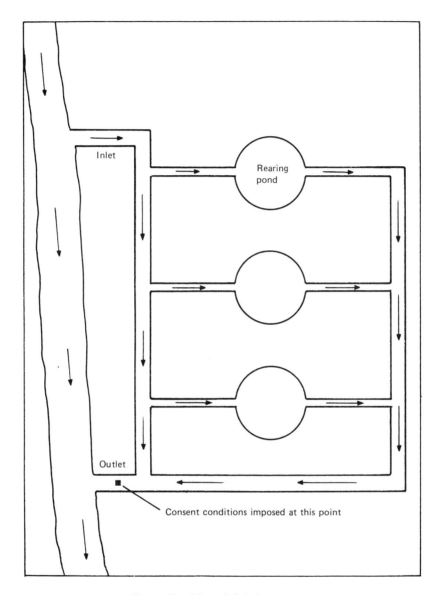

Fig. 2. Land-based fish farm.

In 1988, before the legality of consenting discharges from cage fish farms was confirmed, the Board mounted a number of pilot surveys of water and sediment quality in the vicinity of selected farms. The results were in accordance with those published in the literature [1] and caused the Board concern that uncontrolled expansion of the cage fish farm industry might have a variety of harmful effects on water quality as a whole (Fig. 3). Potential impacts are well reviewed elsewhere [2].

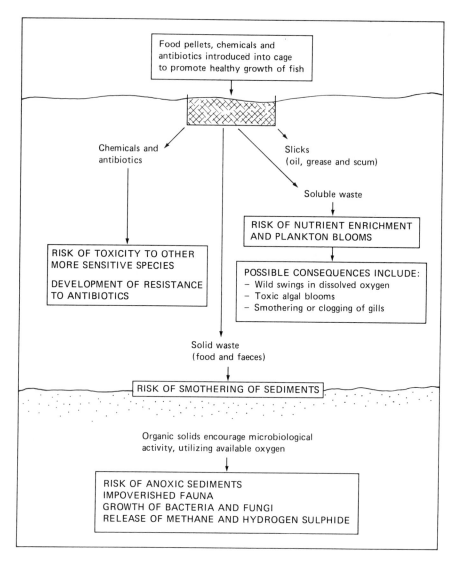

Fig. 3. Potential routes to environmental impact associated with waste from
cage fish farming.

GROWTH OF INDUSTRY

There has been rapid expansion of the industry and the most up-to-date production
figures for trout and salmon are shown in Table 1.

Salmon is the main product in Scotland, and because of the generally very clean,
deep and sheltered sea lochs found along the west coast the industry tends to
be established there and also in Orkney and Shetland and Western Isles. A recent
publication by the Scottish Salmon Growers' Association [3] demonstrates the value

of the industry to the area. The Association states that the industry now has a crop value in the Highlands and Islands greater than the combined output of cattle and sheep. It is estimated to have an annual turnover of £140 m and employs 3600 people on a total of 360 farms which are owned by 170 companies. At one stage production was predicted to reach 70 000 tonnes in 1992, but there was a slump in the industry during 1989/90 because of over-production, and the expansion period is thought to have ended. The current trend is for production increases at existing sites, rather than the establishment of farms at new sites.

Table 1. Production figures for trout and salmon

Trout farming		Salmon farming	
Year	Tonnes	Year	Tonnes
1975	1 300	1980	600
1988	17 000[†]	1989	28 553
		1991	40 000[‡]

[†](13 000 in England and Wales) [‡](estimate)

Many attempts have been made to estimate the population equivalent on the waste produced by fish farming. Solbe [4] estimated that, depending upon the parameter measured, 1 tonne of fish produced waste equivalent to the treated effluent from 122–859 people. In terms of BOD, 1 tonne of fish is estimated to be equivalent to the untreated sewage load produced by 20 people. Although these figures must be treated with caution, the 1991 estimate of production could have a population equivalent of 800 000.

EXISTING LEGISLATIVE CONTROL

Town and Country (Scotland) Planning Act, 1972
The district councils in Scotland have only recently been given control over the development of freshwater fish farms. Fish farming was previously considered to be agricultural development and the General Development Order 1981 allowed the establishment of freshwater farms without the need to apply for planning consent. The Secretary of State for Scotland has amended the Order [5] to remove the permitted development rights, and freshwater fish farms now require planning consent from the district council and the regional councils in the case of Borders, Dumfries and Galloway, and Highland regions.

Crown Estate Commissioners
The district council's planning control does not extend to marine fish farms and such installations are controlled by the Crown Estate Commissioners (CEC). A consultation procedure operates and river purification boards are consultees whose views should be taken into consideration before a licence is granted. A House of Commons Agriculture Select Committee [6] recently criticized the role of CEC in that they were expected to be independent arbiters who at the same time had a commercial interest in the outcome of their decisions. When one considers that the

annual cost of a lease from CEC for a 1600 tonne farm is about £36 000, it would be easy to argue that the financial implications could influence the decision-making process. A further criticism of the CEC was aimed at their failure to establish an advisory committee to assess contentious applications for fish farm licences.

While the Select Committee stopped short of calling for all planning controls to be handed over to local councils, it did recommend that the Scottish Office should draw up planning guidelines to aid the decision-making process.

Control of Pollution Act 1974

Most of the marine cage farms are situated along the west coast, within the areas of the Highland and Clyde River Purification Boards, while a number are located in the Western Isles. Initially there was disagreement as to whether such farms required the consent of the purification authority. All authorities agreed that the industry required to be controlled, but there was considerable debate as to whether discharges from floating installations complied with the definition of trade effluent in section 105 of COPA. The Highland Board argued that 'premises' included land, and that cage fish farms must consist of the land at the bottom of the loch used to anchor the cage. On the basis of their definition, the Highland Board required applications for consent to be submitted under section 34 of the Act.

While the Clyde Board shared the Highland Board's desire to control such developments, it could not agree that the bed of a loch could be land. After all, according to the Act, if a stream bed when dry was still a stream it could not then become land when covered with water in order to satisfy the definition of premises. A Procurator Fiscal also advised the Clyde Board that the Highland Board's definition could not be supported in law.

The Clyde Board, as consultants to the Western Isles Islands Council, had no alternative but to advise that Council of the legal conflict, and therefore the Council decided not to request applications from farms under section 34 of the Act. The Islands Council took the opinion of Senior Counsel [7] who advised 'In relation to cages used for fish farming, consents under s. 34 of the 1974 Act are not required should pollution emanate from them'. This should have clarified the matter but the opinion of the Clyde Board's Counsel [8] was 'I am of the opinion that a system of consents under s. 34 is liable to cover discharges from fish farm cages whether this discharges are classified as effluent or other matter or as trade effluent'.

Upon receipt of this opinion the Clyde Board required applications to be made under section 34, although the National Farmers' Union challenged the opinion of the Board's Counsel. The matter was finally resolved by Schedule 23 of the Water Act 1989, which stated that for the purpose of defining trade effluent, fish farms are deemed to be premises used for carrying on a trade.

Once the matter was resolved, the Board proceeded to impose consent conditions on cage fish farms. These conditions include clauses relating to (a) water body and location, (b) farm size, expressed as a limitation on fish biomass, (c) the prohibition of any discharge of material causing harm to the aquatic environment, (d) the keeping of records of fish biomass and all chemicals used, and (e) advance warning of the use of some chemicals. In the case of the larger marine cage fish farms, there was also a condition requiring the farmer to conduct and report on self-monitoring.

DEVELOPMENT OF A MONITORING POLICY

It will be recognized that pollution is not controlled by simply issuing a consent to discharge but also by regular monitoring to ensure compliance. River boards have a duty to keep public registers which must contain applications for consent, consents granted, analyses of samples and action taken as a result of analyses. The Board has a clear duty to ensure that discharges comply with consent conditions, but there are very real logistical problems in monitoring fish farms.

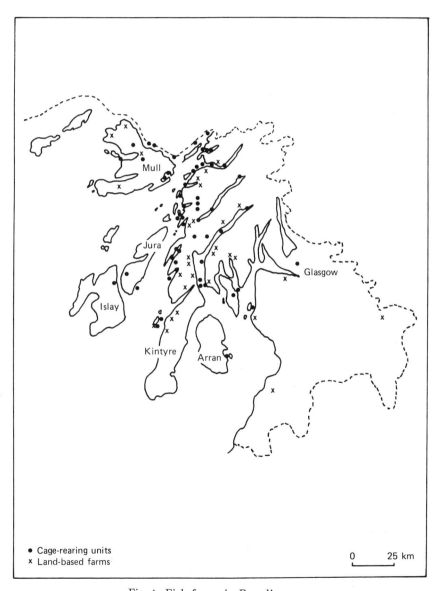

Fig. 4. Fish farms in Board's area.

Fig. 4 shows the location of the farms within the Board's area, and it should be noted that the closest farm is almost 100 km from the Board's headquarters at East Kilbride, while others may involve a round trip of up to almost 500 km. In addition, water and benthic sampling surveys can be very time-consuming. One sampling survey of Loch Sween alone took a survey team two days to complete, during which time other routine work had to be curtailed. Once the Board was empowered to control discharges from cage farms, it immediately improved its monitoring capability by appointing two specialist staff and purchasing necessary sampling equipment, including a survey launch and an underwater TV/video system.

Land-based farms

Apart from certain logistical difficulties, the monitoring of land-based farms is relatively straightforward. The effluent from each farm is discharged via one or more pipes, and the volume and composition of the discharge is consented and monitored in the conventional way. Recently the Board purchased additional 'EPIC-type' automatic water-sampling devices to improve its coverage of the wide diurnal and seasonal variation in quality of fish farm effluents. Similarly, monitoring the impact of the fish farm effluent on the quality of receiving waters presents few problems, involving visual inspections and the regular collection and analysis of water samples and benthos samples from points upstream and downstream of the discharge.

Freshwater cage farms

The provision of these new resources enabled the Board to develop a programme for monitoring the 16 consented cage farms in freshwater lochs on the mainland and islands (Fig. 4). This initially involved twice-yearly surveys of water quality at sites surrounding the farm plus a 'control' site, and an annual biological survey of the loch bed close to the farm. Subsequently, the frequency of water sampling has been increased at those sites where nutrient (especially phosphate) enrichment is regarded as posing a particular risk.

The opportunity is also being taken to compare measured levels of phosphate with the levels predicted using the Dillon and Rigler model [9], which is applied as a management tool to assess the phosphorus loading capacity of a freshwater loch. It is the Board's aim that effluent discharged from a freshwater cage fish farm should not cause the phosphorus level in the receiving water to increase by more than 10% within a trophic band, or move from a lower trophic band (e.g. oligotrophic) to a higher one (e.g. mesotrophic).

Other receiving water quality criteria accord with those for salmonid waters as specified in the EC freshwater fisheries Directive [10]. An attempt is being made to develop an environmental quality standard for the profundal benthos beneath the cage fish farm.

Both the Institute of Fisheries Management (Scottish Branch) [11] and the Nature Conservancy Council [12] have recently published useful reports on the effects of cage fish farming on the Scottish freshwater environment, and the problem of eutrophication has also been addressed [13].

Marine cage farms

Unfortunately, despite the provision of extra staff and equipment, the Board lacked the resources to monitor regularly all 35 of its marine cage farms. Accordingly, it decided to embark on a rolling programme to survey all farm sites over a period of years (5–10 farms/year), beginning with the largest and those located in the most 'sensitive' waters. The sensitivity of sea lochs to hypernutrification was estimated on the basis of flushing time and water volume, and the ranking of the nine most sensitive lochs is shown in Table 2.

Table 2. Derivation of predicted sensitivity indices for nine sea lochs

Sea loch	Crown Estate Commission's classification	Flushing time, T, (days)	Low water volume, V (million m^3)	Predicted sensitivity index, S ($= T/V \times 1000$)
Riddon	B	3	26	115 most sensitive
Caolisport	B	9	146	62
Feochan	A	1	21	48
Melfort	A	9	260	35
Sween	A	6	233	26
Spelve	B	3	152	20
Craignish	A	5	276	18
Creran	B	3	178	17
Etive	A/B	14	940	15 least sensitive

See Crown Estate report [14] for information on classification. Data on flushing time and volume from Edwards and Sharples [15].

To bridge this resources gap the Board decided to impose, as a consent condition, a requirement on the owners of certain farms to carry out self-monitoring, leaving the Board to undertake random sampling of its own to serve as an audit. The smallest farms (i.e. cultivating < 250 t fish biomass) were exempted from self-monitoring unless water quality problems were encountered; medium-sized farms (i.e. biomass between 250 and 500 t) were required to carry out a one-off hydrographic survey and an annual sediment/benthos quality survey; the largest farms (i.e. biomass > 500 t) were also required to carry out twice-yearly water quality surveys. The Board felt that prudent farmers would be gathering such information for their own purposes, and was willing to modify its requirements in the light of any appropriate information supplied. To guide farmers, it provided details of a generalized survey format, focusing on individual cage clusters, with details of sampling and analytical methodologies. Estimated costs are shown in Table 3.

Although accepting the principle of self-monitoring, the salmon-growing industry objected to the Board's detailed requirements, particularly those relating to the benthos quality survey, which it regarded as excessive. The Board subsequently held discussions with the individual appellants and was able, using environmental information supplied by the farmer, to tailor the self-monitoring requirements to suit the characteristics of each farm.

For those farms having many cage clusters and fallow sites within their leased areas, the focus of survey changed from the cage cluster to the leased area as a whole, the aim being to determine whether noticeable effects on water and sediment

quality were confined within the leased area (acceptable) or extended beyond it
(unacceptable). At the time of writing, agreement has been reached with many of
the appellants, who have withdrawn their appeals, and the system of self-monitoring
plus auditing is in operation. The next round of discussions will focus on the
outcome of these surveys, and on the implications of the forthcoming 'charging for
discharges' scheme for self-monitoring.

Table 3. Estimated cost of self-monitoring

Farm's biomass (t)	Cost of Crown Estate Commission lease (£/annum)	Cost of self-monitoring (£/cage cluster/annum)[†]	Value of fish sold (£m/annum)[‡]
1600	36 050	7540	2.1–3.0
600	13 520	7540	0.8–1.1
300	6 760	4940	0.4–0.6
200	4 500	0	0.3–0.4

[†]Calculated as cost to Clyde RPB.
[‡]Calculated at £1.20/lb and £1.70/lb.

Problems requiring resolution

Although good progress has been made in developing schemes for monitoring the
effects on water quality of cage fish farms, it has to be appreciated that the industry
and the controlling authorities are on a 'learning curve' with regard to the potential
impact on controlled waters. The Board will take advantage of section 37 of COPA
if it is demonstrated that consent standards require to be tightened or relaxed. It
is the authors' view that even if standards remain unaltered there are a number of
problems that will require to be resolved in the near future. These are:

(a) Research is required to determine environmental quality standards for phosphorus in freshwaters and nitrogen in saline waters (to avoid hypernutrification and eutrophication).

(b) Aquagard SLT, containing dichlorvos (a 'Red List' substance), continues to
be used under licence to control sea lice on salmon. A suitable substitute
must be found in order that the requisite reduction in loading is achieved by
1995.

(c) Mathematical models require to be developed to determine the capacity of
a water body to accommodate effluent safely and to predict the distribution
and concentration of wastes following release. Existing models for the determination of phosphorus require to be defined for application to long, deep,
poorly mixed freshwater lochs.

(d) The European Community has published a draft urban wastewater treatment Directive stating that all sewage must receive treatment prior to discharge. In addition, industrial effluents of a similar nature to sewage will
require treatment, which may present a serious problem for those companies
operating cage sites. Although some experimental work has been undertaken
to collect waste from the bottom of cages for treatment, there appear to be
very real practical problems to be resolved.

(e) If river purification authorities are to exert total control over discharges from land-based fish farms they require to be given powers to control abstractions. As stated earlier, some rivers have dried up following over-abstraction by fish farmers; and

(f) River purification authorities are soon to be given authority to charge discharges for effluent and environmental monitoring. A decision will need to be made by the authorities whether to continue with the concept of self-monitoring or whether to employ the necessary resources to carry out the monitoring and reclaim the cost from the industry.

CONCLUSIONS

1. Environmentalists claim that the fish farming industry is damaging the environment, while the industry challenges this view and claims that fish farming is dependent upon excellent water quality for survival.

2. The authors believe that there has been so much criticism of the industry that the farmers should see the Board's self-monitoring requirements as a means of demonstrating that fish farming, practised in accordance with the consent conditions imposed, has minimal impact on the quality of the receiving waters.

3. While the Board's requirement for self-monitoring was innovative and initially opposed by the farmers, it is interesting to note that self-monitoring is now considered to part of 'Best Available Technique Not Entailing Excessive Cost' to prevent or minimize pollution [16]. In addition, the National Rivers Authority appears to be actively considering whether a duty to self-monitor should be included in discharge consents in order to overcome the complaint that effluent monitoring is only undertaken during normal office hours [17].

4. It must be stated that it is not the Board's intention to stop the development of fish farming but to ensure that its development proceeds without damage to the aqueous environment.

ACKNOWLEDGEMENTS

The authors thank Mr D. Hammerton, Director of the Clyde River Purification Board, for permission to publish this chapter.

REFERENCES

[1] Gowen, R. J. and Bradbury, N. B. (1987) The ecological impact of salmonid farming in coastal waters: a review. *Oceanogr. Mar. Biol. Ann. Rev.*, **25**, 563–575.

[2] Mainstone, C., Lambton, S., Gulson, J. and Seager, J. (1989) The environmental impact of fish farming – a review. *WRc Report No. PRS2243-M.* 84 pp.

[3] Scottish Salmon Growers' Publicity Brochure (1990).

[4] Solbe, J. F. De L. G. Fish Farm Effluents – Cause for Concern? *Water.* March 1982.

[5] The Town and Country Planning (General Development) (Scotland) Amendment Order 1990, Her Majesty's Stationery Office.

[6] House of Commons Agriculture Committee Fourth Report (1990) Fish Farming

in the UK. Her Majesty's Stationery Office.

[7] Opinion of Senior Counsel for Western Isles Islands Council (1988).

[8] Opinion of Senior Counsel for Clyde River purification Board (1988).

[9] Dillon, P. J. and Rigler, F. H. (1974) A test of a simple nutrient budget model predicting the phosphorus concentration in lake water. *J. Fish. Res. Bd. of Canada*, **31**, 1771–1778.

[10] Council of European Communities. (1978) Council Directive on the quality of fresh waters needing protection or improvement in order to support fish life (78/659/EEC). *Official Journal* L222, 14 August 1978.

[11] Institute of Fisheries Management (1990) Papers from the 1989 Annual Conference on *Freshwater Cage-rearing of Salmonids: Ecological and Environmental Effects*. Institute of Fisheries Management (Scottish Branch).

[12] Nature Consrevancy Council (1990) Fish farming and the Scottish freshwater environment. *NCC Contact No. HF3-03-450*.

[13] Bailey Watts, A. E. (1990) Eutrophication: assessment, research and management with special reference to Scotland's freshwaters. *J. Inst. Wat. & Envir. Mangt.*, **14**, (3), 285–294.

[14] Crown Estate Commission (1989) Marine fish farming in Scotland. Development strategy and area guidelines.

[15] Edwards, A. and Sharples, F. (1986) *Scottish Sea Lochs. A Catalogue*. Scottish Marine Biological Association.

[16] Environmental Protection Act (1990) Her Majesty's Stationery Office.

[17] National Rivers Authority. Discharge Consent and Compliance Policy: A Blueprint for the Future.

21

Assessment and control of farm pollution

J. Seager, BSc, MIBiol, CBiol[†], F. Jones, BSc, MSc[‡], and G. Rutt, BA, MSc, MIBiol, CBiol[§]

INTRODUCTION

Water pollution from farm wastes is a serious problem in the UK and continues to pose a significant threat to river water quality. In England and Wales alone, in 1989 there were 2889 reported pollution incidents caused by farm wastes, of which 18% were classified as 'serious'[1]. This was a significant improvement on the 4141 incidents reported in 1988, although it is not clear whether this improvement was a reflection of improving farming practices or simply due to the fact that 1989 was an unusually dry year with little runoff.

Most reported incidents are caused by the release of animal slurries, silage effluent and yard washings from farm premises into adjacent watercourses. The true extent of the problem may be much wider than the farm pollution incident data would suggest, since they take little account of the possible chronic effects of everyday farming activities and land-use change which are much more insidious in nature and consequently more difficult to detect.

This chapter describes some of the key results of a major national research project which was initiated to provide tools and procedures to be applied in the assessment and control of farm pollution. The research programme is being carried out by the

[†] Environmental Quality and Assessment Officer, NRA Head Office.
[‡] Senior Environmental Appraisal Officer, NRA Welsh Region.
[§] Environmental Scientist, WRc. Medmenham.

National Rivers Authority (NRA) and the Water Research Centre (WRc), principally in Wales, but the project is now being extended to other regions in England.
The approach that has been taken is shown schematically in Fig. 1, and the key
elements of the programme are described below.

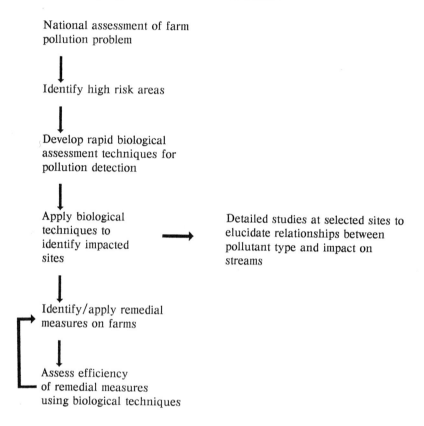

Fig. 1. National research and development programme on farm
pollution assessment and control.

NATIONAL ASSESSMENT OF PROBLEM

Previous national surveys have relied heavily on collating data on reported farm
pollution incidents. Whilst this provides information on the pattern of incidents
which have been dealt with by the pollution control authorities, it is difficult to
intepret the data in terms of the overall impact of farming activities on river quality.
Furthermore, this approach is historical and does not address the important issue
of assessing pollution risk. To overcome some of these problems, data have been
collected concerning not only reported pollution incidents but also any measured
changes in river quality known to be associated with farm pollution. In addition,
national databases have been compiled for those factors which are likely to have a
significant effect on pollution risk. These include animal stocking density and waste
production, soil type and topography. It is possible to overlay these databases using

Geographic Information Systems (GIS) technology to identify high-risk areas. This will provide a valuable tool for the regulatory authorities in targeting resources towards priority areas for the planning and implementation of pollution assessment, prevention and control activities.

RAPID BIOLOGICAL ASSESSMENT

The effective control of surface water pollution from dairy and beef farming is hampered by the large number of farms involved and their distribution over large areas with poor accessibility. The detection of impacted streams by the use of biological indicators has been proposed as part of a proactive pollution control strategy.

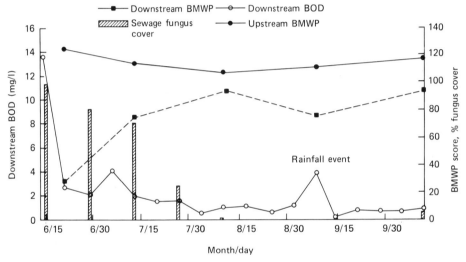

Fig. 2. Water quality at PontfaenBrook and its effects on sewage fungus growth and benthic macro-invertebrates.

The use of biological methods has several advantages over more traditional chemical-specific techniques in assessing river quality. Pollution from farms tends to be episodic in nature and, because of this, may be inadequately detected by conventional 'spot' sampling followed by chemical analysis. Biological communities in rivers, however, will reflect changes in water quality in an integrated way [2]. Their responses can be measured to provide an indication of the recent pollution history at a site. Biological methods, based on the benthic macroinvertebrate fauna, are particularly appropriate in this context since they are relatively cheap, rapid, and have a means of directing pollution control effort to pollution 'hot-spots'.

West Wales was chosen as a region in which to develop a protocol that could be adopted nationally because of the intensity of dairy farming in the area and its impact upon salmonid populations in headwater streams [3]. These streams are (naturally) fairly homogeneous in both physiography and chemistry, and hence in biota, such that observed differences in fauna are likely to reflect the impact of pollution rather than natural variability. Survey work was carried out in late winter/early spring, which is a period when dairy farmers experience the greatest

difficulty in waste containment. The biological methods would then complement
a system devised for use in the summer by Reynolds [4]. Two such systems are
required because of the seasonal variability in benthic fauna.

Fifty-five sites were selected in order to represent a range of pollution impact,
pollutant type (e.g. parlour washings, lagoon overflow, yard washings) and geographical area. Sites were restricted to those likely to support salmonid populations
and unlikely to be affected by other sources of pollution found in rural catchments,
for example pollution sewage-treatment works and light industry. Stream width
was in the range 1–6 m, most sites being between 2.0 and 4.5 m.

Survey work was carried out in the period 27 February to 6 April 1990. At
each site a spot water sample was obtained, sewage fungus cover was assessed,
invertebrate kick samples were taken, and lengths of stream were electrofished semi-
quantitatively. Other environmental data such as substratum composition and
stream width were also recorded. Water samples were subsequently analysed in the
laboratory for a range of determinands, and the invertebrate samples were processed
to species level where practicable.

Information on the abundances of macro-invertebrate species and the percentage
cover of sewage fungus at each of the 55 sites was analysed using the multivariate
classification technique known as TWINSPAN (Two Way INdicator SPecies ANalysis). This analysis split the streams into three groups with significantly different
biological communities [5]. These three groups could be differentiated using a simple dichotomous key involving the abundances of four readily identifiable macro-
invertebrate taxa (Heptagenid and Baetid mayflies, the freshwater shrimp, *Gam-
marus* pulex, and Oligochaete worms) and a crude estimate of the abundance of
sewage fungus. The indicator taxa were found to be animals which have a known
sensivity to organic pollution, and the three groups were found to show significant
differences in a range of variables which reflect pollution from slurry and silage
effluent: BMWP (Biological Monitoring Working Party) score, ASPT (Average
[BWMP] Score Per Taxon), trout density, abundance of sewage fungus and total
ammonia concentration (Table 1). These results indicate that the three groups may
be considered to represent the following levels of pollutant impact: Group 1 – generally unpolluted; Group 2 – mild pollution, pollution further upstream or a recent
history of chronic or acute impacts; and Group 3 – grossly polluted. This classification was supported by direct observation and information from NRA pollution
control staff. For the majority of sites in Groups 2 and 3 it was possible to identify
sources of organic pollution which could account for the observed impacts. These
included chronic sources such as leakages from slurry lagoons and silage clamps or
daily inputs of parlour/dairy washings and more intermittent rain-generated inputs
such as slurry or whey washing off yards, access roads and fields.

The indicator system derived from this survey was subjected to an extensive field
trial in west Wales in February and March 1991. One hundred and forty-six sites in
fifteen catchments (mainly second and third order streams) were examined. Non-
biologists, including pollution control staff and a bailiff, successfully operated the
system after training. Thirty-four farms were found to be polluting watercourses
and fifteen others were suspected. These farms have been visited by NRA pollution
control staff to identify remedial measures. The success of these measures in reducing or eliminating the effects of farm pollution at these sites will be assessed by

a further biological survey in February and March 1992. Fifty sites in Devon have also been sampled to establish whether the present indicator system, or a modification of it, could be applied in this area. The results suggest that the indicator key derived from this survey could be used to identify the farms responsible for pollution in impacted catchments, because the technique is straightforward to use and can be carried out entirely on the bankside.

Table 1. Relationship between TWINSPAN groups and pollution-related variables

Variable	Group 1	Group 2	Group 3	F	p
Group size	20	24	11		
BMWP	129	103	55	31.26	0.0001
	(101–156)	(78–128)	(35–75)		
ASPT	6.36	5.65	5.29		
	(6.15–6.58)	(5.18–6.12)	(4.49–6.08)	20.42	0.0001
Trout density	15.5	4.6	2.6	16.36	0.0001
(No./100 m)*	(9.6–24.9)	(1.4–14.8)	(1.1–6.2)		
% sites with > 10%					
sewage fungus	5	33	91		
Total ammonia	0.02	0.09	0.15	10.71	0.0001
(mg/l)*	(0.01–0.06)	(0.02–0.40)	(0.04–0.62)		

Group values are means with standard deviation ranges in parenthesis. The F statistic and probability values (p) are from analysis of variance. Asterisks denote log-transformed variables.

INTENSIVE STUDIES

One of the west Wales streams identified as severely impacted during the extensive survey phase of the programme was selected as a suitable case study for more intensive investigations. Such studies are required in order to gain an understanding of the effects of particular farm effluents on stream quality and biota. Pontfaen Brook is a tributary of the Afon Gwaun, which discharges to the sea at Fishguard. The principal land use in the catchment is dairy farming and the brook was found to be severely affected by a discharge of silage effluent 2.2 km from its source. Effluent was leaking from two poorly-sealed silage clamps into a road drain which passed beneath a slurry lagoon and into a drainage ditch. The BOD of the water in the ditch at the point where it discharged to the stream was found to be in excess of 3500 mg/l, and the total ammonia concentration was 95 mg/l.

Initial biological assessments in June 1990 revealed that there was a reduction in the BMWP score from 125 upstream of the input to 28 below the mixing zone, whilst trout density was halved from $22/100 m^2$ to $10/100 m^2$. Quantitative sampling of the benthic invertebrates indicated that of 27 taxa found upstream from the input only six could maintain a population below the mixing zone. Of these six, the two most tolerant of organic pollution (oligochaete worms and chironomid midge larvae) had increased their population density by two orders of magnitude below the input.

Continuous monitoring of water quality and spot sampling both upstream and downstream from the input indicated that the observed biological impact could not be interpreted as a direct effect of water quality. Dissolved-oxygen concentrations remained above 7.0 mg/l and the maximum recorded concentration of unionized ammonia was only 0.004 mg/l, that is below levels likely to result in either lethal or sub-lethal stresses to fish or invertebrates. It is suggested that the impact is the result of the growth of a dense cover of the sewage fungus *Leptomitus lacteus*, caused by the organic input and entirely absent upstream of the ditch. Such growths are known to cause localized deoxygenation of the stream bed and smother the habitats of most invertebrates [6].

As the summer progressed, the BOD of the stream was reduced to less than 2–3 mg/l downstream from the input, which resulted in a disappearance of the sewage fungus growth and a progressive recovery of the macro-invertebrate community (Fig. 2). This threshold of 2–3 mg/l supports the findings of other workers who have observed significant growths of fungus at low concentrations of BOD [7,8]. The decline in BOD in the stream was a direct consequence of a steady decline in the BOD load from the ditch, but whether this was a result of remedial action taken at the farm or a consequence of low summer flows remains to be established. A rainfall event in late summer caused an elevated BOD in the stream for a short period, and might have been related to the flushing of animal slurry into the stream.

The results of this detailed work contrast with a previous intensive river quality study carried out in the Eastern Cleddau catchment, west Wales [9]. Continuous chemical monitoring in the Clarbeston Stream within this catchment showed that peaks in ammonia and depressions in dissolved-oxygen concentrations were associated with local climatic factors. This appeared to be caused by the washing of dairy farm wastes from farmyards, ditches and field drains into the stream during periods of rainfall. As with the downstream site at Pontfaen Brook, the macro-invertebrate community in Clarbeston Stream was severely depressed with only pollution-tolerant chironomids and oligochaetes present. Comparison of the chemical and biological data at both sites shows that the nature of the impact of farming activities on river quality may differ depending on the type and pattern of waste discharge and the characteristics of the receiving watercourse. Pollution of the Clarbeston Stream was episodic and rainfall-generated, whereas the impact on Pontfaen Brook was more chronic in nature. Further work is required to investigate a range of different farming practices and waste management problems to improve the understanding of river quality impact.

POLLUTION CONTROL

Although the pollution control authorities have invested considerable effort in farm visits and campaigns with some notable successes in some areas, farm pollution problems are still widespread. River quality investigations have generally been reactive in the sense that they have been confined to examining the extent of the damage after incidents have occurred. Relatively little effort has been placed in the use of river quality assessment techniques proactively to identify farms causing problems and then assessing the efficacy of remedial measures. The results of this study to date suggest that the proposed biological assessment methods would provide a rapid, reliable and cost-effective means of pin-pointing problems in high-risk areas. Further work is now required to assess the applicability of the approach developed in Wales to other regions in the UK.

Clearly, the solution to the farm pollution problem does not lie solely in increasing the vigilance of the regulatory authorities. A more rational approach in catchment planning is required to underpin decisions on acceptable levels of farming activity in terms of their environmental impact. What was lacking in the past were commonly acceptable procedures which can be used to make these judgements. This need is currently being addressed by WRc through the development of a catchment-based pollution risk-assessment procedure called 'FARMS' (Farm Activity and River Management Simulation). This procedure allows the assessment of pollution risk associated with the disposal of farm wastes based on catchment characteristics. Although further validation and refinements are still required, this tool shows considerable potential to serve as the quantitative basis for dialogue between pollution control authorities and the farming community to decide upon acceptable levels of farming activity in accordance with designated environmental quality objectives.

CONCLUSIONS

1. Despite considerable efforts of regulatory authorities, farm pollution continues to be a serious problem in the UK. A major national research and development programme has been initiated to provide tools to be applied in farm pollution assessment and control.

2. An assessment of the extent of the problem nationally has allowed the identification of high-risk areas based on factors such as animal stocking density, climate, soil types and catchment topography. This will be valuable in the targeting of resources in pollution assessment and control.

3. The results of an extensive study at 55 sites in Wales suggest that the proposed biological assessment methods would provide a rapid, reliable and cost-effective means of identifying farms causing river quality problems. Further work is required to assess the suitability of the proposed methods for application in other regions.

4. Intensive studies at selected sites have illustrated the variability in the ways that different types of farm-waste discharge affect the biology and chemistry of receiving watercourses. Further work is needed to examine a range of different farming activities in terms of their impact on river quality.

5. The solution to the problem will depend on effective liaison between the regulatory authorities and the farming community. Quantitative procedures are currently being developed (FARMS) which are aimed at providing the basis for decisions on acceptable levels of farming activity in accordance with designated environmental quality objectives.

ACKNOWLEDGEMENTS

The authors gratefully acknowledge the National Rivers Authority for the funding of this work.

REFERENCES

[1] National Rivers Authority/Ministry of Agriculture, Fisheries and Food. (1990) *Water Pollution from Farm Waste* 1989. England and Wales. Joint Report.

[2] Metcalfe, J. L. (1989) Biological water quality assessment of running waters based on macro-invertebrate communities: history and present status in Europe. *Envir. Pollut.*, **60**, 101–139.

[3] Wightman, R. P. (1989) *The effects of agricultural pollution on headwater streams in West Wales.* Environmental Apparaisal Unit, South Weat NRA (Welsh Region). Report No EAW/89/1.

[4] Reynolds, N. (1989) *The effects of agricultural pollution on the biological quality of headwater streams in west Wales – rapid assessment of agricultural pollution by the use of an indicator key.* NRA Welsh Region. Report No. PL/EAW/89/4.

[5] Weatherley, N. S. and Ormerod, S. J. (1987) The impact of acidification on macro-invertebrate asemblages in Welsh streams: towards an empirical model. *Envir. Pollut.* **46**, 223–240.

[6] Curtis, E. J. C. (1969) Sewage fungus: its nature and effects. *Wat. Res.* **3**, 289–311.

[7] Water Pollution Research Laboratory (1969) *Notes on Water Pollution* No. 47.

[8] Quinn, J. M. and McFarlane, P. N. (1988) Control of sewage fungus to enhance recreational use of the Manawatu River, New Zealand. *Verh. Internat. Verein Limnol.* **23**, 1572–1577.

[9] Schofield, K., Seager, J. and Merriman, R. P. (1990) The impact of intensive dairy farming activities on river quality: the Eastern Cleddau catchment study. *J. Instn Wat. & Envir. Mangt.* **4**, (2) 176–186.

Part IV Catchment and groundwater management

22

Integrated river channel management through geographic information systems

M. J. Clark, BA, PhD; A. M. Gurnell, BSc, PhD; J. Davenport, BSc, MSc, and A. Azizi, BSc, PhD[†]

INTRODUCTION

Environmental impact assessment is usually undertaken at the start of a specific project. In exceptional circumstances post-project monitoring and appraisal may provide information on the degree to which the initial impact assessment was correct. Such monitoring and appraisal may in turn provide refinements for future environmental impact assessments for similar projects. Such a temporally and spatially discrete approach is not ideal in the context of catchment and river channel management because these features are subject to continual change as a result of both natural and man-induced processes. Effective, integrated management of both the catchment and its river network requires continual updates on processes, management of those processes and likely impacts of the processes and their management. One vehicle for achieving such effective integrated management is through the application of Geographic Information Systems (GIS) technology to the efficient storage, retrieval and modelling of the components of the environmental system and their interactions.

GIS technology is increasingly being applied to the management of environmental data because of its flexibility in manipulating data of different types. A geographic

[†] Drs Clark and Gurnell are Deputy Directors, GeoData Institute, University of Southampton; Ms Davenport is GIS Manager and Dr Azizi is a Research Scientist, GeoData Institute, University of Southampton.

information system is 'a system for capturing, storing, checking, integrating, manipulating, analysing and displaying data which are spatially referenced to the Earth' [1]. A fully developed GIS will allow inputs of data from a wide variety of local and remote sources, outputs to a range of devices in a variety of forms and formats, and extremely flexible storage and manipulation of the data within the system. In the context of catchment and river management, the manipulation of information within a GIS has many advantages over the use of non-georeferenced data bases and manual procedures, as follows.

(a) A corporate GIS has all the advantages of any computer data base in that it provides all users with efficient access to consistent, standardized, regularly updated data sets.

(b) GIS offers cost-effective means of interrogating the data base for statistical queries (data summary), for combining multiple sets of data on the same geographical area (thus simplifying the representations of spatial distributions), for identifying gaps or deficiencies in the survey base, for constructing predictive 'what if' questions, and for establishing spatial patterns or causal relationships.

(c) GIS provides an ideal framework within which to combine and compare surveys from different dates. It thus plays an important role in long-term monitoring programmes and post-project appraisal.

(d) The GIS can be designed to relate to costs of different possible works or management strategies, thus permitting extrapolation and estimation of project costs – for example in relation to channel maintenance, dredging, weed cutting, etc.

(e) The ability of the GIS to combine many different types of information (as described in (b) to (d)) renders it ideal for use in environmental assessment and environmental impact assessment studies.

GIS applications in the water industry include asset management (e.g. [2]), multi-purpose land use and natural resources planning (e.g. [3]) and specific, focused process monitoring, modelling and analysis (e.g. [4, 5, 6]). These applications represent a range from data-intensive but limited processing applications to processing-intensive functions. This chapter illustrates how a routine, data-intensive function can be developed into a more complex suite of processing-intensive modelling functions through the interfacing of data sets and their joint analysis within a GIS. This development permits effective management and impact assessment for river channel networks.

In considering GIS as a vehicle for integrated river channel management, it is interesting to explore its potential contribution to the evaluation of flood defence performance. This chapter represents a series of possible extensions of a Levels of Service Model for Urban and Rural Flood Defence originally developed by Thames Water in association with Laurence Gould Consultants Ltd [7]. It demonstrates how such a model can be extended in scope and in its level of interaction with other data sets and models to produce an effective base for storing, retrieving and analysing information on the river flood plain and channel network. This approach permits the optimization of flood control, channel maintenance and nature conservation requirements of channel management in the context of both financial and environmental costs and benefits. The functions described could complement any similar formulation of a Levels of Service flood defence model.

THE LEVELS OF SERVICE MODEL

The Levels of Service (LOS) model considers channel reaches within a catchment (here termed LOS reaches), and for each reach it relates levels of service in terms of flood defence and land drainage to current land use. The model which is illustrated in Fig. 1 provides a structure for assembling and assessing operational information on flood defence performance. The approach is based upon four operational steps applied to the LOS reaches and their associated floodplain segments.

Step 1: Identify target levels of service. This is based on a land use assessment which ascribes each LOS reach to a land use band so as to identify its appropriate target score for level of service. (High value areas require a high target level of flood defence.)

Step 2: Identify actual levels of service. This procedure monitors flood and bank full events for the reach so as to assess the reach's actual flood defence performance through an actual monitoring score.

Step 3: Undertake levels of service assessment. This is achieved through a comparison of the target score and the actual monitoring score. This provides the basis for assessing whether the level of service is adequate, inadequate (monitored score exceeds target score) or excessive (monitored score is substantially less than target score).

Step 4: Provide the right service. This takes the LOS assessment and considers channel maintenance history and environmental implications in order to design and implement an appropriate land drainage management strategy in order to provide a service that is neither excessive nor inadequate.

It is the purpose of this chapter to describe how this simple but very effective conceptual model can be transferred from a relational data base management system to implementation on a GIS. Such a transfer permits a number of enhancements to the model and its implementation to be considered.

Automated output of updated maps of the monitored and target scores for the LOS reaches can be provided as new flood event and land use change information is entered into the system. This is a basic but important operational requirement, and such maps are currently manually produced.

More importantly, **fully geo-referenced input of flood process and land use change information** can be easily and efficiently achieved within a GIS implementation. Although the spatial pattern of some of these data (e.g. flood boundaries and some building, road and bridge locations) are currently held and incorporated in the analysis, other data types, particularly information on other categories of land use, are only attributed to proportions of the LOS reach within the statutory flood boundary rather than being associated with specific geographical locations and boundaries. A GIS could hold all land use classes in a fully distributed fashion and it could also hold detailed flood boundaries for every overbank flood event. In the latter case, a rapid automated procedure for entering such data is required and is described below. The potential of GIS to incorporate these input and output tasks provides an enormous step forward in the efficient implementation of the LOS model.

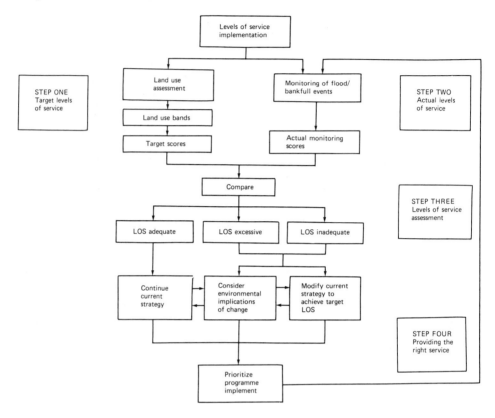

Fig. 1. The Levels of Service model.

By extending the information accessed by the GIS and by developing appropriate modelling modules, it is also possible to begin to address the environmental and financial costs and benefits of maintaining or altering the flood defence maintenance programme. In this way, environmental assessment and environmental impact assessment can be combined with flood defence and land drainage performance in defining a land drainage management strategy for each LOS reach. The great power of GIS technology to address such problems is assessed through a discussion of the interfacing of the LOS model with a land drainage monitoring and post-project appraisal module and an environmental implications module.

These input, output and modelling enhancements offered by GIS technology provide a framework for considering the nature and interactions of four categories of information which are particularly relevant to the character of the channel network and its effective management: flood process information; land use information; land drainage management costs and options, and environmental quality and sensitivity to management.

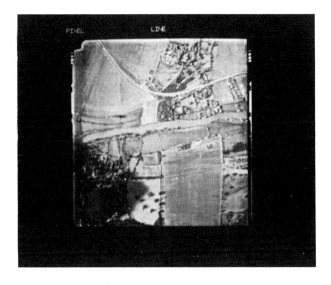

(a) Scanned air photograph.

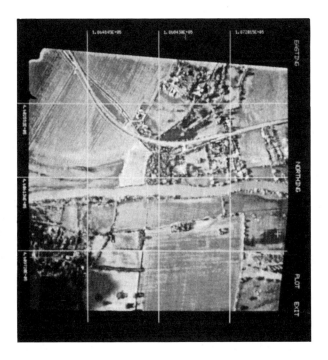

(b) Rectified air photograph registered to the National Grid.

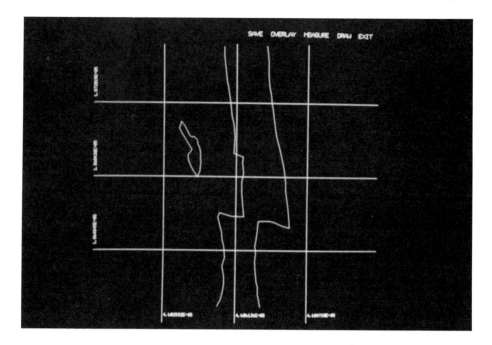

(c) Vectorized flood boundary.

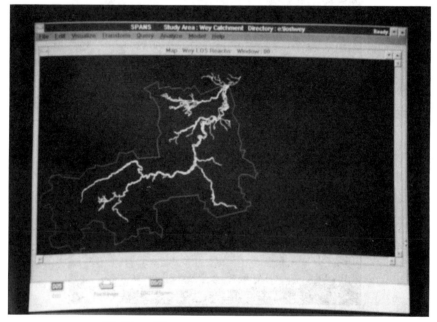

(d) River Wey — statutory flood boundary.

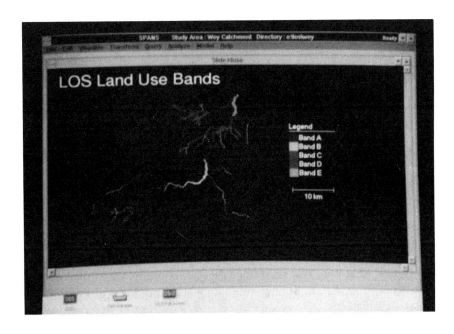

(e) River Wey — land use bands.

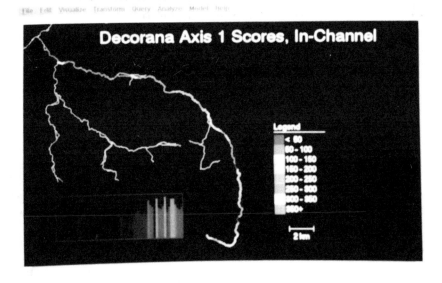

(f) Scores on an environmental gradient estimated for river corridor reaches
using in-channel plant species information.

Flood process information can be improved by streamlining data capture techniques. These improved methods of data capture are outlined in the following section. This is followed by a description of other data sets which can be interfaced to the LOS model to improve the provision of data on the remaining information categories.

IMPROVED DATA CAPTURE: THE INPUT OF FLOOD BOUNDARY INFORMATION

In many cases the extent of flood boundaries is routinely recorded by air photograph or video survey supplemented by sporadic ground data. The translation of such surveys into accurately mapped flood boundaries is time-consuming and the results are of variable accuracy. The difficulties of achieving a consistent level of accuracy result from the great variations in the quality and obliquity of air photographs and the difficulties of accurately extracting information from video images. By implementing the LOS model within a GIS framework it has been possible to develop a flood boundary input module, which takes scanned air photos or video images, and contrasts, stretches, geometrically corrects, resamples and registers them to the national grid. This series of image processing tasks produces a raster image of the flood extent which can be overlaid on vector or scanned map information depicting the river network. In addition, it is a simple task to vectorize the flood boundaries by manually digitizing them over the raster image on screen. This approach has already been implemented for the LOS application. Alternative methods of flood boundary definition include classification of the geometrically correct image using standard image processing algorithms, followed by vectorization of the flood boundary, or the development of pattern recognition routines which will automatically identify flooded areas for boundary vectorization. Both of these approaches could only by expected to provide a partial solution to automatically deriving flood boundaries so that some manual intervention would be necessary. The latter approach, using automatic pattern recognition for flooded as opposed to non-flooded areas, is currently being developed in association with the LOS model.

Registration of all flood boundaries within the GIS provides not only the necessary analysis and reporting of flood area statistics for each of the LOS reaches to input into the LOS model, but it also allows continuous monitoring and updating of the statutory flood boundary and the development of a data set for spatial flood frequency analysis and the definition of boundaries of floods with particular return periods. These floods can then be overlain on maps of land use to assess the likely damage caused by a flood with a particular return period.

INTERFACING THE LOS MODEL TO OTHER DATA SETS

The simple LOS model, as described above, could be interfaced to other data sets and models in order to optimize the financial and environmental costs and benefits of the adopted flood defence strategy. The availability of information on the river channel and corridor environment and on its maintenance history should provide an improved basis for the development of an informed flood defence strategy for each of the Levels of Service reaches.

Land drainage maintenance information

If the assessment of levels of service is viewed as the basis for decision on channel maintenance, then the flood defence performance of a reach needs to be assessed in relation to a record of previous land drainage maintenance activities. Such records may currently be held on a computer data base, in which case interfacing LOS to channel maintenance would depend upon establishing the relationship between LOS river reaches and the corresponding reaches used to archive channel maintenance records. Such interfacing within a GIS would be straightforward. The archive of land drainage maintenance information provides a resource for estimating the costs of land drainage maintenance and its effectiveness, through the frequency and type of maintenance which has been applied. It also provides the baseline from which a changed maintenance schedule can be designed.

River corridor environmental quality and land use information

If the river network is available in digital form (whether as fully digitized maps of bank lines or a simpler representation as channel centre lines), then these can provide a basis for automatically quantifying reach parameters such as channel pattern, slope and sinuosity. Surveyed channel cross-sections provide complementary information on river channel geometry.

River corridor surveys (RCS) are being increasingly widely implemented, often under the Nature Conservancy Council Guidelines [8]. These surveys normally include sketch maps and summary information for river corridor reaches of approximately 500 m in length. The data relate primarily to habitats and land use of the river channel and associated flood plain. However, there are a variety of other types of information which provide a good indication of the geomorphological characteristics of the reach including substrate, channel margin and bank materials, extent of bed features such as pools and riffles, channel and bank size and slope. The summary reach data are easily transferred to GIS and provide a context for assessing the environmental quality and sensitivity of the reach.

The locations and boundaries of areas with statutory constraints (e.g. SSSI and NNRs) are also often available on a GIS. Further potential information sources for land use include digital maps, satellite imagery and census data, but these are not considered here.

AN ENHANCED LEVELS OF SERVICE MODEL

Fig. 2 illustrates a possible enhancement of a levels of service model which contains three components: the LOS module is based upon the previously described model, whereas the monitoring, post-project appraisal, and environmental implications components are new. The model is designed to apply to LOS reaches.

The enhanced LOS model runs in a similar way to the existing LOS system but it formalizes the consideration of the environmental implications of a changed land drainage maintenance programme. In the prototype model (Fig. 2) only two aspects of the environmental implications are detailed: channel/geomorphological sensitivity and conservation/ecological sensitivity. Subsequently, it would be desirable to add other areas of activity which are influenced by or impinge upon land drainage maintenance (e.g. recreation, archaeology, fisheries). In the meantime, the channel/

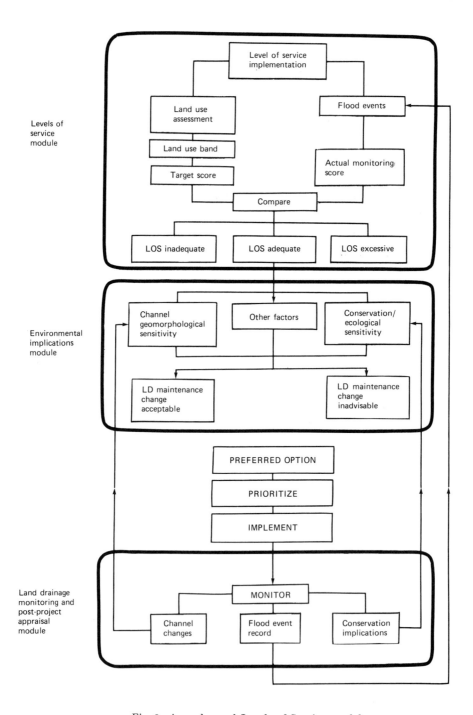

Fig. 2. An enhanced Levels of Service model.

geomorphological and conservation/ecological sensitivity components represent constraints on land drainage maintenance whilst also representing statutory obligations to the environment and a range of amelioration and enhancement opportunities. Monitoring and post-project appraisal would develop from a land drainage maintenance record to evaluate the type and frequency of past management strategies and their levels of success. Thus the enhanced model would assess the level of service for a reach, assess whether a maintenance change is desirable, consider the sensitivity of the reach to particular categories of maintenance, identify any potential for environmental enhancement, and so aid in the selection of a preferred strategy. Once implemented, this strategy would appear in the land drainage record for the reach and its effectiveness would be subsequently judged.

Many of the components of the enhanced LOS model presented in Fig. 2 are already available.

The LOS module is already operational.

The environmental implications module is built upon an analysis of RCS, channel pattern and cross-section, and statutory designation information. The aim is to assess the geomorphological and ecological sensitivity of LOS reaches and their component RCS reaches based upon the above information sources. A research project commenced in 1990 which aimed to develop a river channel typology for British rivers based upon the available information sources, supplemented by any additional survey information that was found to be needed by the study. This research used the typology of channels to classify the river channel system in a way which indicated and defined the sensitivity of the system to management. Although the study considered all aspects of the river channel environment, the typology emphasized the physical characteristics of river channel sensitivity. A complementary study was required to classify reach ecological character and sensitivity from RCS and statutory designation data.

The land drainage monitoring and post-project appraisal module is partly operational since it is based on a combination of the existing land drainage maintenance records (which through frequency and type of maintenance will illustrate the costs as well as levels of success of the existing maintenance programme) and the flood event record (which will be streamlined by the new data capture module). An analysis of within-reach and basin-wide interactions between environmental characteristics, including channel/geomorphological and conservation/ecological status and the maintenance regime, is required to provide guidelines for assessing the sensitivity of the environment to land drainage maintenance. The results of these analyses will complement and aid in the refinement of the geomorphological and ecological sensitivity components of the environmental implications module.

A major strength of GIS in implementing the model of Fig. 2 is that it facilitates extensive data handling so that, for example, basin-wide analysis of the interactions between reach performance, reach characteristics and reach maintenance is straightforward. A basin-wide classification and display of river reaches is also straightforward if the data required to support the classification are encoded.

The major problem facing a GIS approach to the implementation of the model presented in Fig. 2 is not the methodology required to quantify the variables in the boxes for the LOS reaches, but the development of a **reach model** to combine the various data sets that are collected for different types and size of river reach.

The data sets described in this chapter relate to points on the river network and to three different types of reach: the LOS reach, the operational reach, and the RCS reach. If additional data sets were to be included in developments of the prototype model, other reaches would be relevant to the encoding of these additional data types. Although LOS reaches are subdivided into operational reaches so that it is easy to map one onto the other for analytical purposes, this is not true of the RCS reaches. A model is required to locate each reach type (using its start and end co-ordinates) and to map each reach type onto every other type. This is not a trivial task because the method of mapping one reach onto another can have an enormous effect on the derived statistics from analyses of interactions between variables related to different reach types. Because the RCS reaches are small in comparison with the other two reach types in the present model, and because the other two reach types map directly onto one another, the reach model for the current application does not present a major analytical problem. What is needed is an agreed concept and definition of the basic spatial unit for a river system (i.e. the basic river unit) from which all other reach units can be constructed, and a commitment to use the basic river unit or aggregates of it in data collection. In the meantime, reach modelling presents a complex spatial statistical problem.

CONCLUSION

The exploratory enhancement of the levels of service model is well advanced in concept and already possesses some new operational components within a GIS framework. The challenge of producing a completely operational model is dependent upon the success of research programmes. Once the model is operational (and is customized to the operational procedures of a particular user group) there remains the joint challenges of gaining the maximum benefit from the enhanced model and then extending the model to incorporate other relevant information sources. In the context of maximizing the benefit accrued from the model, it is important to explore the ways in which the model can be used not only to improve the fulfilment of levels of service targets, but also to re-order the model so that operational requirements for channel maintenance and environmental enhancement opportunities can in turn become the model's focus. Changing the focus of the model only requires the re-ordering of the boxes so that the requirement implied by any one of the three components can be moderated by the constraints of the other two components. In this way, environmental constraints and opportunities can take an equally important role with levels of service and land drainage maintenance in future river channel management programmes.

ACKNOWLEDGEMENTS

The developments explored in this chapter benefitted greatly from close discussion with colleagues in the National Rivers Authority, Thames Region (D. Mills and C. Candish), but the views expressed are personal and do not represent specific policies of the National Rivers Authority or the NRA Thames Region.

REFERENCES

[1] Department of the Environment (1987) *Handling geographic information*. Report of the Committee chaired by Lord Chorley, Her Majesty's Stationery Office, London.

[2] Elkins, P. (1990) *Mapping awareness*, **4**, 38.

[3] State of Maryland (1990) In: *Introductory readings in geographic information systems*, D. J. Peuquet and D. F. Marble (Eds.), Taylor and Francis, London, p. 65.

[4] Clark, M. J., Gurnell, A. M. and Edwards, P. J. (1990) In: *Proceedings EGIS 90*, First European Conference on Geographical Information Systems, Amsterdam, The Netherlands, **1**, 189.

[5] Meijerink, A. M. J. (1989) In: *Land qualities in space and time*, J. Bouma and A. K. Bregt (Eds.), Pudoc, Wageningen, p. 73.

[6] Moore, I. D., O'Loughlin, E. M. and Burch, G. J. (1988) *Earth surface processes and landforms*, **13**, 305.

[7] Thames Water (1989) *Levels of service for urban and rural flood defence*.

23

Environmental problems in the peat moorlands of the Southern Pennines: reservoir sedimentation and the discolouration of water supplies

D. P. Butcher, BA, CertEd, PhD[†]; J. Claydon, BSc (Eng), DMS, CEng, MICE[‡]; J. C. Labadz, BSc, PhD[†]; V. A. Pattinson, BSc[†]; A. W. R. Potter, BSc[†], and P. White, BSc[†]

INTRODUCTION

The blanket peat moorlands of the southern Pennines have traditionally been viewed as ideal catchment areas for water supply, producing large quantities of water requiring little treament. Erosion of the peat has, however, been of academic interest since the early years of this century. Concern increased in the 1950s and 1960s and workers such as Conway [1] and Tallis [2] used palynology to investigate the timing of peat formation and the initiation of erosion. Their evidence suggested that the peat had begun to develop between 6000 and 8000 years ago, and that dissection began about 4000 years ago but did not become severe until around 1770 AD. Bower [3] divided erosion broadly by process into fluvial and mass movement categories. In 1981 the Peak Park Joint Planning Board published their 'Moorland Erosion Study' [4], which drew together all previous work and included several attempts at the measurement of erosion rates.

[†] The Polytechnic of Huddersfield.
[‡] Yorkshire Water Services Ltd.

The erosion of blanket peat moorlands is significant not only in hydrological and geomorphological terms. Recent emphasis has been placed on the effects of erosion on human activity in the area. Three major problems have been identified.

(1) Environmental degradation, which results in the loss of grazing land and amenity.

(2) The loss of water storage capacity, as the peat removed from the moorlands is deposited in reservoirs; and

(3) Water quality, which is perceived by water managers as by far the most vital and immediate. The link between water quality and severity of erosion is not yet clear, but evidence suggests that water colour is irreversibly raised when eroded peat is desiccated during periods of drought [5].

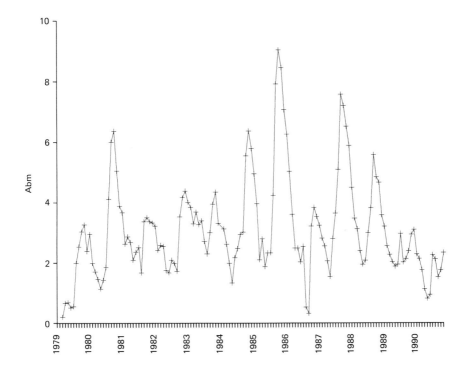

Fig. 1. Thornton Moor raw water colour, 1979–1990.

Direct measurement of reservoir sedimentation has been employed as a means of estimating basin sediment yield, and the technique provides a quicker and more accurate method than does prediction from a short run of stream data. Since erosion is episodic, concentrated in storm events and certain seasons, a stream sampling study may not reveal the true magnitude of the problem, whereas a reservoir preserves a long-term record.

The problem of discolouration of water supplies has attracted attention only in recent years. The chemistry and microbiology of the phenomenon are complex and still poorly understood, but it is believed to involve bacterial activity and the release of long-chain humic substances from the peat. Data showing levels of water colour tend to indicate higher colour in late autumn, especially after a dry summer, and there is a step-like progression in background levels. Fig. 1 illustrates this pattern for Thornton Moor Reservoir near Halifax. Particularly evident are the responses to the drought years of 1976/7 and 1984/5.

(a) Wessenden Head and Blakeley Reservoirs

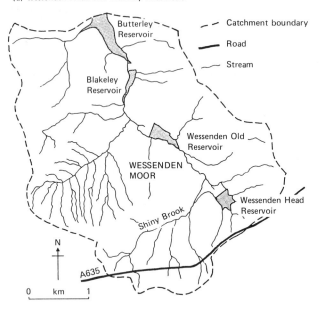

(b) Thornton Moor, Stubden and Doe Park Reservoirs

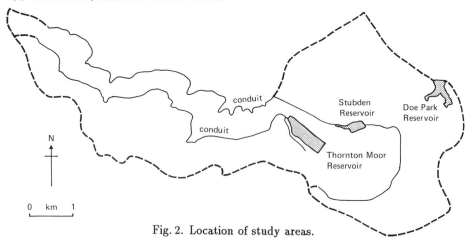

Fig. 2. Location of study areas.

RESERVOIR SURVEYS: METHODS AND CAPACITY LOSS

The use of reservoir surveys in the estimation of catchment sediment yield has been described by Rausch and Heinemann as an 'excellent method' [6]. Ideally, reservoirs should be surveyed using standard techniques, whilst drained for remedial works, and accurate contour plans should be produced.

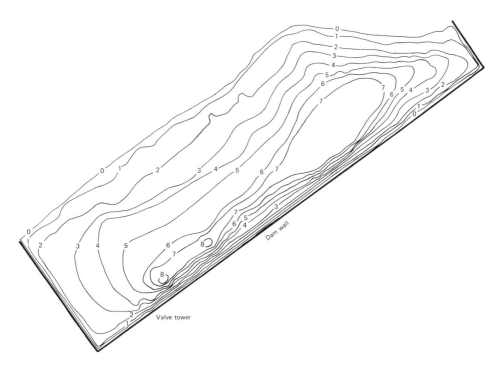

Fig. 3. Bathymetric map of Thornton Moor Reservoir.

In practice these opportunities occur relatively infrequently, though a dry survey was obtained at Wessenden Old Reservoir (National Grid reference (NGR) SE058087) in 1984. These topographic surveys were conducted using a Nikon Total Station (DTM5). More frequently, reservoir capacity is determined using a sonar survey to produce a bathymetric map. In the current study a continuous trace echo sounder (Elliott E100) was used in a series of transects, as proposed by Rausch and Heinemann [6]. All survey data were transferred from the total station to an electronic notepad, which allows input in the form of three dimensional co-ordinates. All the information from topographic and sonar surveys was inserted into a Computervision CAD (computer-aided design) system on a dedicated Prime computer. This computer stores the three-dimensional data and is capable of fitting contours and calculating the resulting capacity of the reservoir.

A large number of reservoirs have been surveyed and analysed (120 to date), but this chapter will concentrate on two main research areas. The first is the Wessenden Valley, on the headwaters of the River Colne to the west of Huddersfield. This has been the site of continuing research at Huddersfield Polytechnic, including investigations on peat hydrology and moorland erosion [7, 8, 9]. The uppermost part of the valley lies on Namurian sandstones and shales overlain by an impermeable sandy clay head deposit, and the pedology is dominated by blanket peat. The peat is 1.5–4.0 m thick over the majority of the catchment, but reaches 6.35 m in depth on the low gradient interfluve in the west. The headwater zones are characterized by a high drainage density (11.15 km/km^2), with steeply incised gullies and areas of exposed bare peat. Water in the Wessenden Valley passes through four reservoirs, constructed between 1836 and 1907 (Fig. 2a). A catchwater diverts a large proportion of the flow from the second reservoir, Wessenden Old, into the uppermost impoundment. Due to the cascading nature of the system it is possible to report capacity loss for individual reservoirs, but erosion rates may only be averaged over the entire catchment.

Table 1. Loss of reservoir capacity

Reservoir:	Wessenden Head	Wessenden Old	Blakeley	Butterley	Thornton Moor	Stubden
Date of construction	1881	1836	1903	1907	1885	1868
Catchment area (km^2)	15.05	15.05	15.05	15.05	5.12	1.94
Original capacity (TCM)	373	486	364	1832	795	451
Revised capacity (TCM)	358	321	137	1724	702	406
Loss of capacity per century(%)	3.7	23	75.2	7.5	11.3	8.2
Annual loss of capacity (m^3)	137	1220	2736	1365	895	318
Annual area-specific loss (m^3/km^2 year)	362.66	362.66	362.66	362.66	174.87	191.58

The second study area, for comparison, is Thornton Moor, to the west of Bradford (NGR SE053332) and shown in Fig. 2b. Here a chain of two reservoirs has been surveyed. Thornton Moor Reservoir itself is at a comparable altitude to Wessenden Head, and again has a blanket peat catchment. Here the peat is shallower, however, around 2 m deep in general on the interfluve areas, with occasional pockets reaching 3 m depth but with none of the intense moorland erosion so evident at Wessenden. Instead, Thornton Moor is drained by a series of larger streams with steeply incised vegetated slopes. These streams are diverted towards the reservoir by means of a Victorian bywash channel which runs along the contour. Stubden Reservoir is below Thornton Moor, and operationally linked to it by the system of catchwaters and also by a facility to pump water between the two. Stubden was constructed in 1868, some 17 years before the upper reservoir. The bathymetric map of Thornton Moor Reservoir is shown in Fig. 3.

Data resulting from surveys of the six reservoirs are shown in Table 1, where loss of capacity is presented in a variety of ways. The annual loss of capacity (volume infilled per year) is by far the smallest at Wessenden Head Reservoir, only $137\,m^3$, and ranges up to $2736\,m^3$ per year at Blakeley. However, the importance of this absolute loss depends on the size of the reservoir and its contributing catchment. Measures enabling more direct comparison between different reservoirs are the loss of capacity per century, expressed as a percentage of the original capacity, and the annual area-specific loss of capacity. From Table 1 it can be seen that the percentage loss of capacity per century varies from 75.2% at Blakeley to only 3.7% at Wessenden Head. Annual area-specific loss of capacity cannot be calculated individually for the Wessenden chain for the reasons previously stated. The value of their combined catchment is $362.66\,m^3/km^2$ year, approximately double the values found at both Thornton Moor and Stubden.

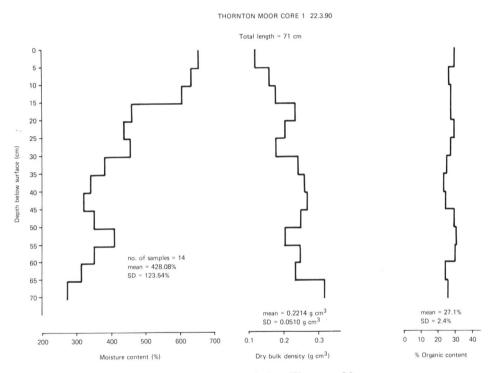

Fig. 4. Sediment characteristics, Thornton Moor.

SEDIMENT SAMPLING AND CALCULATION OF EROSION RATES

If sediment yields are to be estimated from changes in reservoir capacity, it is of vital importance that information is obtained regarding the characteristics of the sediment infill. Rausch and Heinemann [6] discussed the importance of reporting reservoir sedimentation in terms of mass as well as volume.

In this study two methods have been employed to obtain sediment samples; the use of a subaqueous Mackereth [10] corer, and sampling with short cylindrical cores on drained reservoirs. In the laboratory, both types of cores were cleaned and the wet mass and volume of the sediment was determined. Organic content of the samples was measured by the loss on ignition method.

Figs. 4 and 5 illustrate the variation in sediment characteristics with depth obtained by subaqueous coring. The results from both Thornton Moor and Wessenden Head show a tendency towards increased moisture content and decreased density towards the sediment surface, where there is a layer of semi-liquid mud a few millimetres thick.

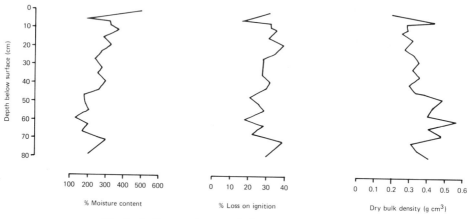

Fig. 5. Sediment characteristics, Wessenden Head.

Sediments obtained at Stubden and Thornton Moor were all composed of fine-grained dark-brown silts, which were very cohesive when oven-dried. At Wessenden Old the silts had a more reddish coloration, and were not nearly so cohesive on drying. At Butterley the samples were of reddish-brown non-cohesive sands mixed with silt, closely resembling deposits found in some of the inflowing streams [11]. The most complex mixture occurred at Wessenden Head reservoir, where the silts and fine sands were combined with dark brown material of obviously organic origin with some retained plant structure.

Table 2 summarizes data for all the sediment samples obtained, giving mean values for moisture and organic content and for dry bulk density. It should be noted that moisture content depends on the final dry mass of a sample, and is therefore not a measure of the absolute mass of water contained. However, it probably does reflect the fact that at Butterley and Wessenden Old the sediments had been exposed to air for some time, with desiccation and possible compaction occurring. Organic content ranges from only 7% at Butterley to over 38% at Wessenden Head, confirming the visual evidence mentioned above. The former value is more usual for UK soils and sediments; the very high organic contents in the other reservoirs are clearly a result of erosion of peat from the catchments. All these reservoirs have very acid waters, with correspondingly low autochthonous production.

Mean dry bulk densities for the reservoirs discussed here range from 0.145 t/m^3 at Stubden to 0.925 t/m^3 at Butterley. For comparison, the average dry bulk density of blanket peat on Wessenden Moor has been established at 0.1 t/m^3 [11], whilst typical values for mineral soil might be around 2.0 t/m^3. The variation in reservoir sediment density reflects the composition. The mixture of low density organics with sands at Wessenden Head produces a higher mean value than at Thornton Moor and Stubden, where the silts are lower in organic content but more uniform in texture.

Table 2. Reservoir sediment analysis

Reservoir:	Wessenden Head	Wessenden Old	Blakeley	Butterley	Thornton Moor	Stubden
Number of samples	34	13	0	5	14	99
Mean moisture content (%)	244.5	157.5	N/A	50.4	428.1	810.2
Mean dry bulk density (t/m^3)	0.3391	A S S U M E D 0.4438	0.4438	0.925	0.221	0.145
Mean organic content (%)	38.2	A S S U M E D 27.2	27.2	7	27.1	29.6
Total sediment infilling (t/year)	46.44	541.68	1214.78	1262.63	198.23	53.89
Sediment yield (t/km^2 year)	203.7	203.7	203.7	203.7	38.72	27.78
Organic infilling (t/year)	17.73	147.45	330.66	88.38	53.72	15.95
Organic yield (t/km^2 year)	38.82	38.82	38.82	38.82	9.52	8.22

The mean dry bulk densities and organic contents presented in Table 2 have been used (a) to calculate total gravimetric sediment infilling for each of the reservoirs, (b) to apportion this as an area-specific sediment yield over each catchment, and (c) to calculate the organic fraction of such infilling. Total infilling is greatest at Blakeley and Butterley (both over 1200 t/year) and lowest at Wessenden Head. When only the organic fraction is considered, the picture is rather different, with Butterley achieving only a relatively low 88 t/year, and Thornton Moor being comparatively high at 54 t/year. These figures take no account of the size of the reservoir or its catchment, and the difference between the two areas becomes clear when such a correction is made. Mean sediment yield for the Wessenden Valley chain is almost an order of magnitude greater than that at Stubden, and the organic-only yield at Wessenden is comparable to the total yield at Thornton Moor (39 t/km^2 year).

TRAP EFFICIENCY

Given the rapid rates of accumulation of sediment in these reservoirs, some measure of their trap efficiency is of paramount significance if a true erosion rate is to be calculated. A number of empirical studies have been carried out using such a research design, and generalized statistical relationships have been found between trap efficiency and catchment parameters. Of these the most well-known and commonly used is that of Brune [12], who drew up envelope curves (Fig. 6) for normal ponded reservoirs; the capacity-inflow (C/I) ratio provides an index to the type of reservoir, that is, reservoirs with C/I ratios of 1 or less are seasonal storage reservoirs, and those with a ratio greater than unity are hold-over storage reservoirs.

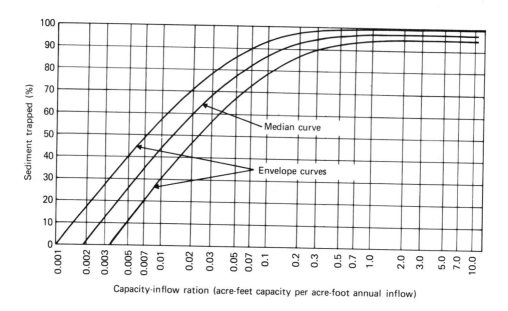

Fig. 6. Brune's (1953) envelope curves for trap effeciency.

EMPIRICAL STUDIES OF TRAP EFFICIENCY

Trap efficiency has been monitored at Wessenden Head and Blakeley Reservoirs (Table 3). The two reservoirs differ in their overflow characteristics: overflow occurs infrequently at Wessenden Head, usually during winter when precipitation and snowmelt keep reservoir levels high. Blakeley overflows continuously, directly into Butterley Reservoir and only ceases to do so for short periods in the summer. For each reservoir, a number of storms have been monitored [8] so that both the input and output of sediment could be determined. The theoretical trap efficiencies of a number of authors have also been calculated. The results for both reservoirs are shown in Table 3. It may be clearly seen that where trap efficiency is measured on a storm basis it is a highly variable value, not one of the empirical relationships being generally appropriate. For each storm the trap efficiency is matched fairly closely by at least one of the estimates, but one method does not provide similar results for all storms. When the results of the whole period are totalled for each reservoir, the measured trap efficiency is close to all the estimated values, except for that of Brune and Allen. However, the methods of Churchill [13] and Brune and Heinemann [14] really depend on annual inflow, so the intensive sampling period of six weeks is an insufficient length of time to establish a reliable trap efficiency value. A simple evaluation of the estimated trap efficiency values shows that, for Wessenden Head, the best methods to use are those of Brune and Allen [15] and of Brown [16]. For Blakeley, those of Brown and of Churchill are most appropriate.

Table 3. Theoretical and measured trap efficiencies
at Wessenden Head and Blakeley Reservoirs

Wessenden Head		Storm[†]			
		1	2	3	4
Sediment loss (kg)		654.9	2.16	100.8	39.6
Measured TE[‡](%)[§]		0	97.3	50.7	73.8
Area-weighted TE (%)		0	98.2	69.75	83.9
Theoretical TE (%)					
(i)	Brune & Allen (1941)	78	78	78	78
(ii)	Brown (1943)	96	96	96	96
(iii)	Churchill (1948)	65	70	50	60
(iv)	Brune (1953)	99(97–100)	99	99	99
(v)	Heinemann (1981)	100	100	100	100
Carver (1986) (measured)		65.17	—	—	—

[†]Storm: 1. 20.1.86–23.1.86; 2. 5.3.86–6.3.86; 3. 26.3.86–27.3.86; 4. 2.4.86–4.4.86.

[‡]TE = trap efficiency

[§]Shiny Brook catchwater only.

Blakeley		Storm[†]					B[‡]
		1	2	3	4	5	
Sediment loss (kg)		116.0	114.0	3528.0	414.7	1969.9	632.2
Measured TE(%):		0	97.3	50.7	73.8	67.6	0
Theoretical TE (%)							
(i)	Brune & Allen (1941)	37	37	37	37	37	37
(ii)	Brown (1943)	88	88	88	88	88	88
(iii)	Churchill (1948)	96	99	65	86	88	100
(iv)	Brune (1953)	99.9	—	92	98	99	—
(v)	Heinemann (1981)	100	—	87	94	95	—
Total Flow 4.3.87–6.4.87							
Measured TE (%)		= 88%					
Estimated TE (%):							
(i)	Brune & Allen (1941)	= 37%					
(ii)	Brown (1943)	= 88%					
(iii)	Churchill (1948)	= 87%					
(iv)	Brune (1953)	= 86%					
(v)	Heinemann (1981)	= 80%					

[†]Storm: 1. 4.3.87–7.3.87;   2. 13.3.87–16.3.87;   3. 27.3.87–30.3.87;   4. 1.4.87–2.4.87;
5. 4.4.87–7.4.87.

[‡]B = baseflow period 11.3.87–13.3.87

SEDIMENT YIELD CORRECTIONS USING TRAP EFFICIENCY

In the previous section, sediment yields were presented for reservoirs in the southern Pennines without any consideration of trap efficiency. This section carries out that correction using an amended version of Brown's (1943) equation. There were a number of considerations in the decision to use Brown's relationship: (i) the empirical data described earlier for two of these reservoirs suggested that Brown's relationship was appropriate to both of these quite different reservoirs; (ii) Brown's method is based on a capacity : watershed ratio rather than capacity : inflow. This is easier to calculate for small reservoirs where no inflow data are available; (iii) the reservoirs used in Brown's study would appear to be of a hydrological regime more comparable to Pennine reservoirs than those described by other authors. They have a high percentage trap efficiency due to a smaller and more variable runoff; (iv) whereas Brune and Allen used measures of erosion in the catchment, Brown used measures of sediment delivered to the reservoir. This is more appropriate firstly because it compares with the technique adopted in the empirical studies, and secondly because it overcomes any difficulty with sediment delivery ratios.

Since trap efficiency depends on reservoir capacity, it is clear that high rates of sedimentation will cause it to vary over time, usually declining from a maximum at construction to the present day. This variability must be taken into account in any correction of sediment yield:

$$Cap \times n = Original\ cap - (annual\ loss\ of\ cap \times n) \qquad (1)$$

where annual loss of capacity is the total loss of capacity divided by the number of years since construction. Using this delay function, the trap efficiency for each year since construction may be calculated:

$$CTn = 100\left(1 - \frac{1}{1 + 0.1SRn}\right) \qquad (2)$$

where CTn = trap efficiency at time n; SRn = reservoir capacity at time n.

Fig. 7 shows the decline in CT for a number of reservoirs in the southern Pennines. Blakeley from this study is illustrated, showing the rapid rate of decline in CT for reservoirs with high rates of infilling. Tunnel End is another small reservoir which has silted up from an original capacity of 103 Ml to only 8.2 Ml in 1988, whilst Redmires Reservoir is more typical of Yorkshire Water's direct supply reservoirs in remaining close to a trap efficiency of 100%. Where reservoirs are linked in a cascading system, only losses from the lowest reservoir can be appropriately treated in this way. Sediment lost from the upper reservoirs is accounted for in the sedimentation survey of the next reservoir down the chain. The final coefficients in the current study were therefore 96.06% at Butterley, 96.72% at Thornton Moor, and 97.80% at Stubden. In all three cases this is a decline of less than 0.4% from the original figure, indicating that these reservoirs remain effective sediment traps with minimal loss over the spillway. This lends credibility to the appropriateness of the technique of reservoir surveys for estimation of catchment sediment yield.

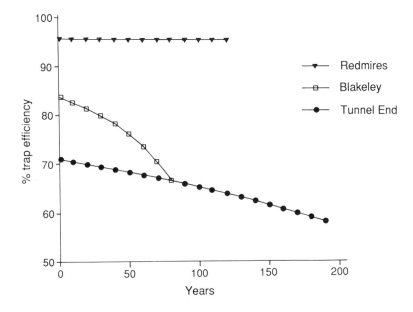

Fig. 7. Trap efficiency over time.

Using the trap efficiency data for each year since construction, it is possible to recalculate total sediment yield for each reservoir:

$$Total\ yield = \sum_{i=1}^{n} annual\ yield \left(1 + \frac{100 - CTi}{100}\right) \qquad (3)$$

Using such calculations here would suggest that more realistic sediment yields would be 212.0 t/km² year at Butterley, 28.4 t/km² year at Stubden, and 40.0 t/km² year for Thornton Moor.

DISCUSSION ON SPATIAL PATTERNS OF MOORLAND EROSION AND WATER DISCOLOURATION

Moorland erosion

Walling and Webb [17] have suggested an average rate of sediment supply to British reservoirs of around 30 t/km² year for catchments of 10 km², based on evidence from published studies. The current research at Stubden and Thornton Moor has given results in accordance with this suggestion. For the Wessenden Valley the estimated sediment yield is very much higher, confirming the visual evidence of blanket peat erosion combined with active incision through mineral deposits. Labadz *et al.* [9] have summarized the results of recent reservoir studies in the UK, and have found only one study comparable to the values obtained here for Wessenden. This is a survey of the North Third Reservoir, Stirlingshire, described by McManus and Duck [18]. Here a yield of 205 t/km² year was recorded, from a catchment described as 'thickly peat-mantled upland', and it was suggested that such areas produce much higher yields than those dominated by till soils with subordinate peat cover.

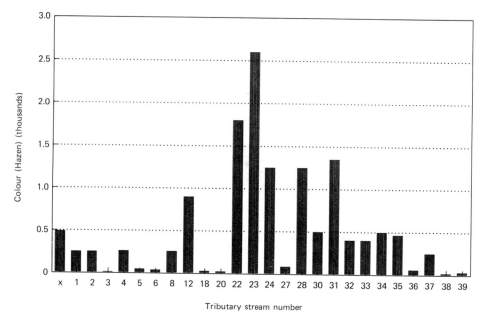

Fig. 8. Range of colour in tributary streams, Thornton Moor.

WATER COLOUR

One of the six reservoir catchments, Thornton Moor, has been studied with regard to water colour. Within the Thornton Moor catchment there is a conduit system which captures some 40 tributary streams and diverts this water into the reservoir. Water colour has been sampled in each of these tributaries on a weekly and bi-weekly basis over a period of two years. Fig. 8 shows the range of water colour measured over a two-year period. This diagram clearly shows the extreme spatial variability of water colour. Thornton Moor is not a catchment with a general problem of discoloration, but rather one with small areas generating extreme discoloration. A number of authors such as McDonald and Naden [18] have described a technique of tributary turnout or exclusion based on colour and this policy is now being practised at Thornton Moor.

Colour storage and release from reservoir sediments

Recent research has suggested that the sediment within the reservoir may act as both a sink and a source for colour. Research has been carried out at Thornton Moor Reservoir, into the long-term effects of the reservoir sediments on colour entering the reservoir from a catchwater conduit.

Since construction in 1885, Thornton Moor Reservoir has accumulated over 93 Ml of sediment within the reservoir basin. The sediments have a mean dry bulk density of 0.221 t/m^3 and a 27% organic content. The 11% capacity loss at Thornton Moor is comparable with other reservoirs of similar size, at this altitude, and the rate of erosion in the catchment would not appear to be excessive. However, the nature of the indirect catchment is such that the trap efficiency of the catchwater/reservoir system is lower than in other reservoirs. Significant amounts of sediment are likely to be transported past the catchwater during times of high flow.

The sediments within the reservoir are primarily located in a flat delta form adjacent to the inlet and below the active area of the reservoir. The shape of the reservoir is such that the sediments are likely to be spread out over the whole surface of the reservoir bottom. Given the shallow nature of the water body, and the broad expanse of sediment, there is a considerable area of interface between the raw water and the sediments. The reservoir is also subject to considerable wind disturbance.

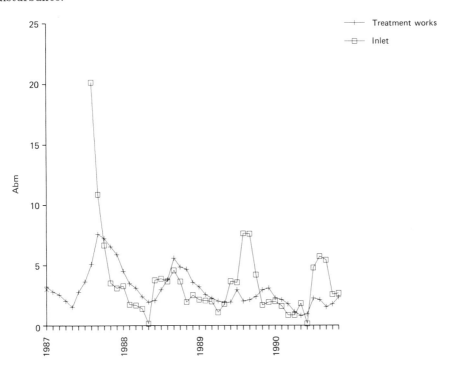

Fig. 9. Variation in colour between inlet and treatment works.

A detailed study was initially made of water company records of colour for both the conduit inlet, where water enters the reservoir, and the draw-off for supply. These two sampling points lie at either end of the reservoir. It would appear that there are periods when the reservoir body is a net importer of colour, and periods where the colour leaving the reservoir is greater than that entering it. This is shown in Fig. 9. It may be seen that in the summer months, when catchment colour is high, the reservoir acts as a buffer, effectively diminishing raw water colour. In winter, it would appear that colour is released from the reservoir basin since colour is higher at the treatment works than at the inlet. However, the data-set lacks temporal resolution despite its great length, and there is a considerable need for storm-based sampling on the inlet stream.

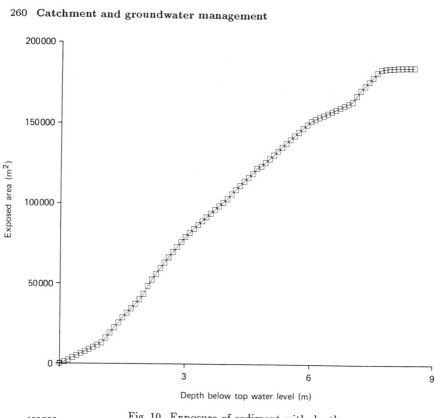

Fig. 10. Exposure of sediment with depth.

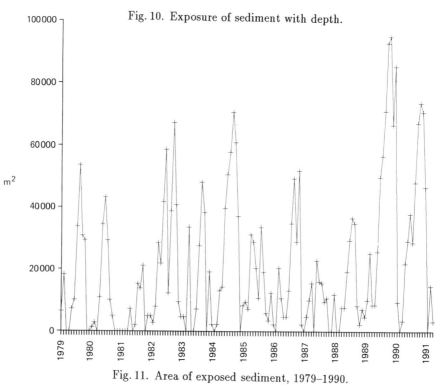

Fig. 11. Area of exposed sediment, 1979–1990.

Some initial research has been carried out into the relationship between colour in the reservoir and the area of sediment exposed during the summer months. In order to calculate the area exposed, a relationship has been calculated between reservoir level and the area of exposed sediments. This is possible through the CAD model of the original survey of the reservoir in 1989. The relationship with exposed area of sediment and depth below top water level is shown in Fig. 10. Reservoir levels, on a daily basis since 1979, were obtained from Yorkshire Water and a conversion to exposed area by month calculated (Fig. 11). There is a significant negative relationship between the area of exposed sediment and the difference in colour between the inlet and the draw-off.

Furthermore, it is possible to discern a cyclical pattern in which the colour would appear to be removed within the reservoir system during the period May to October, when the reservoir is low and sediments are exposed, and released from the reservoir in the winter months. Research is currently being carried out into the effects of wind turbulence on colour release from reservoir sediments.

CONCLUSIONS

1. As a result of their topgraphy, high rainfall, remote location, and low-intensity land use, the UK uplands are well-suited to the gathering and storage of water. The relatively low cost of reservoir water, and the limited amount of treatment required, has promoted their use in preference to groundwater supplies.
2. The treatment works adjacent to upland reservoirs were often small in size and ill-equiped to cope with significant changes in water quality.
3. A combination of increased usage of upland reservoirs during drought years, and changes in water quality associated with increased erosion and reservoir sedimentation, have reinforced the need to improve treatment works.
4. All the single-stage works treating water from the Wessenden and Thornton catchments have been replaced by multiple-stage works, either completely new or by extension of works which were of recent construction.
5. The question of loss of yield from sedimentation is now being considered by Yorkshire Water, using the updated depth/capacity data derived during the study.

ACKNOWLEDGEMENTS

The reservoir surveys were carried out for Yorkshire Water Services Ltd, who also granted access to catchments and original information. The initial work in the Wessenden Valley was partially funded by an NERC Research Studentship (GT4/83/AAPS/23, J. C. Labadz, nee Oldman). The authors are grateful for the contribution of Tiffany Pemberton-Jewitt. Many members of staff and students of Huddersfield Polytechnic have assisted with fieldwork, and Steve Pratt drew the diagrams.

REFERENCES

[1] Conway, V. M. (1954) Ringinglow Bog near Sheffield. *J. Ecol.*, **34**, 149–181.

[2] Tallis, J. H. (1965) Studies on southern Pennine peats IV – evidence of recent erosion. *J. Ecol.*, **53**, 509–520.

[3] Bower, M. M. (1960) Distribution of erosion in blanket peat bogs in the Pennines. *Trans. Inst. of British Geographers*, **29**, 17–30.

[4] Phillips, J., Yalden, D. and Tallis, J. (1981) *Peak District moorland erosion study*. Phase 1 report. Peak Park Joint Planning Board. Bakewell.

[5] Edwards, A., Martin, D. and Mitchell, G. (eds) (1987) Colour in upland waters. *Proceedings of a workshop held at Yorkshire Water*, Leeds, Yorkshire Water.

[6] Rausch, D. D. and Heinemann, H. G. (1984). Measurement of reservoir sedimentation. In: Hadley, R. F. and Walling, D. E. (eds) *Erosion and sediment yield*. Geobooks, Norwich, pp. 179–200.

[7] Burt, T. P. and Gardiner, A. T. (1984). Runoff and sediment production in a small peat-covered catchment: some preliminary results. Ch. 8 133–151 In: Burt, T. P., and Walling, D. E. (eds), *Catchment experiments in fluvial geomorphology*. GeoBooks, Norwich, pp. 133–151.

[8] Pemberton, T. J. L. (1987) The use of reservoir sedimentation surveys for the prediction of sediment yield in upland catchments. Unpublished MPhil thesis, The Polytechnic of Huddersfield.

[9] Labadz, J. C., Burt, T. P. and Potter, A. W. R. (1991) Sediment yield and delivery in the blanket peat moorlands of the southern Pennines. *Earth Surface Processes and Landforms*, **16**, 255–271.

[10] Mackereth, F. J. H. (1969) A short core sampler for sub-aqueous deposits. *Limnol. Oceanog.*, **14**, 145–151.

[11] Labadz, J. C. (1988) Runoff and sediment production in blanket peat moorland: studies in the southern Pennines. Unpublished PhD thesis, CNAA (Huddersfield Polytechnic).

[12] Brune, G. M. (1953) Trap efficiency of reservoirs. *Trans. Am. Geophysical Union*, 34, (3), June.

[13] Churchill, M. A. (1948) Discussion of: Analysis and use of reservoir sediment data, by L. C. Gottschalk. *Proceedings, Federal Inter-Agency Sedimentation Conference*, Denver, Colorado (US Bureau of Reclamation), 139–140.

[14] Heinemann, H. G. (1984) Reservoir trap efficiency. Chapter 8. In: *Erosion and Sediment Yield*. Hadley, R., and Walling, D. E. (eds) Geobooks, Norwich.

[15] Brune, G. M. and Allen, R. E. (1941) A consideration of factors influencing reservoir sedimentation in the Ohio Valley region. *Trans. Am. Geophysical Union*, **22**, 649–655.

[16] Brown, C. B. (1944) Discussion of: Sedimentation in reservoirs, by J. Witzig. *Trans. Am. Soc. Civ. Engrs.*, **109**, 1080–1086.

[17] Walling, D. E. and Webb, B. W. (1981) Water Quality. In: Lewin, J. (ed.) *British Rivers*, George Allen and Unwin, 126–169.

[18] McManus, J. and Duck, R. W. (1985) Sediment yield estimated from reservoir siltation in the Ochil Hills, Scotland. *Earth Surface Processes and Landforms*, **10**, 193–200.

[19] McDonald, A. T. and Naden, P. S. (1987) Colour in upland sources: variations in small intake catchments. *Wat. Serv.*, **91**, 121–122.

24

Diversion of the River Calder at Welbeck: an engineering and environmental challenge

C. McDonald, BSc, PhD, MRTPI,[†] and C. E. Rickard, BSc, FICE, MI-WEM[‡]

INTRODUCTION

The diversion of the River Calder forms an integral part of the Welbeck reclamation and landfill project. This scheme[1] is being implemented jointly by the City of Wakefield Metropolitan District Council and the West Yorkshire Waste Management Joint Committee.

It is an imaginative and far-reaching project, the purpose of which is to reclaim and restore an area of land covering about 400 ha, much of which is currently derelict, largely inaccessible and, in certain parts, potentially dangerous.

The Welbeck project area is located in the Yorkshire coalfield, north-east of Wakefield, and lies in greenbelt between the outer fringes of the city and the nearby town of Normanton. One of the main aims has been to avoid tipping on greenfield sites elsewhere in the region through the creation of a major new landform in an already despoiled stretch of river valley.

The project is a complex and controverisal one, potentially including residual mineral extraction, the disposal of power station and sewage sludge, the recovery

---

[†] Head of Environment, Wakefield Metropolitan District Council.
[‡] Divisional Director, Mott MacDonald Group

and utilization of landfill gas and road, rail and canal developments, besides its basic reclamation and spoil and waste disposal objectives.

The principal motivation for the river diversion is to create sufficient landfill capacity to accommodate the co-disposal of domestic refuse and colliery spoil from a wide area over many years. The diversion will enable voidspace volumes on the site to more than double, at little extra cost to the local authorities in respect of site preparation, access provisions, etc. Studies have shown that, given the terms on which spoil has become available from the Selby Coalfield, the diversion scheme offers significant savings in the unit costs of waste disposal.

The existing river channel is heavily encroached upon by the legacy of past mineral working. Diversion of the river will not only increase landfill capacity but will also give early access and after-use potential to a restored flank of the site. The creation of a new river corridor in this manner is seen as an opportunity not only to deliver early reclamation and screening achievements, but also to help turn a tide of public disaffection with landfill in general and the Welbeck project in particular.

Fig. 1. Aerial view of project area.

ENGINEERING ASPECTS

Proposed river diversion

The site of the river diversion is illustrated in Figs 1 and 2. The reach of the river which is to be diverted extends form just downstream of Kirkthorpe weir to just upstream from Stanley Ferry, a channel length of 3.3 km. The western boundary of the site is defined by the Aire and Calder canal, which runs in a northerly direction,

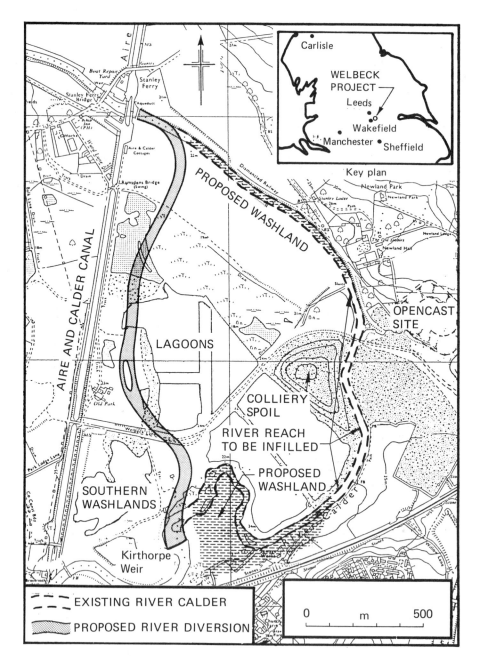

Fig. 2. Site of proposed river diversion.

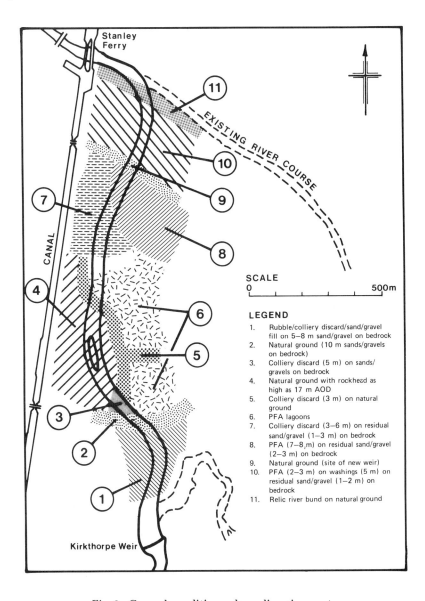

Fig. 3. Ground conditions along diversion route.

leaving the river at Broadreach lock above Kirkthorpe weir, and passing over the river at Stanley Ferry aqueduct. The river will be diverted in order to cut off the broad loop which currently divides the site. The new channel will follow a shorter route much closer to the canal, reducing the river length by about 1.3 km.

Site topography, geology and hydrogeology

The site is dominated by large mounds of colliery spoil and by the remnants of old sand and gravel works in the form of lagoons, many of which have been filled or are being filled with pulverized fuel ash (PFA) from the local power station, or with colliery waste. An area of old gravel pits near Kirkthorpe weir has been successfully converted into washlands for flood storage as part of the Welbeck project, whilst providing recreational facilities and environmental enhancement. Otherwise the site is relatively level at 18–20 m AOD.

The natural superficial geology of the site is dominated by a layer of sand and gravel which varies in thickness from 1 m to 7 m, depending on the degree of past commercial extraction. This stratum overlies directly the mudstone/siltstone bedrock, except for an occasional intervening thin layer of clay. Rockhead varies across the site from 8 m to 17 m AOD.

The route of the diversion has been investigated by 38 boreholes in order to define the extent and nature of the materials through which the channel will pass, particularly the substantial volumes of infill materials. Fig. 3 summarizes the findings in the form of eleven zones defining the predominant superficial materials found therein. Zones 2, 4 and 9 are largely natural ground. Zones 1, 3, 5, 7 and 11 comprise fill material of either indigenous (sands, gravels, clays) or imported (colliery discard) materials. Zones 6, 8 and 10 have substantial deposits of PFA, typically 7–8 m thick, overlying 2–3 m of residual sands and gravels. In Zone 10 the situation is more complex, with 2–3 m of PFA overlying about 5 m of very fine material, assumed to be washings from the sand and gravel industry. It is Zones 6, 8 and 10 which present the greatest challenge for the engineering design of the river diversion.

The hydrogeology of the site has been investigated to ascertain present water quality and piezometry. Groundwater levels generally reflect the local topography, with higher levels to the west, but locally the groundwater levels are influenced by the river or those lagoons which still have standing water. No widespread or significant contamination of groundwater was evident from the tests which were undertaken.

Hydrology and hydraulic design

The River Calder rises in the Pennines and is subject to flashy floods. The 100 year return period flood is estimated at about 400 m^3/s whereas the normal flow varies from 10 m^3/s to 20 m^3/s, depending on the season. The channel slope is about 0.40m/km, although it is locally steeper immediately downstream from Kirkthorpe weir where the channel follows a somewhat tortuous course.

For a river diversion of this type, where the channel length is to be shortened substantially, it would be normal practice to provide a weir for grade control. In the case of the Welbeck site this is not essential, because the whole length of the diversion channel is to be lined with rock to preserve channel stability through the weak strata described above, and a series of riffles could be provided in the channel. However, it is proposed that a weir is constructed near the Stanley Ferry end, as one of the means of enhancing the environmental status of the river corridor.

Hydraulic performance of the new channel has been checked using a simple backwater analysis for a range of flows and varying channel roughness characteristics. The main aim has been to ensure that flood levels are not increased. The new weir will therefore be a relatively low structure which, nevertheless, will have a drop of about 1.5 m at low flows. Substantial provision has been made for flood storage within the site, more than enough to compensate for that lost due to the channel shortening. The two main storage areas are indicated in Fig. 2, being the Second Southern Washlands and the Newland Washlands, both incorporating some of the existing river channel, which will be made redundant by the diversion.

Erosion protection

It was realized from the early days by the joint promoters of the scheme that the new reach of river channel would require substantial erosion protection to ensure that, once diverted, the channel would not stray from its designated course and thereby endanger the landfill site. The prospect of large quantities of local rock becoming available, at no cost to the local authorities, from the residual opencast coal operations within the Welbeck site led to the selection of riprap as the obvious medium for channel stabilization. Should it become necessary to obtain rock from elsewhere for any reason, the designs will be reviewed and adjusted accordingly. The project economics are robust enough to bear any resulting increase in the cost of erosion protection.

Fig. 4. Temporary works to stop erosion of the existing river bank. On completion of the diversion this will form part of the Southern Washlands.

The local rocks include a fine-grained sandstone (The Warmfield Rock), which comes from the coal measures between the Sharlston Low Coal and the Sharlston Yard Coal. This rock is not particularly strong and has micaceous bands which form planes of weakness. The longevity of the local rock is therefore questionable in that it may tend to degrade over the years. However, there are no rigid requirements for the quality of riprap stone, and the local rock has already been used successfully to arrest local bank erosion on the existing channel, albeit only in the recent past (Fig. 4). The proposed channel lining will comprise a minimum thickness of 1.0 m of rock on slopes no steeper than 1 : 2 (vertical to horizontal), and stone sizes will be correspondingly larger than those required to resist the forces of erosion, so that any tendency for the stones to degrade has been anticipated, and will be catered for. In the most vulnerable areas, for example downstream of the new weir, the riprap thickness will be increased to 1.5 m, that is three times the calculated requirement. Where the exposed strata are highly erodible, the riprap will be laid on an underlayer of gravel with a geotextile where appropriate.

Geotechnical design

The geotechnical design has addressed the two main issues of stability of the channel excavation during construction (and appropriate construction methods), as well as the long-term stability of the river channel and associated landfill site. The main problems are caused by the PFA and the sand and gravel washings. Both have a high moisture content, low permeability, and low strength, making excavating the channel a difficult operation. Dewatering the washings has been shown to be impracticable by pump tests carried out in one of the boreholes. The underlying gravels provide an unrestricted source of water, and the low permeability of the washing makes conventional dewatering impossible.

The use of sheet piling or a jet grouted wall to cut off the source of water and to stabilize the strata adjacent to the excavation is to be avoided on the grounds of cost. The favoured solution is to excavate in stages and to form a retaining bund of colliery discard as the excavation progresses. Large quantities of coarse colliery spoil will be available on the site, which should provide an economic answer to the problem, requiring only a major earthmoving effort to achieve the desired effect. It is intended to include indicative proposals in the tender documents, but tenderers will be required to provide a detailed method statement.

Proposed weir structure

The site of the weir is conveniently located at the downstream end of the diversion where, in the midst of the problematic fill materials, there remain the remnants of a narrow strip of virgin land – a main access route in the former sand and gravel workings. Nevertheless, the weir design has been chosen to give a low-weight structure incorporating a cut-off to prevent destabilizing under-seepage. The weir design has also been influenced by the desire to avoid a rigid geometrical structure which would detract from the aim of environmental enhancement. The overall concept for the weir is very simple – a double-steel sheet pile wall forming the basis for the superstructure as well as the cut-off. The weir itself will have a gently curved planform, and will be constructed from concrete to accurate line and level to ensure even flow conditions. The weir will not have vertical abutments, which would detract from the natural appearance: instead the weir crest will merge into the sloping sides of the channel. The weir will incorporate a fish pass on one side, and the weir pool downstream will be deepened to provide a region of tranquil water.

Other engineering considerations

Diversion of the river involves a number of other engineering works beyond the for-mation of a stable channel. The proposed washland areas require inflow and outlet structures. The inflow structures will take the form of overspill side weirs, with reinforced grass, designed to overtop for low-order floods. The outlets are expected to comprise culverts with an upstream weir inlet set at the normal retention level, and an outlet box in the river bank.

The southern limb of the old river channel will require a constant supply of water for the foreseeable future because the local Sharlston colliery currently pumps water from the river. This will be achieved by opening up an abandoned 17th century canal, parts of which are still evident on the site, which used to bypass the tortuous Kirkthorpe bends. The discharge from the existing Kirkthorpe sewage-treatment words would also have had to be re-routed, but current proposals are to decommission this works and divert the sewage elsewhere.

ENVIRONMENTAL ASPECTS

Underlying principles and approach
The river diversion studies have addressed the design of the adjacent landfill only to the extent necessary to satisfy river-corridor design issues. However, all these studies have been frank in recognizing that the basic justification for the river diversion is the creation of additional voidspace for landfill.

The opposition which such recognition invites is countered by ensuring that high environmental standards are maintained in the landfill design and management, and by advancing notions of express compensation for the local disturbances necessarily associated with such a regional landfill facility. In other words, the acceptance of colliery spoil and controlled wastes as fill materials at Welbeck not only enables the basic reclamation of the area, but is also intended to achieve and finance *(inter alia)* a faster restoration to a higher standard.

In that context, the river diversion is effectively evolving as an environmental project which stands in its own right. As will be seen below, immediate benefits are to be derived from the river diversion and improved river corridor, in terms of countryside recreation and tourism and leisure potential. The radical nature of this transformation is increasingly being seen as a potential initial spur to economic and environmental regeneration of a much longer stretch of the Lower Calder Valley.

It developing this approach, it is significant that the emphasis is as much upon the creation of a river corridor as on the diversion of a river channel. The frame of reference has been consciously broad, to enable the design of a consolidated habitat and recreation area, linking the river and its banks with adjacent landscaped landfill, to maximize the environmental and after-use potential of the investment.

Existing river corridor
The loss of river corridor as a result of the diversion will be limited to a stretch of about 1 km which is of little environmental merit in terms of landscape, habitat and potential public access (Fig. 2). It is a stretch of regular cross-section, the banks are steep-sided and largely devoid of vegetation, the channel width shows little variation and for a good part pre-existing colliery spoil tips encroach virtually to the edges of the channel. This reach of channel supports very little wildlife (Fig. 5).

There is an interesting historic parallel at Welbeck insofar as there exists an area of recessive ecology, known as the Half Moon, now prized for its fishing and birdlife, which was created by Victorian engineers who cut off a smaller loop of the River Calder during the construction of a railway. One of the key realizations in approaching the river diversion proposition was that the same principle could be applied on a greater scale in the case of the new river diversion. Significant lengths of the former river will be retained as wetland areas, therefore the net effect is a greater area of habitat that promises to be richer, more diverse and more accessible.

Re-creation of river corridor

The diversion of the river presents a significant opportunity to enhance the environmental value of the river corridor, by creating a channel which has a wide variety of sectional form and features. Typical cross-sections of the diversion channel are shown in Fig. 6, and the following are expected to be incorporated into the new channel: (a) a gently curving alignment in planform; (b) varying width, depth and side slopes; (c) shallow water margins at the channel edges; (d) shallow banks/beaches; (e) deep pools at certain points in river bed; (f) an angler's berm; (g) an island; and (h) a second weir.

Fig. 5. Section of the River Calder to be lost, looking upstream between two colliery coal tips.

Access berms for inspection and maintenance are to be provided on both banks, in accordance with the wishes of the National Rivers Authority (NRA). The primary access berm is to be located on the left (western) bank, at a level just above the annual flood level. A secondary, occasional access reservation on the right bank will have a more variable level and will be allowed to develop a more natural appearance.

Looking beyond the river channel to the wider river corridor, it was recognized that there was scope, within the scale of the reclamation proposed at Welbeck, for the creation of an exciting and diverse new landscape. The landscape design produces a blend of views of wide open spaces, with intimate and potentially complex ecological areas, as have already been achieved in the Welbeck Southern Washlands.

In the river channel itself, damp margins will be created at the water's edge. Thickets of bramble, blackthorn and other low-growing scrub will be established from place to place along the river bank. A large island, with its protected habitat, will be created within the new channel. The edge of the river will be indented and deep pools will be incorporated in the river bed. The overall character of the new channel will be as diverse and complex as possible.

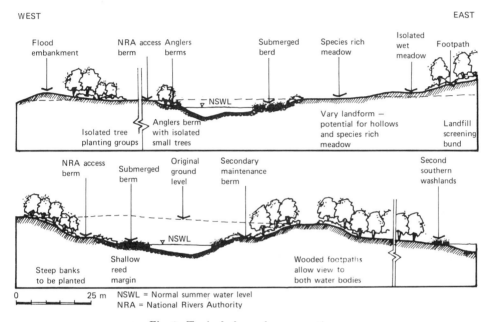

Fig. 6. Typical channel cross-sections.

The dominant after-use for most of the river corridor will be for passive leisure and recreational activities, although the size of the area permits the development of active recreation at specific locations. A new footpath, bridleway, cycleway network, including a limb of the proposed cross-Pennine route, will cross the river valley at various points. The focus for more active recreation and tourism is likely to extend around the existing marina development and historic aqueduct at Stanley Ferry. Studies are being undertaken with British Waterways on the possible extension of marina facilities. The possibility of a canoe slalom is under investigation.

Environmental testing points

Any project of this nature tends to involve trade-offs between engineering and environmental priorities, and the way such trade-offs are resolved is an incisive test of whether the environmental aspirations have been fundamental, or are just a surface gloss. The Welbeck River diversion proposal can readily pass such test.

Firstly, no constraint was set in terms of a minimum landfill area or volume which was to be sacrosanct, nor was the river corridor to be designed to maximize landfill potential. If these had been over-riding criteria, the end-result would be a river channel and engineered washlands, rather than the holistic river corridor concept described above.

A measure of the habitat lost, retained and created is provided by comparison of the lengths and surface areas of river corridor and wetland at Welbeck, before and after the diversion works. Taking into account both the new channel and retained river arms, the scheme is likely to provide an increase of almost 1 km in the total length of riverside habitat. A net gain of about 10 ha of wetland will result, the new wetlands being planned to provide a wide range of habitat types, much richer than most of the existing gravel lagoon areas.

The provision of a new weir on the diverted river course gives another instance of the balance of priorities. The cost of the weir could approach £0.5 million and, as indicated above, there is no overriding hydraulic reason for its provision in this particular situation. However, the weir offers the following potential environmental benefits: (a) providing an immediate focal point with visual interest, thereby enhancing the amenity of the area; (b) by backing up the water level in the new river channel, the weir helps avoid a deeply incised appearance which would detract from the visual impact of the river; (c) increasing the recreational potential for the river corridor, e.g. by facilitating the incorporation, if only at some future date, of a canoe slalom course; and (d) increasing the aeration and the downstream water quality of the River Calder. These advantage are considered to be significant, but only if a reasonably long-term view is taken about the crucial role that rivers can play in the environmental inheritance left for future generations.

Finally, there is the issue of water quality. The River Calder at Welbeck is of poor quality, being designated Class III. No additional discharges to the river will occur as a result of the river diversion, and the scheme will have no detrimental effect on river water quality, which is dominated by discharges upstream from Welbeck. Inevitably a major issue of potential concern is the possible contamination of the river due to leakage of leachate from the proposed landfill. While the detail is beyond the scope of this chapter, given the design and operational regime intended for the landfill, it is considered that the risk of contamination is minimal, and that if any were to occur there is unlikely to be a significant effect on the river. Nevertheless, the microbiological quality of the River Calder at Welbeck will not permit immersion-type watersports in the short term. To some extent this limitation may be overcome by feeding certain of the retained water features from alternative sources of less-polluted water. Significant improvement in the water quality of the River Calder itself can only be expected in the longer term. However, the approach at Welbeck has, again, been not to use that as an excuse to discount facilities like canoe slaloms, but as an obligation to preserve such possibilities for future generations.

Information and public consultation

Much emphasis has been given, not only on the river diversion element but throughout the Welbeck project, to underwriting project information and public consultation obligations by a sound technical base of studies and documentation. The river-corridor proposals have been preceded and paralleled by basic survey including existing riverine habitat, water quality, flora and fauna and heritage. A full environmental impact assessment followed, covering ecological, visual, noise, access, and other impacts. The assessments were structured around comparative environmental evaluation of the old and new river corridors at Welbeck and considered not only the benefits that will accrue as the creation and restoration of the river corridor is completed, but also the implications of the disruption from construction works meanwhile.

A major phase of public participation on the Welbeck project was built around the river diversion planning submission in early summer 1990. This involved not only a series of publications and public presentations, but launched an ambitious and ongoing programme of river corridor and site tours utilizing a purpose-built visitor centre and a specially recruited Interpretation Officer. The level of effort and professionalism which has been seen as appropriate to this aspect of the work is evidenced by the quality of the public broadsheet [2] and the development of interactive computer graphics displays.

IMPLEMENTATION PROGRAMME

The local authorities are committed to a major reclamation and co-disposal scheme at Welbeck whether or not the river is diverted, and many aspects of the project are already being undertaken. The new landform is being completed in stages, with areas being returned to use progressively over a period of 20–25 years. Notwithstanding this extended timescale, significant improvements will be achieved early in the project. The final contoured landform is being developed from the outside working inwards, to provide screening of landfill activities and phased restoration of the perimeter of the site. The sponsors of the project were (in 1991) promoting a Parliamentary Bill which would give them the necessary powers to proceed with the diversion works. The river diversion could be undertaken within two years and wetland habitats created within four to five years of work commencing. The new river corridor is expected to be open to public access many years earlier than would have been possible if the river was not to be diverted away from the landfill works.

ACKNOWLEDGEMENTS

The foresight and resolve of the City of Wakefield Metropolitan District Council and the West Yorkshire Waste Management Joint Committee, and the skills and dedication of professional colleagues in advancing the concepts and achievements described in this chapter, are acknowledged with gratitude.

REFERENCES

[1] Mott McDonald (1990) Welbeck Reclamation and Landfill Project – Diversion of the River Calder: Summary Report; Technical Report, Vol 1 – *Environmental Review*; Technical Report, Vol 2 – *Engineering Proposals*.

[2] Wakefield Metropolitan District Council (1990) *Welbeck – the Creation of a River Corridor*. Broadsheet, August.

25

Design and implementation of the River Sheaf comprehensive flood-alleviation scheme

D. R. Young, BSc, CEng, MICE, FIWEM[†], **and J. G. Cross, BSc, CEng, MICE, MIWEM**[‡]

PRELIMINARY STUDY AND DESIGN

Catchment

The River Sheaf drains an area of about $68 \, km^2$ to the south-west of Sheffield, including the city centre where it outfalls to the River Don. The upper part of the catchment is suburban, becoming urban and increasingly industrialized as it progresses towards the centre of the city. The lower catchment is subject to much redevelopment.

Over the length of river covered by the scheme the only large tributary is the Porter Brook, which joins the Sheaf 1 km upstream from the confluence with the River Don. This part of the river system is almost entirely culverted. The catchment plan is shown in Fig. 1 and schematic layout of the scheme is shown in Fig. 2.

HISTORY

Since the 18th century the Sheaf Valley and its immediate surroundings have been subject to extensive industrial development. Most of the lower river runs between

[†] Principal Engineer, Balfour Maunsell Limited, Sheffield.
[‡] Project Engineer, National Rivers Authority (Yorkshire Region).

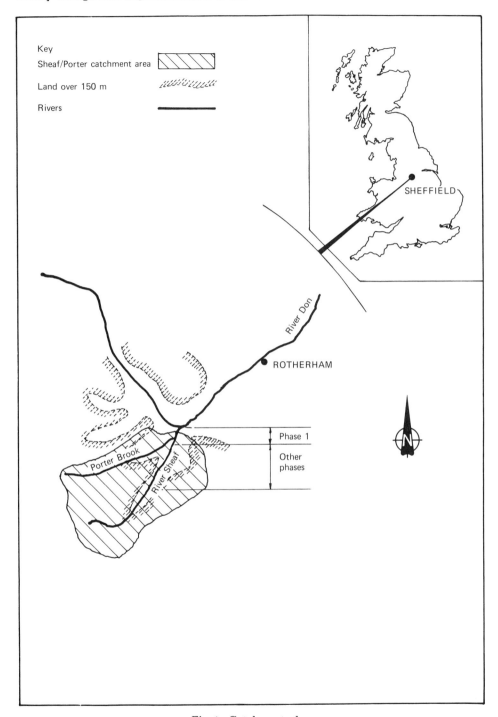

Fig. 1. Catchment plan.

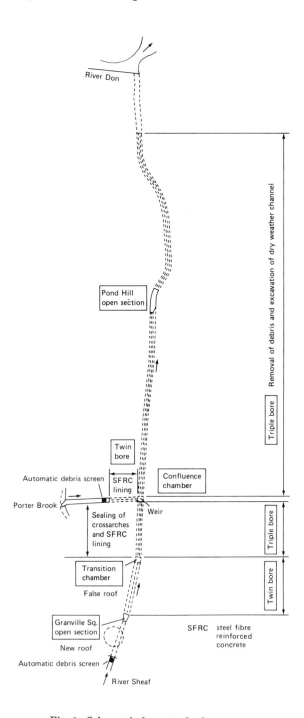

Fig. 2. Schematic layout of scheme.

walls and buildings or in culverts, and the bed is lined in several sections. There are several weirs where water has beeen abstracted in the past for industrial purposes. Typically, any culvert or bridge is made up of several structures of differing ages and construction.

Powers to carry out scheme

The National Rivers Authority (NRA) has powers to carry out flood-defence improvement works granted to it under section 147 of the Water Act 1989 and section 17 of the Land Drainage Act 1976. These powers are limited to designated main rivers and the River Sheaf was designated as such in November 1978. All schemes are approved internally at feasibility, outline and detailed design stages. The finalized scheme is approved by the Regional Flood Defence Committee and the Ministry of Agriculture, Fisheries and Food who also make a contribution of approximately 15% to the construction costs.

Flooding

The worst floods occur on the river when a summer-type thunderstorm occurs over the catchment. The resulting flash runoff from the urban part of the catchment is exacerbated if the upland peaty areas are already saturated from previous rainfall. In this case, these areas tend to behave as an 'urban' catchment with surface water quickly running off into the river channel.

Flooding occurred at a threshold of about 1-in-5 years, with major floods of recent years having occurred in 1958, 1973 and 1982. Major flooding also occurred in 1922. The flooding causes considerable traffic disruption in the city centre, and could cause closure of Sheffield's main British Rail station and the flooding of many commercial and residential properties in the city.

Flood water escaped from the channel at five discrete points, each with its own unique problems. Water which left the channel flowed towards the city centre, using London Road as a flood channel. Houses and roads on both sides of London Roads were flooded and traffic disrupted on this and other adjacent commuter routes.

Hydrology

The minimum design standard for protection of urban areas in the Yorkshire Region is to withstand a 1-in-50 year event, with a 300 mm freeboard.

Flood flows were determined from the Flood Studies Report [1], having due regard to the limited historical data available. Some level data were available from the flood events of 1958 and 1982, and four years of data were available from Highfields gauging station. For low-order floods, the flow frequency relationship was established from the Highfields data. For floods of greater magnitude, the Flood Studies growth curve was used to predict the levels at Highfields gauging station. The flow frequency curves at other points on the catchment were determined using the Flood Studies ungauged catchment method.

The 1-in-50 year flow at the Don confluence is $75.2\,\mathrm{m^3/s}$ with Porter Brook contributing $16.1\,\mathrm{m^3/s}$ of this total.

Hydraulics

The flood levels were calculated by a mainframe computer program based on the

Manning roughness coefficient and subcritical flow. Existing weirs were used as control points with losses at bridges, culverts and constrictions being calculated manually. From these it become apparent that critical lengths could not be modelled numerically, and it was decided to model physically four sections of the river.

Because of the lack of reliable data it was impossible to calibrate the models from historic events, but the flow phenomena observed in real life were reproduced accurately in the models. The models tested various improvement options and the optimum solutions produced were the result of very close liaison between the modellers and the NRA engineers. Results show that this liaison is essential, as frequently the best solutions were not those originally envisaged.

The models, which cost £60 000, have avoided unnecessary work estimated at £1 500 000.

Benefit analysis

Benefits were calculated based on the works of Penning-Rowsell and Chatterton [2] combined with an assessment of the out-of-channel flow and its depth. No allowance was made for the kinetic effects of the flood water. Traffic benefits were calculated using the MROAD program described by Parker, Green and Thompson [3]. There were no allowances made for loss or disruption of life or public utility losses. When compared wih the costs, the benefit/cost ratio of the overall scheme was 1.8 at 1991 values.

Environmental considerations

During the outline design of the alignment and location of flood defences, and the regrading works to the channel, a river-corridor survey was carried out under a joint funding by the NRA and the Nature Conservancy Council. This survey identified the sections of bed and banks which provided a good habitat or had other significant features. Riverside structures which were of interest from an industrial archaeological point of view were also identified. Both temporary and permanent works were avoided or minimized in these areas. The local authority and conservation bodies were also consulted and their comments incorporated in the design. Following completion, an in-house assessment was made of the environmental impact of the permanent and anticipated temporary works. The decision not to publish an 'environmental statement' was then made, and this decision was publicized in the local press.

Early liaison

In order that the hydraulic design options could be assessed for both physical and economic viability, it was essential to liaise not only with public utilities but also with Sheffield City Council (for alterations to highway and footbridges) and all riparian owners. The most important riparian owner was British Rail, who owns most of the lower culvert system and who has an on-going programme of maintaining and strengthening the culverts.

The NRA has consent powers over development within 8 m of the river and, where possible, developers have been required to upgrade the defences to the 10-in-50 year standard.

Transfer from NRA to Balfour Maunsell

Because of limited in-house resources, the NRA decided that the detailed structural design, contract preparation and construction supervision should be undertaken by an outside consultant. A brief was prepared detailing the required flood-defence positions and levels, culvert improvements, and channel regradings and realignments. Any future work to be carried out by the NRA was also identified. Balfour Maunsell were interviewed with three other consultants and then appointed.

Table 1. Summary of phases involved in scheme.

Phase	Title of phase	Length of phase (km)	Date of construction	Outline cost estimate (£000s)	Brief description
1	River Don confluence to Farm bridge	1.3	1990–1	985	Cleaning and hydraulic improvement of culverts, mechanical screens, roof to open chamber
2	Farm Bridge to Myrtle Road	0.5	1992–3	188	Lowering of weir and new floodwalls
3	Myrtle Road to Arnside Road	1.7	1993–4	855	Raising of services and footbridges, new floodwalls, lowering of weir, new weir, regrading and realignment
4	Arnside Road to Archer Road	1.6	1994–5	370	Access ramp and defence wall, regrading and culvert improvement
			Sub-total	2 398	
	Site investigation			20	
	Hydraulic models			77	
	Topographical survey			5	
			Total	2 500	

Costs are at September 1990 prices.

IMPLEMENTATION

Project phasing and estimates

The improvements and capital works identified by NRA in the early design period led them to split the scheme into four phases, as summarized in Table 1. By early 1991, Phase 1 has been designed in detail, contract documents prepared and sent out to tender.

Phase 1 of the scheme involved the section of the River Sheaf from the Don confluence to Farm Bridge, a distance of 1.3 km. About 76 m of the Porter Brook, from its confluence with the River Sheaf, has also been improved as part of the scheme.

The main components of Phase 1 were:

(a) debris cleaning and partial regrading of the River Sheaf culverts, including a dry-weather flow channel in the left-hand culvert bore;

(b) application of fibre-reinforced sprayed concrete to some sections of the culverts to improve hydraulic performance;

(c) infilling to the cross arches upstream from the confluence chamber;

(d) false roof infilling to the River Sheaf transition chamber to reduce turbulence;

(e) provision of a roof and cutwater to Granville Square chamber to reduce turbulence;

(f) hydraulic improvement of the existing storm-sewer inlet to Granville Square chamber;

(g) lowering of the weir at the entrance to Porter Brook culvert; and

(h) provision of two automatic screens at the two entrances to the culvert system.

Survey work

Physical and topographical

The NRA had carried out a large amount of survey work, consisting of a physical inspection, topographical survey, and detailed photographic record of the deposited silt and debris.

The first task of the consulting engineers was to follow up this work with a further detailed survey to identify missing details and levels, carry out a structural assessment, and update the photographic record.

A team comprising staff with appropriate confined space and breathing apparatus training and experience was selected. A method of working for underground survey work was established and strictly adhered to by the team.

The survey work called for considerable physical strength in the staff for carrying safety equipment and the appropriate surveying equipment. During the work there was only one emergency escape incident due to methane stirred up in the bed sediments. Fortunately an exit manhole in the railway station was only 20 m away.

From the results of the survey it was apparent that excavation and removal of consolidated silt and stone in the river bed would be a major task, and expensive to excavate in all culverts. Since much of the culvert system consisted of either double or triple bores, if one was selected as a 'dry-weather flow' channel excavation could be minimized. Furthermore, future deposition would also be reduced. The remaining culverts could then cleared of loose debris and silt accumulation (Fig. 3).

What were originally thought to be scour holes at the downstream end of the weirs near the confluence chamber were found to be purpose-built stilling pools, with varying amounts of siltation. The survey of these 'pools' involved taking underground, by hand, a lightweight boat to plumb the depths of the pools.

The structure of the culverts was closely inspected to record their condition, given that the intended use of a sprayed concrete coating for hydraulic smoothing would seal off water ingress and cover up cracks and other deficiencies. As a result of this inspection, special measures had to be taken in the design of the lining.

Site investigation

To enable detailed design of the screen structures a site investigation contract was let which revealed 'made ground' overlying mudstone from the Carboniferous series.

Liaison with appropriate authorities
British Rail
Much of the culvert system runs beneath Sheffield's Midland Station, and two of the three proposed structures were to be constructed on British Rail land.

British Rail's major concerns were: (i) the structural effect on the culverts which are their property; (ii) loss of car parking spaces at Porter Brook; and (iii) disruption during construction.

Fig. 3. Debris in culvert.

The earliest meeting with British Rail concerned the lining specification, and much assistance was given by their laboratory staff at Derby. Staff of the Area Civil Engineer's office were concerned with covering structures with the lining, and the construction of the new roof at the transition chamber. The solution adopted for the lining was a layer thin enough to allow cracks to be seen as a result of structural movement rather than a more substantial thickness with inherent structural strength. The roof of the chamber was designed as a removable false 'ceiling' to allow future inspection.

The site of the Porter Book screen was initially agreed on the basis that an electricity sub-station would require moving to another site which minimized the loss of car parking bays. However, an alternative location proved hard to find and was very expensive, and eventually a compromise was sought.

By moving the screen upstream, alternative car parking was created over the river by the construction of a roof to the stream, and additional parking was provided at the Granville Square site.

Public Utilities
The NRA approached all the public utilities at an early stage, and extensive records and plans were made available to the consulting engineer.

The existing Granville Square open section had a 600 mm diameter high-pressure gas-main crossing the river, the hence close liaison with British Gas at the design stage was necessary.

At the same site an existing 1.2 m diameter storm overflow outlet and flap valve required cutting back, and hence liaison with the local authority, as agent for Yorkshire Water, was essential.

Local Authority

At an early stage of detailed design, meetings with officials of Sheffield City Council were convened to map out the various permissions/approvals required as a result of the scheme.

The Water Act 1989 allows the NRA to construct any necessary works to alleviate flooding 'on, in or under the watercourse'. It was construed by the Council that permission for the river works was unnecessary, but the skip compound alongside the banks would require full planning permission.

The scheme also impinged on the Council's roads and bridges, and required footpath diversions to be sought.

Initial planning problems stemmed from concern with respect to the visual impact of the overhead runway for the screen grab mechanism and the associated fencing to the skip compound. The highway engineers were concerned about skip lorries turning off the busy Granville Square roundabout.

The negotiations of an aesthetically acceptable solution took twelve months, with approval given subject to various agreed conditions. These included tree screening, natural stone compound wall of the River Sheaf screen, colour pigmented concrete, and acceptably coloured paviors and metalwork.

Detailed design

All the hydraulic design had been carried out by NRA staff before the consultants were commissioned. The detailed design mainly concerned the river structures, culvert improvements and lining, and the screens' installations and compound.

Contracts

The Phase 1 work was divided into two contracts:
    Contract No. 1 – Civil-engineering works;
    Contract No. 2 – Screens and associated works.
The contract period for the civil works was 52 weeks with a stage completion to allow screen installation at week No. 26. The mechanical/electrical contractor had 10 weeks to install and commission the plant.

Lining to culverts

Due to the thin coating requirement a thickness of 50 mm was adopted using steel-fibre reinforced concrete (3% by weight) spray applied by the 'dry' process. The fibres were 25 mm long and 0.4 mm diameter melt extract stainless steel. The coating was applied in two layers, the first 40 mm being reinforced and the second 10 mm unreinforced, applied whilst the first was still 'green', and trowelled smooth (Fig. 4). Before application, the substrate was cleaned and repairs/voids filled using heavy-duty repair mortar.

Fig. 4. Sprayed concrete lining.

Due to water leaking through the existing structure, drainage was required be-hind the sprayed lining. The design allowed for a drain comprising high-density polyethylene dimpled sheet with integral woven-mesh for lining adhesion, fixed to the substrate before spray application. As the distance between the drains was fixed at about 3 m intervals, this was considered suitable for the contraction joint locations and thus a triangular crack inducer was incorporated with the liner drain.

Granville Square Roof
The roof required at the open section of the River Scheaf at Granville Square had a maximum span of 13 m, and the stability of the existing retaining walls was difficult to assess. It was considered unwise to excavate to about 2 m below bed level to obtain a sound foundation for the supporting structure. The required cutwater could act as a central support, and thus the concept of an *in-situ* culvert was developed. This form of structure would easily resist uplift forces under surcharge conditions. The structure was designed to be constructed in two halves to maintain river flows.

The backfilled structure yielded the dividend of parking space to alleviate British Rail's loss of space at the Porter Brook site.

False roof at transition chamber
The false roof, which was to be removable to satisfy British Rail's requirement, consisted of GRP units installed within the existing jack-arched roof. The GRP units were filled with rigid foam for lightness and stiffness, and were fixed using resin-anchored stainless-steel bolts.

Screens

Initially the NRA had identified the type of screen to be installed at the entrances to the culvert system, and the consulting engineer was required to confirm the suitability and viability.

In the urban area, with no immediately available operational staff, the automated grab-type screen suspended from an overhead gantry/runway was the most attractive system. It had been installed in a similar environment in London, though on a smaller scale, and was able to handle the varied debris deposited from an urban river (Fig. 5).

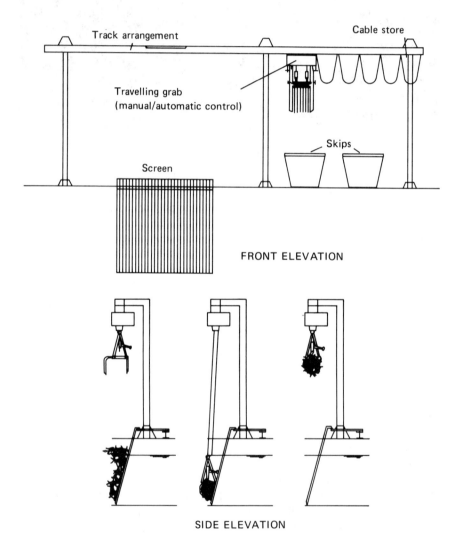

Fig. 5. Typical arrangement of automatic screen.

The screen widths were 8.5 m in each location, although the height of the Sheaf screen was 3.6 m whilst the Porter Brook screen was 2.1 m.

The foundations and river walls to the screen were designed to be separate structures, with the wall foundations being located at the level of the mudstone, about 2 m below bed level.

The highway authority initially stipulated that none of the screen installations should be attached to the adjacent bridge. This was subsequently relaxed so that the right-hand wall was dowelled to the bridge abutment and to the adjacent weathered dry-stone wall. The dry-stone wall was then grouted up prior to the new wall being poured against it.

The screen grab mechanism can be automatically initiated by river level or timer, but some concern had been expressed regarding the safety of such operations should anyone be in the river. Initially, therefore, the river levels operated a dial-out alarm during a storm to summon the duty operator. Otherwise, the regular operation was anticipated to be weekly using manual initiation.

Future phases

Construction work for Phase 1 started on site in February 1991, with design work for Phase 2 already in hand. In this way contunuity was achieved, and by 1995 the entire scheme will be implemented, giving flood protection to the lower Sheaf Valley to a standard return period of once in 50 years.

CONCLUSION

The production of a technically and environmentally sound solution to the severe flooding problems in the Sheaf Valley has been very reliant on close and careful liaison with engineers, planners and other concerned authorities, starting at the feasibility stage and continuing through into the construction phases.

ACKNOWLEDGEMENTS

The authors wish to thank their colleagues at the National Rivers Authority (Yorkshire Region) and Balfour Maunsell who assisted in the design of this scheme.

REFERENCES

[1] Institute of Hydrology. (1975) *Flood Studies Report.*
[2] Penning-Rowsell, E. C. and Chatterton, E. B. (1977) *The Benefits of Flood Alleviation – Manual of Techniques.*
[3] Parker, D. J., Green C. H. and Thompson, P. M. (1987) *Urban Flood Protection Techniques – A Project Appraisal Guide.*

26

River flow failure at Sleaford: context, causes and cure

P. J. Hawker, BSc, CEng, MICE, MIWEM[†], K. J. Harries, BSc, FGS, MI Geol[‡], and N. F. Osborne, BSc, MIFM[§]

INTRODUCTION

The River Slea runs through the middle of Sleaford in Lincolnshire and is a significant feature of the town. Footpaths follow one or both banks through most of the urban area, and weirs in the town centre and on the eastern outskirts form potentially attractive water features. The weirs also cause water to pond for a considerable distance upstream due to the shallow channel gradient. The river is fed principally by springs which issue from the top of the Lincolnshire Limestone aquifer, at the contact with overlying clays just to the west of the town in the 'Boiling Wells' area. The aquifer also serves public water supplies by means of pumped abstraction.

In recent years, the springs have failed periodically. When this occurs, river flow ceases and water quality in ponded sections of the channel through the town deteriorates. Understandably, there is considerable concern among the local community about this problem.

An investigation into the problem was carried out in late 1989 by Sir William Halcrow and Partners Ltd for the National Rivers Authority (NRA), Anglian Region. The work included historical research, hydrological and hydrogeological studies to

[†] Principal Engineer, Sir William Halcrow & Partners Limited.
[‡] Senior Hydrogeologist, Sir William Halcrow & Partners Limited.
[§] Assistant Engineer Water Resources, NRA Anglian Region.

establish how much of the flow problem may be attributed to groundwater abstraction, and how much to other factors such as natural seasonal and climatic controls. The value of the river corridor in ecological and amenity terms was also assessed in the vicinity of Sleaford, and consultation within the local community undertaken, to help determine targets for flow rehabilitation.

A programme of field trials was undertaken to determine the feasibility of seasonal river support using groundwater as a solution to the problem. In association with the trials, the groundwater resources of the aquifer were evaluated in order to see whether such an augmentation scheme was sustainable in the long term. The findings from these studies and the field trials are presented.

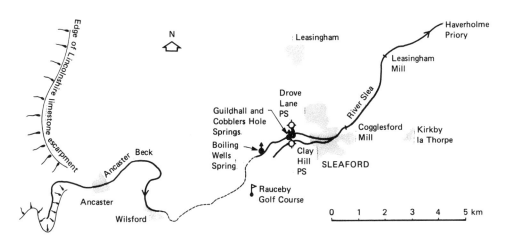

Fig. 1. Location map of study area.

CATCHMENT DESCRIPTION

General

Outcropping in the Sleaford area, the Lincolnshire Limestone forms a distinctive escarpment feature through Lincolnshire; the steep scarp slope rising to a maximum elevation of some 120 m AOD 10 km west of the town. Ground elevation declines gradually eastwards on the dip slope of the escarpment to reach 15 m AOD at Sleaford. A location map of the study area is shown in Fig. 1.

The surface catchment area to the Slea at Sleaford is approximately 40 km^2, with the western watershed being defined by the escarpment edge running through the head of the Slea Valley just west of Ancaster. However, far more important for river flow is the groundwater catchment: this is the extent of outcropping limestone, from which groundwater drains into the River Slea, and is estimated to be 80 km^2 in area. Arable farming dominates the land-use in the catchment.

Geology

The strata forming the escarpment are wholly of Jurassic age; the broad geological divisions and structure are shown as a cross-section in Fig. 2. Lias Clay (Lower

Jurassic) represents the oldest strata in the study area, and these beds outcrop at the base of the escarpment. The overlying strata of the Inferior Oolite Series (Middle Jurassic) comprise thin basal Northampton Sands and Lower Estuarine Beds, followed by some 30 m of Lincolnshire Limestone. The succeeding Great Oolite Series (Middle Jurassic) comprise the Upper Estuarine Beds, Great Oolite Limestone, Blisworth Clay and Cornbrash. Finally, the youngest Jurassic strata are represented by the Oxford Clay (Upper Jurassic), and these beds are present in the east of the study area. The dip of Jurassic strata is remarkably low and consistent, being approximately 1° to the east.

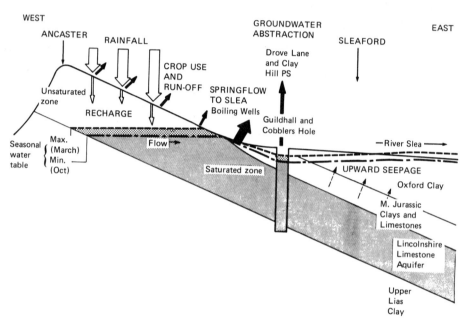

Fig. 2. Conceptual hydrogeological section.

The River Slea

The headwaters of the Ancaster Beck, a tributary of the River Slea, are fed by springs issuing at the base of the Lincolnshire Limestone escarpment, where it is underlain by impermeable Lias Clay. The beck then flows easterly through a gap cut into the escarpment Downstream of Wilsford, the river loses flow by bed leakage back into the limestone, and by Rauceby Golf Course the channel is frequently dry.

Substantial flow reappears just west of Sleaford due to three major springs issuing at the contact between the Lincolnshire Limestone and Upper Estuarine Clays. All three springs, however, are currently ephemeral. Boiling Wells Spring (3 km upstream of Sleaford) is typically dry between June and December nowadays; Guildhall and Cobblers Hole Springs (1.5 km upstream of Sleaford), are typically dry between September and December. When the latter two springs are dry there is no flow in the River Slea as far as Haverholme Priory, 6 km downstream of Sleaford.

From Sleaford, the River Slea flows north-easterly to join the River Witham near Coningsby.

THE HYDROGEOLOGY OF THE LINCOLNSHIRE LIMESTONE AQUIFER

The Lincolnshire Limestone constitutes a major source of groundwater for public and private water supplies in the county. The Lias Clay forms the effective, impermeable base to the aquifer, whilst the clays of the Upper Estuarine Beds provide a confining cover. A conceptual hydrogeological cross-section through the aquifer at Sleaford is shown in Fig. 2.

Recharge to the aquifer occurs at outcrop; average annual recharge to the Slea catchment is estimated to be 195 mm, 90% of which occurs during the months November to March inclusive. In response to recharge, groundwater levels in the aquifer rise and attain a maximum typically in March. High aquifer transmissivity (2000-$3000\,m^2/d$), caused by fissure development, results in rapid groundwater flow in the saturated zone. Groundwater flow is predominantly easterly towards the major spring discharges and pumping centres. A groundwater contour map (with the catchment defined) for a typical March is shown in Fig. 3; the map is based on data from the NRA Anglian Region observation borehole network.

During the summer months, groundwater levels decline due to reduced recharge and continued discharge; minimum groundwater levels are usually reached in October, before the onset of winter recharge. When groundwater levels drop below spring level, the springs cease flowing.

Current groundwater abstraction from the Slea catchment amounts to some 3400 Ml/a. Of this total, 2800 Ml is pumped at two public water supply sources at Drove Lane and Clay Hill, both near the Guildhall and Cobblers Hole Springs (see Fig. 1). The remaining 600 Ml is pumped from a number of private sources. Analysis of NRA river spot flow gauging data shows that in an average year 1900 Ml of groundwater is discharged at Boiling Wells Spring, whilst 10 800 Ml is the combined discharge of Guildhall and Cobbers Hole Springs.

To summarize, the average annual water balance for the Slea groundwater catchment is:

Inflow	Outflow	
Recharge	Springflow + Abstraction	
16 100 Ml	12 700 Ml	3400 Ml

Current abstraction, therefore, amounts to 21% of the average annual recharge to the catchment.

Groundwater quality in the Lincolnshire Limestone aquifer is characterized by relatively high hardness at outcrop, dominated by calcium and bicarbonate. Groundwater flowing into the confined zone undergoes softening where calcium is replaced by sodium. About 10 km east of the eastern edge of the limestone outcrop there is a distinctive saline interface, beyond which the groundwater is relatively high in total dissolved solids – dominated by sodium and chloride (see Fig. 3). Groundwater here is mostly of connate origin, reflecting limited circulation of recent, meteoric waters.

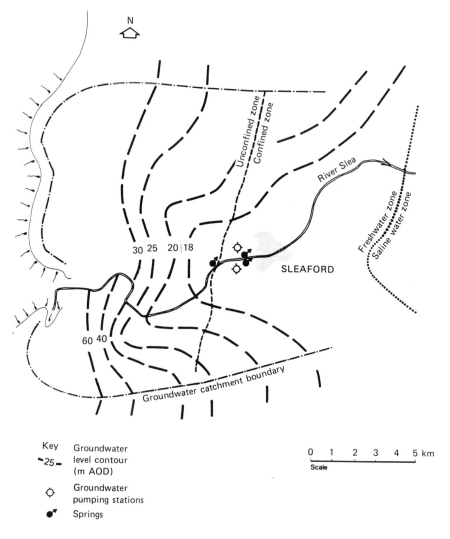

Fig. 3. Groundwater level contours: a typical March.

INVESTIGATIONS

Historical research

Research was undertaken to set current river flow patterns at Sleaford in context historically. This aspect of the work involved: (i) discussion with local people who had a long association with the river; (ii) a literature search concentrating on activities relating to water, such as navigation, milling, water-cress farming and fisheries; (iii) examination of council records, particularly during notable dry weather periods.

The literature search embraced local newspapers, books and previous reports of river maintenance. A Master of Philosophy thesis by M. Hunt of Boston provided a fascinating and informative account of the Slea as a navigable waterway between 1794 and the advent of the railways to south Lincolnshire in about 1860.

The main findings from the research were:

- in the 18th century, eight mills operated on the Slea within two miles of each other at Sleaford;
- navigation along the Slea in the nineteenth century, from Sleaford to the River Witham, was never halted through lack of flow, but was limited in the droughts of 1844 and 1854;
- in the first half of the twentieth century, commercial water-cress growing, which needs an all-year-round flow of spring water, took place at Guildhall Springs;
- the Slea was locally renowned for its trout. These fish bred naturally within the town of Sleaford until the 1960s, but have not done so since;
- there are no reports of the river drying out in Sleaford itself until the drought of 1976, though since then it has become a regular occurrence.

Thus, the historical evidence suggests that regular failure of river flow through Sleaford is a problem of relatively recent origin.

Ecology, amenity and public consultation

The ecological assessment showed that the main interest in the river corridor is floristic. Winterbourne habitats now extend downstream of the formerly perennial Guildhall Spring by some 300–400 m; restoration of perennial flow to this reach is thus undesirable in ecological terms.

Because of its high degree of public access, the river is important in the town and there is much concern about it among the community. The town is at present growing rapidly so that both demand for public water supplies and pressure for river improvement are likely to increase.

Sleafordians are keen to see flowing water of good quality returned to the river through the town all year-round, of sufficient depth to ensure unsightly mud-banks are not exposed. There is also interest in restoring the former navigation along the river between the town and the River Witham.

Analysis of hydrological data

To examine factual evidence for causes of low flows, data on aquifer recharge, groundwater abstraction and river flows have been analysed.

Recharge data as calculated by the Meteorological Office were used. Groundwater abstraction data were obtained from licence returns (and estimated where necessary) and a graph illustrating abstraction changes since the late 1940s is shown in Fig. 4. Abstraction was steady up to the mid 1950s at about 2.0 to 2.5 Ml/d. A rapid increase then occurred, reaching 5.5 Ml/d (equivalent to a rise of 240 Ml/a) by the early 1960s, mainly at Drove Lane pumping station. Since then, there has been a steady increase up to a current rate of about 9 to 10 Ml/d (3400 Ml/a).

Earliest flow data come from spot gaugings at Leasingham Mill (3.5 km downstream of Sleaford) between 1960 and 1965. Unfortunately these data post-date the dramatic rise in abstraction in the late 1950s. From the early 1960s, there are flow data from spot gauging sites at Boiling Wells Spring, and at Castle Causeway in the town. A gauging weir was installed at Leasingham Mill in 1974, and an almost continuous flow record is available since then. At all three gauging sites the data show that the river ceased flowing in the notable dry periods in 1964/65 and 1976. In addition, low (less than 0.1 Ml/d), or no flow periods, have frequently occurred at all three sites since the early 1970s.

Comparison of recharge, abstraction and flow data has clearly shown the dominant influence of recharge upon flows. However, what is not clear from the data is the effect of abstraction. It is evident from historical research that the low flow conditions, which occurred frequently during the 1970s and 1980s, did not happen in the past.

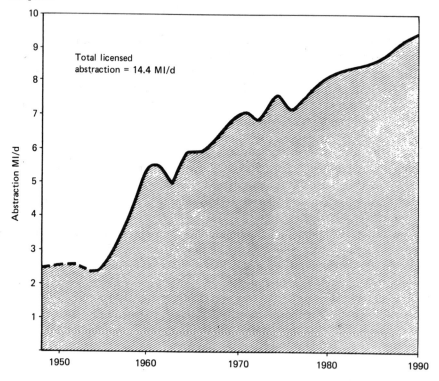

Fig. 4. Total groundwater abstraction from Slea Catchment.

Since available data are inconclusive, analytical methods were employed to estimate natural flow conditions with no groundwater abstraction. Two methods were employed.

(i) Springflow correlation
Current groundwater abstraction in the Slea catchment amounts to 9.3 Ml/d. If abstraction ceased, there would be an extra 9.3 Ml/d (or 3400 Ml/a) on average available for spring discharge to the River Slea.

Due to the recharge-discharge control on groundwater flow in the aquifer, this volume would be distributed seasonally to the springs according to groundwater level fluctuation. To determine such a distribution, the seasonal behaviour of two springs issuing from the Lincolnshire Limestone Aquifer in the vicinity of Sleaford have been analysed; the particular springs chosen are relatively unaffected by abstraction.

It was concluded that, if there was no groundwater abstraction, then the seasonal minimum flow in the River Slea at Sleaford would be approximtely 3 Ml/d in a year with average recharge.

(ii) *Drawdown Analysis*

In a typical year, when significant recharge effectively ceases in April, groundwater levels decline in the aquifer due to the combined effects of spring discharge and groundwater abstraction. When groundwater levels drop below the spring points, the springs cease flowing. Calculations have been carried out to predict what groundwater level recessions would have been at both Guildhall and Boiling Wells Springs without groundwater abstraction at Dove Lane and Clay Hill. In this way it has been possible to estimate the effects of abstraction on springflow cessation.

It was concluded that, in an average year, without abstraction: (a) the flow from Guildhall and Cobblers Hole Springs would be low, but would not cease altogether; (b) the flow from Boiling Wells Spring would be ephemeral, with cessation of flow from the end of September to mid-December.

Conclusion

Recharge to the Slea groundwater catchment is the dominant control on the flow regime of the River Slea at Sleaford, with high flows in winter months and low flows in late summer/autumn.

Historical evidence strongly suggests that groundwater abstraction over the last 30–40 years, notably at Drove Lane pumping station, has further reduced flows to critical levels (and frequently stopped flow altogether) during the late summer-autumn months.

It is inferred, by analytical methods, that in an average year, under natural conditions (no abstraction):

(i) the flow through Sleaford would be maintained by the springs at Guildhall and Cobblers Hole and would be continous – falling to a seasonal minimum of about 3 Ml/d;

(ii) Boiling Wells Spring would be ephemeral – with no flow occuring from late September to mid-December.

Nowadays, under average conditions, groundwater abstraction has resulted in failure of the Guildhall and Cobblers Hole Springs for about 90 days, between September and December/January. Abstraction has also resulted in earlier cessation of flow at Boiling Wells Spring, which now typically occurs in July.

RIVER AUGMENTATION TRIALS

Introduction

Field trials were carried out to determine whether river augmentation using groundwater was a feasible solution to the Slea flow failure problem. Essentially the trials comprised pumping groundwater into the river upstream of the town from a redundant well at Drove Lane Pumping Station.

Extensive monitoring was carried out before, during and after the trials, as shown in Table 1.

Table 1. Field Trials Monitoring Programme

Determinand	Monitoring Method
Groundwater levels	Chart recorders on two existing boreholes, and 3 new boreholes drilled as part of the investigation
Pumped augmentation flow	Weir tank at well head
River flows	Temporary v-notch weirs installed at strategic locations downstream of the augmentation
Water quality	Analysis of samples from well head, and the river through and downstream of Sleaford
Public reaction of environmental impact	Tour of works and meeting with selected community representatives during trials; feedback from amenity/ecological/historical research consultations and informal contact with local people as trials proceeded

The trials commenced on 27 November and ended on 18 December, 1989; apart from heavy rain in the last few days, the period was dry. There was no natural flow in the river during the period of the trials.

Pre-trial works

Works carried out in advance of the trials included:
- three observation boreholes, fitted with chart recorders;
- wellhead works at the redundant well at Drove Lane, comprising pump installation and provision of a weir tank to measure flow;
- 750 m temporary pipeline (300 mm diameter, laid overground);
- construction of three temporary v-notch weirs in the river channel.

The location of these works is shown on Fig. 5. As the trials proceeded, bunds were constructed across the River Slea above the augmentation point, and across the Nine Foot River, to prevent backflow to more permeable sections of channel and consequent leakage losses.

Observations

Groundwater levels and aquifer characteristics

Before the augmentation commenced all observation boreholes showed a gradual groundwater level recession of about 20 to 30 mm/day due to pumping 8–8.5 Ml/d for supply at Drove Lane and Clay Hill Pumping Stations. Once augmentation began, the recession increased in response within about four hours of start-up. The steeper recession trend continued relatively steadily until 15 December, when heavy rainfall (starting 13 December) promoted groundwater level recovery.

At Drove Lane Pumping Station drawdown caused by extra pumping from the redundant well for augmentation was about 1.5 m in the well itself, and estimated to be less than 0.3 m in Production Borehole No. 1 (20 m from the well). There was, therefore, no significant effect on production for supply at Drove Lane Pumping Station (or at Clay Hill, because of the distance) due to the augmentation trial.

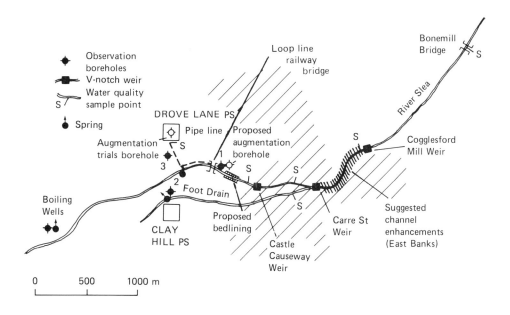

Fig. 5. Augmentation trials and recommended scheme.

The groundwater level data obtained from the trials was used to calculate transmissivity and storativity of the limestone aquifer. Generally, values of transmissivity are in the range 1700–3000 m^2/d, reflecting the highly fissured nature of the limestone. Values of storativity are in the range 10^{-3} to 10^{-1}, which reflect a semi-confined aquifer. These values were used in the drawdown analysis to assess the effects of abstraction in springflow, described above.

Flows and leakage
Augmentation commenced at a rate of 1 Ml/d, but this was increased after one day to 3 Ml/d (28 November), and this rate was steadily maintained until the end of augmentation on 18 December.

Because of the very shallow gradient, backflow occurred as the channel filled up, eventually extending some 300 m upstream towards Guildhall Spring. Flow over the first v-notch weir at Castle Causeway, 450 m downstream of the discharge point, initially stabilized at about 0.7 Ml/d, thus showing bed leakage to be occurring at a rate of 2.3 Ml/d. To reduce this a sandbag dam was built across the river immediately upstream of the augmentation discharge. This eliminated the backflow problem, and reduced leakage over the section between the discharge and the Castle Causeway weir to 0.9 Ml/d.

At the next temporary weir downstream (Carre Street) a steady-state flow of about 1.1 Ml/d was achieved. Some improvement was sought by limiting backflow in the Nine Foot River which joins the Slea here; however, this had negligible effect. The inferred rate of leakage of 1 Ml/d between Castle Causeway and Carre Street is puzzling, since the rate of loss from the ponded section upstream of the permanent weir at Carre Street (into which there had been no inflow for many weeks before the trials started) suggested minimal bed leakage over an extended period. This conflict of evidence has yet to be fully explained.

Downstream of Carre Street a steady-state flow of about 1 Ml/d was observed at Cogglesford Mill. Thus, leakage over this section is apparently negligible. The rate of fall in the water level in the ponded section upstream of Cogglesford over the weeks preceding the trials supported this finding.

Water quality

As expected, the augmentation resulted in a great improvement in surface water quality. Before the trials, the ponded sections of river sampled had low dissolved oxygen (down to 12%), high ammonia, and high chlorophyll-a levels. This would place the river in Class 3 of the NWC/DoE classification system. By analogy with lakes and reservoirs, water in the ponded sections could be described as eutrophic.

As augmentation continued, water quality in the river generally improved. Dissolved oxygen increased to at least 64%, and the chlorophyll-a concentration decreased such that a mesotrophic state was established. At the end of augmentation, water quality through Sleaford improved sufficiently for the river to fall within Class 1B. In amenity terms, the rate of flow achieved was effective in moving floating debris from the town centre, and in controlling pollution caused by ducks.

Public reaction

Generally, the local community were highly supportive of the trials, and satisfied with the results. The transformation of the watercourse from a dry, grassy channel to a river flowing with clear well water upstream of Castle Causeway was dramatic, and eloquently demonstrated why Sleafordians are so concerned about flow failure in their river. Consultation suggested that the rates of flow through the town, and the water quality improvement achieved during the trials, were acceptable to the local community.

FLOW ALLEVIATION

The requirement

In the past the River Slea has served a number of purposes including cress growing, milling, navigation and angling. The amenity value of the river in Sleaford is important both to the local community, and also in respect of current initiatives to promote tourism in the area. The river is a prominent feature, with footpaths following one, or both, banks almost throughout the town.

There is no doubt that groundwater abstraction has had a significant effect on river flow through Sleaford, particularly during the late summer and autumn months.

Abstraction has certainly had an adverse impact on the amenity value of the river; notably in the cessation of fisheries activities which require a perennial flow.

However the use of the water is important for public supply; furthermore local demand is likely to increase as a result of planned growth around Sleaford.

The primary objective in alleviating the low flow problem must be to develop a sensible strategy for overall groundwater abstraction, which ensures that the amenity value of the river through Sleaford is protected. Following the investigations described above, four objectives have been identified for amenity preservation:

(1) flowing water through Sleaford from a point just west of the town throughout most years;

(2) sufficient volume of flow to counteract the pollution load imposed by the resident duck population;

(3) sufficient depth of water to fill the channel between banks throughout most years (i.e. no seasonal exposure of mud); and

(4) sufficient flow velocity to prevent accumulation of floating debris in the town.

Whilst there is considerable local interest in restoring the Slea navigation, the amount of water needed for this purpose would be an order of magnitude greater than that needed to achieve the above objectives. In view of the quality and resource value of local groundwater, therefore, restoration of the navigation was not included as an alleviation target.

On this basis, the required flow to support the river when springs are dry is assessed as 3 Ml/d. This flow equates with both the estimate of the average seasonal minimum under natural conditions, and the rate during the trials, which was deemed sufficient by those community representatives whose views were canvassed. This flow would also satisfy the above four amenity objectives, although judicious channel modifications downstream of Carre Street Bridge would be needed to satisfy the third objective.

Evaluation of alleviation options

Along with the detailed investigation of groundwater augmentation, various alternative options for low flow alleviation were considered. These included: (i) diverting mains water into the river; (ii) diverting treated effluent into the river; and (iii) intercepting perennial flow in the Slea at Wilsford, in the upper catchment, and piping it to Sleaford. It was concluded that these alternative schemes were either impractical, or too costly (or both), compared with the groundwater augmentation.

The groundwater augmentation scheme would involve pumping from a new borehole adjacent to the Slea for discharge to the river. In terms of the impact on groundwater resources, the results from the trials showed that the additional abstraction for augmentation had no significant effect on the two adjacent public supply sources. Furthermore, the timing of the trials coincided with a period when groundwater levels were at the seasonal minimum. To place December 1989 in context, groundwater levels were about 3–4 m below average seasonal minimum, and 4–5 m above levels reached at the end of the 1976 drought. Therefore, in most years, a 3 Ml/d pumping rate for augmentation is not unreasonable in relation to the effect upon existing supply wells.

In terms of the total resources of the Slea groundwater catchment, less than 5% of the average annual recharge to the catchment would be abstracted for augmentation of the River Slea. By pumping in the summer/autumn months, some storage deficit would be built up in the aquifer, which would then be made up rapidly by the following winter's recharge. Thus, the effect would be simply to delay (slightly) the natural onset of springflow, which would be compensated for by extending the augmentation period by a few days. Such a redistribution of storage, therefore, would have no overall adverse effect upon the resources of the catchment.

The recommended alleviation scheme

A groundwater augmentation scheme was proposed to support the river when the springs which feed it are dry. A new borehole adjacent to the river near Observation Borehole 1 (See Fig. 5) would be pumped at a rate of 3 Ml/d, and water from it piped to the river for discharge just downstream of the loop line railway bridge. To limit the leakage observed during the trials, the river bed would be lined for about 200 m in the vicinity of the discharge point. In addition, a demountable barrage would be installed upstream of the discharge to prevent backflow and consequent leakage. Thus the scheme should afford a minimum flow approaching 2 Ml/d at Carre Street/Cogglesford Mill, which would fulfil public expectation as assessed during the trials, and meet the objectives set out above.

Downstream of the town at East Banks, channel narrowing and enhancement works were suggested, including the reformation of banks, and the planting of reeds and other marginal plants (see Fig. 5).

The augmentation should commence when flow to the Slea from Guildhall Spring ceases, and, in typical years, should continue until such spring flow resumes. It is suggested that, in view of the competition for groundwater resources in the Slea catchment, augmentation should cease on 31 January if springflow recovery has not occurred by then. This control rule is somewhat arbitrary, but could be reviewed once the scheme has been in operation for a few years.

The recommended scheme for augmentation was accepted by the NRA Anglian Region and is currently being implemented, along with a licence reallocation of groundwater abstraction amongst the two existing sources at Drove Lane and Clay Hill, and the new source at Kirkby La Thorpe, 3 km east of Sleaford.

CONCLUSIONS

No evidence has been found relating to flow failure of the River Slea through Sleaford prior to 1962. If it did fail the event was extremely rare and of short duration. Historical usage, such as cress growing, milling, navigation and angling, testify to the general reliability of flow.

However, during the 1970s and 1980s, flow failure has been a common occurrence between the months of September and December. The failure is considered to be due to increased groundwater abstraction since the 1950s (for public water supply) from boreholes near to major springs feeding the River Slea, just upstream of Sleaford. Without this abstraction, it is likely that the River Slea through Sleaford would be perennial; though in summer months flow would recede to a minimum of about 3 Ml/d, on average.

The River Slea is potentially a prime amenity of Sleaford, but nowadays it is an eyesore after an extended period without flow. The amenity value of the River Slea is significant not only to the local community, but also in respect of the tourist potential in the area. Thus, there is an obvious requirement to alleviate the flow failure problem.

A programme of field trials, together with a resource assessment of the Slea catchment, has demonstrated the feasibility of seasonal augmentation of the River Slea using groundwater. Consequently, a scheme has been recommended which essentially involves pumping about 3 Ml/d from a new borehole with discharge to the Slea upstream of Sleaford during periods when springflow ceases. In addition, the river channel would be lined near to the discharge point to limit leakage losses. Downstream of Sleaford, at East Banks, a series of channel enhancement measures is suggested.

The recommended scheme for augmentation was accepted by the NRA, and was implemented in 1991.

27

Investigations into the proposed water abstraction from the River Spey

G. D. Watt, CEng, BSc, FICE, FIWEM[†] and B. Gill, CEng, BSc, MEng, MICE, MIWEM[‡]

INTRODUCTION

The need for water

At the completion in 1975 of the North East of Scotland Water Resources Study [1], the River Spey was identified as a potential source for the supply of 38 Ml/d to the Lower Moray and Banff coastal area. Following local government reorganization in 1975, Grampian Regional Council was formed and adopted the scheme from the North East of Scotland Water Board.

Discussions with the Spey District Fishery Board and other interested parties resulted in the Council investigating other sources including abstraction from the alluvial gravels on the west bank of the River Spey to the south of Fochabers. The latter appeared to find favour with the fishing interests provided that some form of protection would be given during times of low flows, a matter which was still under discussion in 1991.

In 1975 the Consultants predicted a demand of 38 Ml/d on the Spey scheme by the year 2001. By the mid-eighties it was evident that North Sea oil and gas was not likely to generate any major water users in Grampian Region and expected demand figures were substantially reduced to 8.7 Ml/d and 18.5 Ml/d for the years 2001

[†] Senior Depute Director, Grampian Regional Council, Department of Water Services.
[‡] Regional Associate, Mott MacDonald Scotland.

and 2011 respectively. These figures were reviewed again in 1988 because of the Drinking Water Directive (80/778/EEC) which indicated that maximum admissible concentrations (MACs) would not be measured against annual or seasonal averages but against individual samples. This strict interpretation meant that water from many sources within the area to be covered by Spey water would not meet the standards and in time these sources would have to be abandoned. The net result was that demand forecast had to be adjusted upwards to give a predicted demand of 27 Ml/d by the year 2011.

Choice of the River Spey

The River Spey with a catchment area of some 2960 km^2 is by far the largest river in the area and has been identified as a more suitable option than any of the other rivers in the area.

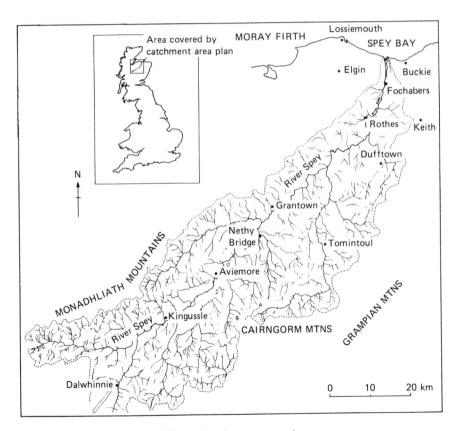

Fig. 1. Catchment area plan.

Other sources have been rejected, mainly because none of them, as a single source, could meet the long-term demand forecasts and all would require at least two-stage treatment to meet the latest water quality regulations.

The choice of the River Spey was therefore accepted by Grampian Regional Council and subsequently a testing programme has been ongoing since 1984 to confirm the suitability of the alluvial gravel abstraction scheme as an alternative to a direct river intake scheme in terms of water quality, economic grounds and environmental considerations.

Summary of earlier work

A paper entitled *Groundwater investigations in the Lower Spey Valley, near Fochabers* [2] outlined the series of comprehensive studies and investigations undertaken to first of all identify potential groundwater abstraction areas, and then to confirm the suitability of the area of the proposed wellfield in the Dipple area to the south of Fochabers on the west bank of the River Spey as an abstraction site.

The paper concluded that the studies had shown that the alluvial gravel aquifer in the Dipple area could provide the quantity of water required, and that the quality of the water could meet drinking water standards after limited treatment.

RECENT CHANGES IN LEGISLATION AFFECTING THIS SCHEME

The Water (Scotland) Act 1980 consolidated certain provisions of the Water (Scotland) Acts of 1946, 1949 and 1967 as well as the appropriate sections of the Local Government (Scotland) Act 1973. Furthermore a number of Directives have impacted on the water scene in Scotland and in particular the Directive relating to the quality of water intended for human consumption (80/778/EEC).

This Directive has now been embodied in Scottish legislation by amending the Water (Scotland) Act 1980, as scheduled in the Water Act 1989. The latter also empowered the Secretary of State for Scotland to make Regulations which set defined standards for water quality. The Water Supply (Water Quality) (Scotland) Regulation 1990 came into being on 1 July, 1990. Under these Regulations the Council are required to abandon all unsatisfactory sources, or to undertake to treat them to meet the new standards, or to seek dispensations where the deficiencies are naturally occurring.

The Regulations also require a different approach to treatment according to whether the raw water is classed as surface water or groundwater. This classification is in accordance with Directive 75/440/EEC concerning the quality of surface water intended for the abstraction of drinking water. The alluvial gravel waters have, in terms of this Directive, been classified as groundwater.

ADDITIONAL INVESTIGATIONS

Introduction

Following the changes in the interpretation of the Directive 80/778/EEC, Mott MacDonald was requested to review the scheme and to make recommendations for any further investigations required to confirm the suitability of the scheme.

Previous recommendations had been to develop the scheme in two distinct phases, using the first phase almost as a large scale pilot scheme to test previous conclusions and to use these results in the design of the second phase.

However, following the review, demand figures had risen to 22 Ml/d by 2001 and to 27 Ml/d by 2011. It was therefore deemed impractical to develop the wellfield as a two-stage operation. Furthermore, as the EC water quality parameters were now more stringent, it was considered prudent to have more field data available in order to predict more accurately the range in water quality and quantity likely to be achieved when the full wellfield had been developed.

Pump tests and water quality monitoring

Two new test wells were installed 100 metres apart in the summer of 1989 with a number of associated piezometers. In order to try to confirm the colour removal ability of the alluvial gravels, the test wells were located where the river could be expected to have its greatest influence on the wellfield, that is where the ratio of river water to groundwater abstracted from the wells would be highest.

Weekly sampling for water quality showed one of the test wells to be consistently good, confirming the results of the previous long-term test. Water quality was below MAC levels throughout the sampling period, except for one occasion when flooding of the site occurred causing direct entry of river water into the well, resulting in unacceptable levels of coliform and E. coli.

Unfortunately the water quality from the second test well was not as good, with colour, iron and manganese levels often above MAC.

The results from this second test well were obviously a great disappointment and led to a significant reappraisal of the scheme and the way in which it will be implemented. This is further discussed under the section 'Scheme as now proposed'.

Noise levels from borehole pumps

The River Spey is a famous salmon fishing river and a number of concerns from the fishing fraternity were expressed on the effect a water abstraction scheme could have. One of these concerns was the effect sound pressure levels from the wellfield pumps may have on salmon.

To study this, the Department of Agriculture and Fisheries for Scotland Marine Laboratory in Aberdeen was commissioned by Mott MacDonald in 1988 to undertake a series of sound pressure recordings in three rivers and to report one their findings.

Soundings were taken on the River Spey to record background river noise levels; on the River Lochy, where a pumping test was being carried out, to record both background river noise levels and the river noise levels when the pump was in operation; and on the River Dee at the site of a bed intake to record both background river noise levels and river noise levels when the pumps, pumping to the adjacent bankside reservoir, were switched on.

The results from this study showed that the noise levels associated with the pumps on both the Lochy and the Dee were low in relation to the levels of background noise and tend to be masked. Natural background levels on the Spey were rather higher than those measured for the Lochy and the Dee, suggesting that an abstraction scheme on the Spey would also tend to be masked.

However, as the noise levels were taken when all three rivers were in moderate spate, the conclusions given in the report were subsequently questioned as to their validity for low summer flows. Further work was therefore considered to check if the low frequency noise levels generated by the wellfield pumps would exceed the ambient noise levels in the river at low summer flows.

Monitoring of sludge disposal areas

Sewage sludge from Rothes sewage treatment works is spread on several fields between 1.5 and 3.0 km from the proposed wellfield site. In order to identify any likely pollution to the wellfield caused by these activities a series of four boreholes were installed to monitor water quality.

Two boreholes were located outside the sludge disposal areas but at positions likely to intercept groundwater flow from the sludge disposal areas. Two boreholes were situated within the disposal areas.

Water quality from the samples collected indicated that nitrate levels in one of the boreholes outside the sludge disposal areas exceeded MAC on one occasion. Iron and manganese levels in boreholes outside the sludge disposal areas also exceeded MAC. Results from within the sludge disposal area indicated that high nitrate and nitrite concentrations were to be found within the groundwater immediately underneath.

The conclusion drawn from this monitoring exercise was that while there was no evidence of a deterioration in groundwater quality in the monitoring boreholes outwith the areas receiving sludge, it was recommended that the future spreading of sludge be restricted and that movement of groundwater towards the wellfield should be closely monitored.

Chlorine dosing

Previous recommendations proposed that chlorination of the raw water should take place at the wellfield before being pumped to Badentinan. However, the purpose of the operational storage reservoir at Badentinan was to provide a volume of water available for backflushing the wellfield should a pollution incident occur to contaminate the wellfield.

Previous water quality studies concluded that it was a bacteriological process which contributed to the oxidation of organic matter in the River Spey water which led to a removal of colour. In order not to risk upsetting this process it was considered prudent to backflush with untreated water and therefore chlorination would take place at Badentinan after the water had passed through the operational storage reservoir.

SCHEME AS NOW PROPOSED

Introduction

Due to the uncertainty over the final quality of the water to be abstracted from the wellfield, a cautious approach was adopted for the implementation of this scheme.

The conclusions drawn from the pumping test and water quality investigations indicated that a significant number of wells would produce water of a quality which would only require disinfection and pH control before going into supply.

It was therefore proposed that all wells be installed and tested before a decision was taken on full treatment requirements. Wells with water requiring partial treatment only would be allowed to supply the distribution system and sufficient

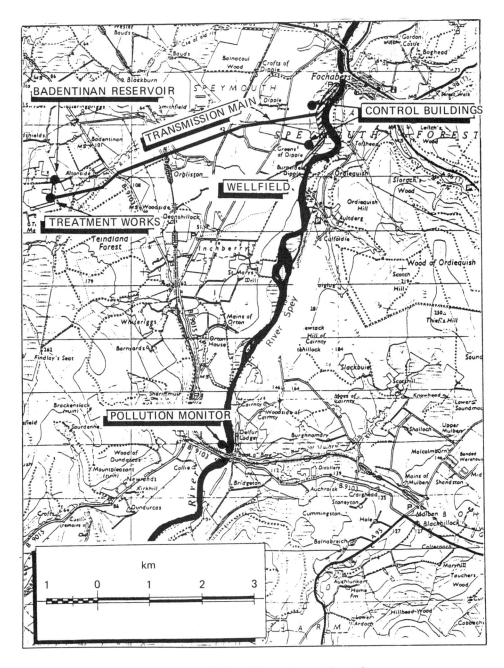

Fig. 2. Layout of River Spey abstraction scheme.

water would be available from those wells to meet the mid-1993 demand of 10 Ml/d.

By the end of 1995 it is expected that any treatment works which would be required to bring water from wells with poorer quality up to the required standard will have been constructed.

The layout of the scheme is shown in Fig. 2.

Wellfield

Some 36 wells would be constructed along a 3 km stretch of the west bank of the River Spey upstream of Fochabers. Wells would be spaced at approximately 80–90 metre intervals, and would be set back generally 50–55 metres from the river bank.

The depths of wells would be between 10.5 and 12.5 metres. On average each well was expected to produce some 0.85 Ml/d.

Well chambers to house individual wells would be watertight to allow for occasional overtopping during periods of flooding. All electrical equipment within the wellhead would therefore need to be waterproofed.

The River Spey is a very mobile river and certain stretches of the river bank in the vicinity of the wellfield are vulnerable to erosion. It was therefore proposed that bank protection works be undertaken to safeguard the integrity of the wellfield at those locations.

Road access to the wellfield would be via Fochabers to the north and the existing metalled road running north-south parallel to the river. Two access points from this road were proposed, each existing farm tracks, which would be upgraded to allow access to the control buildings.

Due to the uncertainty over the final water quality being abstracted from individual wells, it was proposed to have a twin mains collecting system along the wellfield. Once water quality from each well had been assessed, those wells requiring only partial treatment (i.e. pH control and disinfection) could be separated from any wells producing water of a poorer quality which would need a greater degree of treatment.

Two control buildings would be required to house pump controls and substations for the electrical supply system.

Pumping mains from wellfield to Badentinan

Twin pumping mains, each with a capacity of 75% of the total demand, would deliver water from the wellfield to Badentinan. A twin 450-mm diameter ductile iron pipeline with internal mortar lining was recommended. A resistivity survey assessed the need for any external protection requirements.

Should it be that, after testing of all wells for water quality, some wells require additional treatment, one of the pipelines could be used to transfer such water to the full treatment works.

Treatment facilities at Badentinan

The initial treatment required for water abstracted from the wellfield would be limited to: (a) disinfection by super-chlorination to a 1 mg/l residual followed by dechlorination by sulphur dioxide; (b) pH correction by lime dosing with lime slurry at 7% concentration. However, in planning these facilities, account needed to be taken of the treatment works which could be required if the quality of the blended well water was worse than expected. For this reason a conceptual design, layout and hydraulic profile were completed for the full treatment of the water.

Pollution monitors

In the design of the alluvial abstraction scheme, account was taken of the risk to the scheme from a pollution incident in the river which could affect the integrity of the wellfield. In order to identify pollution incidents in the river as quickly as possible, an upstream pollution monitor was proposed at Boat o'Brig.

A telemetry outstation would deliver signals to a control station at Elgin. The pumps would be capable of being remotely shut down and backflushing of the wellfield initiated.

Monitors for pH, iron and manganese levels would be installed at Badentinan for both raw and treated water.

ENVIRONMENTAL CONSIDERATIONS

Fishing

The Council, recognizing the great importance and value of the salmon fishings on the River Spey, sought the advice of two highly respected Fishery Consultants, Dr D. H. Mills and Mr N. W. Graesser. Their considered view was that the infinitesimal reductions in flow, even at the lowest daily discharges, would have no effect on water quality or the flora and fauna of the river, and would have no influence either on the movements of migratory fish or on angling and netting success.

The alluvial gravel scheme allows for water to be drawn indirectly from the river over a length of some 3 km. There would be no permanent structures within 50 metres of the river bank and consequently there would be no interruption to fishing during or after construction. Lastly, the quantity drawn from the river would be reduced because of the landward element of abstracted water.

The Fishery Board sought an agreement with the Council that when the flow drops below the 95 percentile an equivalent quantity of water to that being abstracted would be released into the upper catchment.

Visual impact of wellfield

The wellfield is within an area of great natural beauty and tourism plays an important part in the economic prosperity of the area. The wellfield was therefore designed to create the minimum amount of visual impact. The electrical supply to individual pumps would be by means of underground power cables from two sub-stations.

The surrounding buildings are mainly farm houses and farm buildings. The two control buildings on the wellfield have been designed to complement the local architecture with landscaping and tree planting providing suitable screening. Fencing would be kept to a minimum and is of a type already in use in the area.

Pipeline route from wellfield to Badentinan

The pipeline route was selected following discussions with the local land owners, tenant farmers, and other interested parties. A compromise had to be made between the best 'engineering' solution and the wish to avoid upsetting the local community. A flora and fauna survey is to be carried out along the pipeline route before construction commences, and appropriate reinstatements included in the construction contract in an attempt to reduce the impact of these works as quickly as possible.

Reservoir and treatment works site

The area around Badentinan is mainly coniferous forest and plantations and is some distance from the nearest public road. It was therefore considered that the visual impact of this section of the works would be extremely low.

Chemicals in water treatment

Initial sampling of the test wells showed the water to be of variable quality within the wellfield. It was not practical to test the quality of water throughout the length of the wellfield, but the results suggested that the majority of the wells would yield a satisfactory water. By duplicating the pipe connecting the wellheads and the pumping main to the treatment works it would be possible to treat only that water which fails the MAC before putting it into supply.

As previously stated the initial treatment would be by lime dosing and chlorination only. This has a significant advantage over conventional treatment of river water obtained from a direct intake scheme in that very little in the way of chemicals is added to the water. With increasing public concern being expressed regarding the continued use of some chemicals in the water treatment industry, any reduction in the amount of chemicals used can be a significant bonus and one which could be especially beneficial in this particular scheme, if the final quality of the well water proves satisfactory for partial treatment only.

CONTROLS PROPOSED TO SAFEGUARD WATER QUALITY

Farming practices

The quality of the water within the wellfield area did not exhibit the effects of agricultural pollution and in order to maintain that condition it was decided to purchase the two farms, extending to some 180 hectares, within which the wellfield is sited. The farms are now operated on a limited partnership basis with the Regional Council and are subject to very severe restrictions on the farming practices. The following are totally banned: pig rearing; horticultural productions; sheep dipping; application of organic fertilizers and slurries; chemical control of vermin; and burial of carcasses. Silage production is permitted but its storage is rigorously controlled. The application of inorganic fertilizers and herbicides is strictly controlled in terms of the rate of application and its timing.

All of these controls are designed to protect the treatment process, which takes place within the alluvial gravels, and to maintain the high quality of the raw water. Prevention rather than cure is the main reason for restricting farming practices.

Pollution monitor

As stated previously it was proposed to install a pollution monitor at Boat o'Brig.

This monitor was to be placed as far upstream of the wellfield as possible, but its location took into account the risk of pollution between the monitor and the wellfield. The site agreed was at Boat o'Brig, with the intake positioned immediately downstream of the railway bridge.

Backflushing of the wellfield
If a major pollution incident affects the river and threatens the wellfield, (hopefully a very rare event) the raw water reservoir at Badentinan would be required for backflushing of the wellfield to prevent contaminated water from entering the wells.

The reservoir capacity of $9000\,m^3$ would allow backflushing at an average rate of 8.7 l/s through each well for a period of 8 hours. It has been estimated that this will be a sufficient duration to allow any pollution incident to pass and to flush out any pollutants which may have entered the aquifer.

Monitoring boreholes
A possible threat to the wellfield caused by a deterioration in groundwater quality entering the aquifer has been identified, and investigations carried out to assess this threat have been discussed. Although there has been no evidence to suggest deteriorating groundwater quality, a series of monitoring boreholes on the western and southern boundaries of the aquifer have been installed to monitor the quality of the groundwater in the alluvial gravels. If there is a significant build-up of concentrations of contaminants over a period of time, then steps can be taken under the Water Act 1989 or the Control of Pollution Act 1974 to prohibit or restrict the activities which are causing concern.

ACKNOWLEDGEMENT

The authors are grateful to Mr J. M. T. Cockburn, Director of Water Services, Grampian Regional Council, for permission to submit this chapter.

REFERENCES

[1] Sir M. MacDonald and Partners (1975) *Water resources study*, North East of Scotland Water Board.
[2] Watt, G. D., Mellanby, J. F., van Wonderen, J. J. and Burley, M. J., (1987) Groundwater Investigations in the Lower Spey Valley, near Fochabers. *J. Instn. Wat. & Envir. Mangt.* **1**, 1, 89.

28

Groundwater abstraction and river flows

M. Owen, PhD, C. GEOL, FGS, MIWEM[†]

INTRODUCTION

Since the passing of the Water Resources Act in 1963 it has been the explicit duty of the appropriate regulatory authority of the day to 'secure the proper use of water resources'. With the passing of the Water Act 1989, this duty now rests with the National Rivers Authority.

Within the detailed provisions of the 1963 Act was a requirement to take into account the needs of the natural environment for water. This requirement was embodied in the concept of 'minimum acceptable flows' in determining which 'character of the inland water and its surroundings' had to be taken into account. This requirement was reinforced successively by the Water Act 1973, the Wildlife and Countryside Act 1981, and ultimately, and most strongly, by the Water Act 1989. There is now a requirement for any relevant body 'so far as may be consistent with the purposes of any enactment... so to exercise any power... to further the conservation of natural beauty, flora, fauna etc.'.

Combining this duty with the duty to manage water resources leads to the conclusion that the NRA should strike a balance between the use of water resources for water supply and for the support of amenity, wildlife and fisheries in the environment. With the recent legislation there is once again the distinction between responsibility for the overall management of water resources (the NRA), and the development of resources (mainly the water undertakers).

[†] Groundwater Resources and Licensing Manager, National Rivers Authority, Thames Region.

The legislation applies equally to the management of surface water and ground-water resources and recognizes the relationship between the two. It is with certain aspects of this relationship and its consequences for the river environment that this chapter is concerned.

TRENDS IN GROUNDWATER DEVELOPMENT

By the 1960s, after nearly a century of traditional groundwater development for direct supply, concern was starting to grow about the consequences of such abstractions on the depletion of river flow. The Water Act of 1945 had provided for some control of abstraction, but many sources had been established through earlier statutory rights. Although the coming into force of the 1963 Act provided the first comprehensive legislative and organizational framework to manage water resources, it did not provide for any curtailment of pre-existing abstraction rights without paying compensation, and that remains the case in the current legislation. These rights are now embodied in 'Licences of Right' under the 1963 Act.

Many of these existing rights included entitlement to quantities which were far in excess of need at the time they were obtained. As demand for water has grown substantially over the last forty to fifty years the actual quantities abstracted have grown considerably and are now often approaching the authorized limits. In some catchments, where the total reserved quantity is a large proportion of recharge to the aquifer and the water is largely exported after use, marked depletion of river flow has occurred. Whilst it is generally impossible to develop water resources without having some environmental impact, the balance between the needs of supply and the needs of the environment is not being achieved in these cases.

The development of such major river depletion was foreseen 30 years ago by Ineson and Downing [1] who set about examining new ideas for groundwater abstraction which utilized the delay effects of aquifer storage in a more imaginative way. Numerous schemes which utilize this characteristic in various ways have been implemented since then [2] and have resulted in groundwater resource development which is environmentally acceptable.

RELATIONSHIP BETWEEN GROUNDWATER ABSTRACTION AND RIVER FLOW

Classical theory of groundwater flow and well hydraulics has long recognized the effect of pumping a well on the flow of a river in an extensive continuous aquifer, for example Glover and Balmer [3], Jenkins [4]. Accepting the assumptions that surround these theories, the depletion of river flow due to groundwater pumping can be calculated.

Depletion builds up with time and increases more rapidly as the distance between the well and the river decreases (see Fig. 1); it is also dependent on the aquifer properties of transmissivity and storativity. Given enough time, the depletion will eventually equal the pumping rate; for wells within a few hundred metres of a river the bulk of this effect will usually happen within a year or two.

Depletion consists of two components, interception and induced recharge. Interception is groundwater flow that was on its way to the river and is diverted into the well. Induced recharge is water in the river which is induced to infiltrate the aquifer through the river bed by a hydraulic gradient to the well generated by pumping. A well close to a river (case 'A' in Fig. 1) will obtain its yield from a combination of these components, a well further away (case 'B' in Fig. 1) will obtain its yield only by interception.

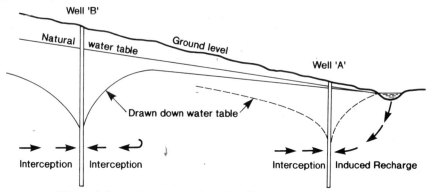

Fig. 1. Schematic cross-section of wells pumping along the valley
of a groundwater-fed river.

In many typical chalk valleys the wells are situated towards the head of the valley, near to the source of the perennial river (Fig. 2). The effect of pumping a well in such a location is to divert into the well the groundwater flow that previously fed the springs at the head of the river. As a result these springs tend to dry up, particularly when the water table is low, and the upper part of the river changes from perennial to intermittent flow or, in the extreme, may become virtually dry.

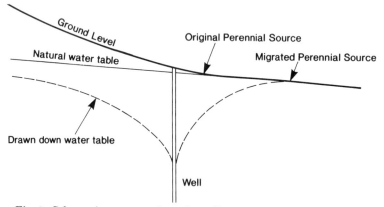

Fig. 2. Schematic cross-section of a well pumping near the source
of a groundwater-fed river.

These theoretical examples illustrate simplified conditions. Often the relationship between aquifers and their dependant rivers is hydrogeologically complex. The depletion effects may be much delayed by marked vertical components of flow in the aquifer as in the Permo-Triassic Sandstones [5], or may be selectively generated on certain sections of rivers due to variable lithology and geological structure as in the Cotswolds [6].

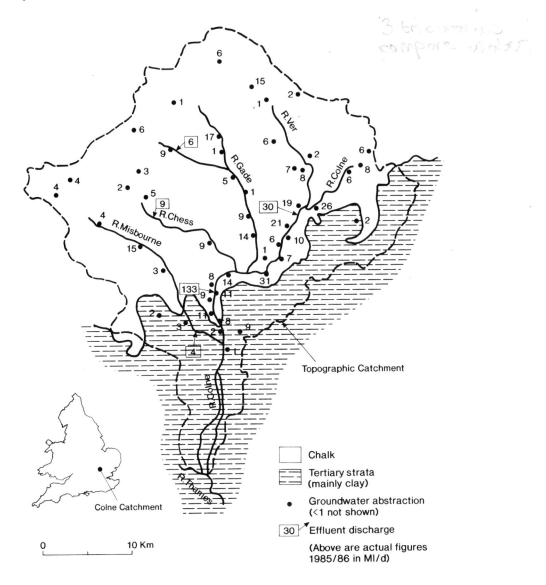

Fig. 3. Abstractions and sewage effluent returns in River Colne catchment.

THE COLNE CATCHMENT

The substantial growth of residential areas on the outskirts of London in the first half of the twentieth century was supplied with water mainly by extensive traditional direct supply abstraction from the chalk valleys of the Chilterns and North Downs. This development has had marked depletion effects on certain rivers exemplified by the Ver and the Misbourne, tributaries of the River Colne in the Chilterns (Fig. 3), and on parts of the Colne itself. However, other tributaries and other

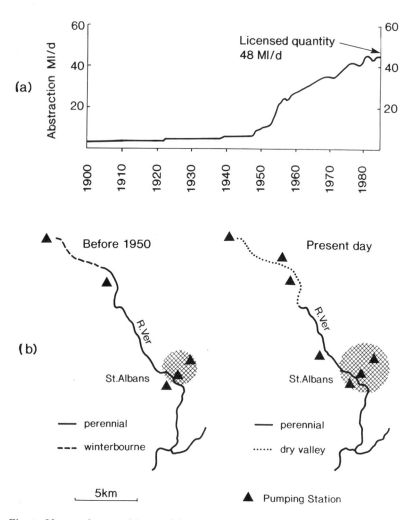

Fig. 4. Ver catchment: history (a) of abstraction, and (b) effects on river flow.

parts of the main river show little evidence of these effects and serve to illustrate how groundwater abstraction can be made in a more environmentally sympathetic way.

In the Ver groundwater catchment, approximately 75% of the average annual recharge to the chalk aquifer is allocated to licensed abstraction which is now almost totally taken up (Fig. 4a). Approximately half the water abstracted is used to supply areas outside the catchment; the remainder is used within the catchment and generates effluent which is returned to the river system much lower down the Colne. There is a major pumping station at the original perennial head of flow and others further downstream alongside the river.

The effects of pumping near the head of a perennial river, explained in the previous section, are well illustrated by the Ver. The upper 10 km of originally perennial or regular bourne stream are now normally dry (Fig. 4b). Further downstream, the remaining perennial section suffers much reduced flows and under low flow conditions, accretion gauging shows the depleting effects of riverside groundwater abstractions. This once typical chalk stream has suffered substantial environmental degradation with major shifts of habitat, loss of naturally sustained fisheries, loss of its watercress farming and reductions in the general amenity value of the river as an integral part of the rural and urban landscape.

In contrast, the River Chess, another typical chalk stream tributary of the Colne shows little of the problems of the Ver. Abstraction does take place but at a lower level, approximately 30% of average recharge to the groundwater catchment. Abstraction near the perennial head of the Chess is modest, used within the catchment and balanced by return of consequent effluent a short way downstream. The quantity of effluent is augmented by net imports of supplies into the upper catchment from external sources. Therefore reasonable flows are sustained from the head of the valley and in combination with natural accretion progressing down the valley, river flows are sustained throughout and there is no discernable environmental loss due to downstream abstraction.

The main valley of the River Colne itself is an object lesson in the effects of heavy groundwater abstraction on river flow. Over a distance of about 13 km between the confluences of the Gade and the Ver tributaries some 120 Ml/d are abstracted, resulting in substantial reduction of flow which becomes very obvious at low flows. The Blackbirds sewage treatment works was constructed in the early 1970s to augment flows by discharging approximately 30 Ml/d to the depleted section, an early attempt to alleviate low flows and improve the river environment. Further downstream, inputs from the major sewage treatment works at Maple Lodge, which serves the bulk of the Colne Valley, and from tributaries, produce a river with much higher base flows. In this area, abstractions totalling up to 80 Ml/d on average and up to 120 Ml/d at peak times over a distance of 9 km produce no adverse environmental effects even at low flows.

LOW FLOWS AND THEIR ALLEVIATION

For the last thirty years or so the effect of groundwater abstraction on river flows has been widely recognized and taken fully into account in new groundwater developments. With the implementation of the Water Resources Act 1963, these have been permitted only where the effects on river flow are insignificant or where the mode of operation or deliberate compensation discharges protect flows. There remains however the legacy of rivers like the Ver and Misbourne, markedly depleted by groundwater abstractions inherited from antecedent rights (Figs 5 and 6).

This issue was brought out into the open by the erstwhile Thames Water Authority in a study of six catchments around the London basin (Halcrow [7]). On privatization of the water industry the subject clearly came into the environmental arena and was taken up by the National Rivers Authority. It rapidly became clear that the condition was not restricted to the chalk streams around London. The Authority [8] carried out a preliminary survey which identified similar problems in up to forty catchments, mainly spread over the groundwater-bearing areas of the

Fig. 5. River Misbourne (Warren Water) at Great Missenden Abbey in February 1936.

Fig. 6. River Misbourne (Warren Water) at Great Missenden Abbey in March 1990.

south eastern half of England. This survey was carried out independently by the ten regions of the Authority and has recently been taken further by a national research project. The purpose of the project has been to produce commonality and consistency in assessing the cases nationally to aid in the formulation of a programme of alleviation schemes in justified cases.

Three approaches to alleviation can be considered:
(1) engineering schemes which allow the offending abstractions to continue by artificially maintaining flows;
(2) direct source replacement linked with revocation or reduction of the licences for offending abstractions;
(3) abstraction charges which persuade operators to use other sources through loaded tariffs.

The Halcrow study in the Thames Region considered the first two approaches. Direct source replacement was invariably the more expensive with £100 million identified as the approximate cost of replacing all the 180 Ml/d of sources contributing to problems in the six study areas. The recommended solutions in all cases were engineering schemes with a total capital cost of £10 million, but with a net present value of £26 million at 1987 prices when lifetime operating costs were included.

The schemes all involved manipulation of the use of groundwater locally but, in combination with other measures, in ways which would depend on some degree of artificiality in the system of restored river flow. In discussion with local interest groups, with professional conservation and fisheries people, and with NRA committees, this approach has (in Thames and other regions) come under increasing criticism for attacking the symptoms of the disease rather than the disease itself. There is thus a change of opinion developing generally towards a greater element of the second approach of finding alternative sources.

However, artificially sustaining flows may be appropriate in some cases. Two schemes, which do not directly use groundwater for augmentation, have been operating satisfactorily in the Thames Region for many years. One is Blackbirds STW, already mentioned in this chapter, the other is the Carshalton Ponds on the River Wandle at Sutton, Surrey, where recirculated river water is used to maintain springs which are affected by groundwater abstraction.

The third approach is speculative at present. The NRA is working towards a new abstraction charging scheme with a common national structure for implementation in April 1992. There is the possibility of including within this scheme an environmental impact factor whereby abstractions which have a serious impact on river flow would be charged at a higher rate than those which do not. In some cases there may be sufficient differential to persuade operators to move partly away from environmentally-sensitive abstractions through amendments to licences.

In the author's view there is no reason to say that any of these approaches is the only correct one. The most appropriate one will depend on the circumstances of each case and sometimes the answer may be a hybrid. The aim should be to produce the 'best practicable environmental option'. Whatever approach is taken there is the need for an objective method for evaluating the environmental requirement in setting prescribed river flows. Such a tool is needed in order to set meaningful targets for the restoration of flows; it is also needed to deal with new abstractions.

The provision of such a tool is being addressed through research projects entitled 'Ecologically acceptable flows' and 'Landscape appraisal methodology' in the NRA national R and D programme. These projects illustrate the interdependence of water resources management and a range of conservation factors. The former project, involving close collaboration between hydrologists and biologists, will take several years to complete and an interim subjective approach will be necessary to deal with some proposed cases for low flow alleviation. Through this and other work one of the fundamental concepts of the Water Resources Act, 'minimum acceptable flows', is now being tackled though perhaps an alternative term, such as 'river flow objectives', would be seen as less rigid and would allow for seasonal variation. In due course this should produce a consensus view on a reasonable environmental and economic balance in the proper use of water resources.

CONCLUSIONS

The interaction between groundwater abstraction and river flow and its implications for acceptable abstraction management are well illustrated in the major catchment of the River Colne in the Chilterns. In particular, the tributaries of the Rivers Misbourne and Ver show the serious flow depletion and consequent environmental damage caused by high levels of abstraction.

The issue of alleviating low flows in these and other catchments suffering a similar condition has been taken up by the NRA. It is the author's view that the solutions lie in a mix of three approaches: (1) engineering schemes to artificially maintain flows; (2) alternative sources; (3) loaded abstraction charges.

Whatever solution is chosen, it should recognize that a reasonable balance must be struck between the completing claims of water supply and the environment on limited water resources. In this context, alleviation schemes cannot be aimed always to restore pristine flow conditions. Arriving at a reasonable balance will ultimately need objective methods of assessing the environmental value of rivers.

REFERENCES

[1] Ineson, J. and Downing, R. A. (1964) The groundwater component of river flow discharge and its relationship to hydrogeology. *J. Instn Wat. Engrs*, **18**, 519–541.

[2] Owen, M., Headworth, H. G. and Morgan-Jones, M. (1991) Groundwater in basin management In: *Applied Groundwater Hydrology*. (R. A. Downing and W. B. Wilkinson, eds) Oxford University Press, Oxford.

[3] Glover, R. E. and Balmer, G. G. (1954) River depletion resulting from pumping a well near a river. *Trans. Amer. Geophys. Union*, **35**, 468–470.

[4] Jenkins, C. T. (1968) Techniques for computing rate and volume of stream depletion by wells. *Groundwater*, **6**, 37–46.

[5] Salmon, S. (1990) Flow mechanisms within a sandstone aquifer and the implications for groundwater resources development. *Proc. Water and Waste Water Conf.*, Barcelona, Spain.

[6] Rushton, K. R., Owen, M. and Tomlinson, L. M. (1991) The water resources of the great Oolite Cotswolds aquifer (in press).

[7] Halcrow, (Sir William Halcrow & Partners) (1988) *Study of Alleviation of Low River Flows Resulting from Groundwater Abstraction.* Thames Water Authority, Reading, UK.

[8] National Rivers Authority. (1990) *Extent and Impact of Over-Abstraction on River Flows.* London.

29

The protection of urban groundwater from pollution

D. N. Lerner, PhD, CEng, MICE, MIWEM[†] and J. H. Tellam, PhD[†]

INTRODUCTION

Groundwater has the same roles in urban areas as elsewhere, principally as a water resource and water supply, and as an environmental agent. In developed countries there are large volumes of groundwater beneath cities such as London, Liverpool, Manchester, Birmingham and Coventry. Many cities in the UK draw upon their own groundwater for public and industrial supplies, although these are often supplemented by surface-water schemes. These groundwater resources should be growing more valuable as the pressure on surface-water resources and rural aquifers increases, but their value is being diminished by actual or potential pollution.

In many other countries, urban groundwater is a vital water supply. In arid and semi-arid climates there may be no (affordable) surface-water source. In developing countries, the rate of growth of cities can outstrip the rate of provision of public water supplies. Many of the urban populations will then rely on shallow wells for their supply.

Groundwater resources are often higher in urban aquifers than equivalent rural areas, provided that water supplies are imported into the city. There are then additional sources of recharge which more than counterbalance any reduction of infiltration which might occur in covered areas. These sources include leaking water mains and sewers, soakaways for storm runoff and over-irrigation of parks and gardens [1–3].

[†] Lecturers, Hydrogeology Research Group, School of Earth Sciences, University of Birmingham.

All groundwaters have a role in the aquatic environment. Groundwater supported baseflow keeps rivers flowing, supports wetlands, and dilutes effluents. Urban groundwater is potentially as valuable as any other, as an environmental agent and as a water resource.

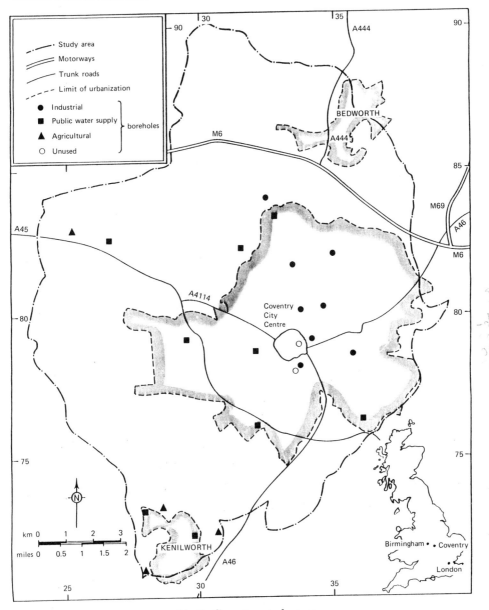

Fig. 1. Coventry study area.

IMPACT OF CITIES ON GROUNDWATERS

General review

Cities have impacts on both the quantity and quality of their groundwater. Discussions of changes to recharge, over-abstraction, and rising groundwater levels are left for another time. This chapter concentrates on groundwater quality. Table 1 reviews a few of the many studies which have shown that there is widespread deterioration of quality below cities. (A fuller review is given by Somasundaram et al. [4].) All the likely pollutants are involved: bacteria and other micro-organisms, major inorganic ions (NO_3, Cl, SO_4, etc.), trace ions such as heavy metals, and a wide range of organic chemicals. Of the organics, chlorinated hydrocarbon solvents (CHSs) are particularly common as they are widely-used, mobile, and resistant to degradation.

The remainder of this section is a brief description of three case studies – Coventry, Madras and Birmingham. These will illustrate the extent of urban pollution and the sources of pollutants as a prelude to a discussion of groundwater protection.

Coventry, UK Midlands

The quality of groundwater under Coventry is the subject of an ongoing study funded by the European Community (EC) and involving teams from the University of Birmingham, the Geological Survey of Denmark, and the Bureau de Recherches Geologiques et Minieres of France [8, 17]. The study area (Fig. 1) includes rural and urban areas, and is drawn to follow geological and topographical divides.

Table 1. Some urban groundwater pollution case histories

Author	City/region	Country	Determinands
Cruickshank et al. [5]	Merida	Mexico	Bacteria
Eisen and Anderson [6]	Milwaukee	USA	Cl, SO_4, bacteria
Ford [7]	Birmingham	UK	Majors, metals, B, P, Si, CN
Gosk et al. [8]	Coventry	UK	Org.
Kimmel [9]	Long Island	US	Metals, NO_3, Cl, Org.
Marton and Mohler [10]	Bratislava	Czechoslovakia	Oil
Razack et al. [11]	Narbonne	France	SO_4, NO_3
Rivett et al. [12]	Birmingham	UK	Org.
Roman et al. [13]	(Various)	Poland	Metals
Shahin [14]	Cairo	Egypt	NO_3, majors, metals
Sharma [15]	Bhopal	India	Majors, metals
Thomson and Foster [16]		Bermuda	Micro-organisms, Cl, NO_3

Abbreviations:
   Org. = organic determinands
   Metals = heavy metals
   Majors = Na, K, Ca, Mg, Cl, SO_4, HCO_3

The area is mainly composed of Permian and Carboniferous strata, overlain in places by glacial drift. The sequence is alternating sandstones and marls (mudstones) with occasional conglomerates. The discontinuous, lenticular units form a complex aquifer system extending down to 500 m in places. The aquifer is well-used for water supply with an average abstraction from the boreholes (Fig. 1) of 37 000 m^3/d. About 24 000 m^3/d of this is used for public supply and makes up 24% of Coventry's supplies; the remainder is surface water from Oldbury Reservoir and the River Severn.

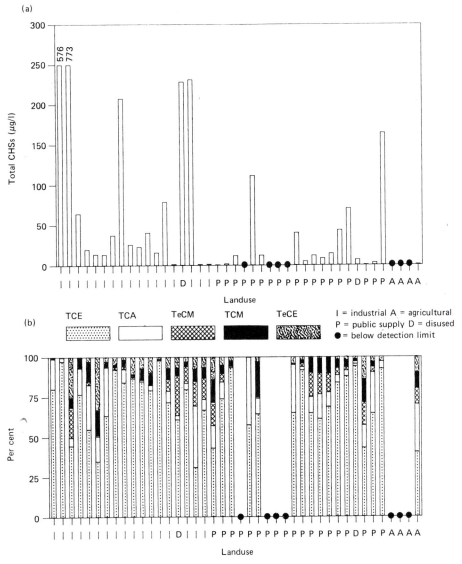

Fig. 2. Chlorinated solvents in Coventry groundwater: (a) total concentrations; (b) proportions of five major solvents.

The extensive marls in the system might be expected to protect the groundwater from major pollution, and it is true that there is little sign of serious inorganic pollution. Virtually all determinands (NO$_3$, Cl, SO$_4$, Ca, Mg, metals) are at higher concentrations in the urban area, but the levels are not excessive [18]. For example, the average nitrate concentration in rural areas is 7.6 mg N/l, increasing to 9.3 mg N/l in the industrial boreholes.

However, there is widespread and major pollution by CHSs, and Fig. 2 (a) shows that almost all urban boreholes are polluted with over 10 μg/l of total solvents. The five solvents listed in Table 2 have been found, and their relative distributions are given in Fig. 2 (b).

Table 2. Commonly found chlorinated hydrocarbon solvents

Abbreviation	Alternative names	Drinking water limit [19] (μg/l)
TCM	Trichloromethane or chloroform	—
TeCM	Tetrachloromethane or carbon tetrachloride	3
TeCE	Tetrachloroethene or perchloroethylene	10
TCA	1,1,1-trichloroethane	—
TCE	Trichloroethene	30

Although trichloromethane has been used by industry in the past, its widespread occurrence is probably a result of mains leakage, where it is found as a by-product of chlorination. Tetrachloromethane is surprisingly common, considering that it is hardly used today. The remaining three, tetrachloroethene, 1,1,1-trichloroethane and trichloroethene, are used in thousands of tonnes per year. They have some uses as solvents, but are mainly used for cleaning – drycleaning in laundries, metal cleaning and degreasing in Coventry's metal-working industries, and component cleaning in many new 'high-tech' industries.

The legal limits for these compounds in drinking water, where set, are given in Table 2. These limits are exceeded in one public water supply borehole (where the water is treated before going into supply), and at many industrial boreholes.

The ultimate source of the CHSs is the industrial sites themselves. The actual routes into groundwater are harder to determine, but visiting any site soon gives an impression of slapdash management of chemicals. Spillages, leaking pipework, deliberate disposal, leaking sewers and drains are all probable routes – and the losses of stock will be attributed to evaporation. There are several hundred sites using CHSs in Coventry, ranging from major motor manufacturers to 'under-the-arches printshops'. As every industrial site sampled has polluted groundwater below, it is probable that all the unsampled sites are also actual pollution sources. This story of widespread CHS pollution is repeated in Birmingham [20], Milan [21], and in the US [22] (to give some examples).

Madras, Southern India

Madras is a city of 3.8 million people on the eastern coast of India (Fig. 3(a)). It is built on thin marine deposits and estuarine and fluvial alluvium. Most of the water supply is imported surface and groundwater, providing about 70 1/head d, but a proportion of the population relies on shallow wells. These draw about 10 Ml/d from a water table aquifer in the alluvium within a few metres of the surface.

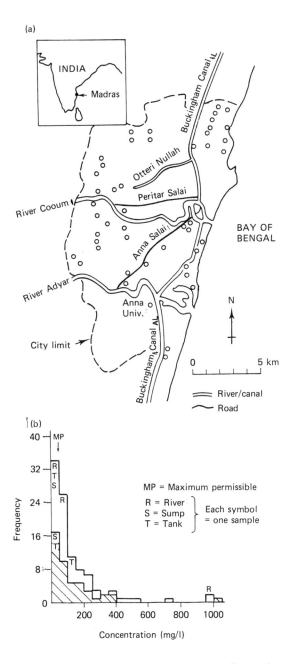

Fig. 3. Groundwater survey in Madras: (a) location of sampling points; (b) comparison of nitrate concentrations with Indian drinking water limits [4].

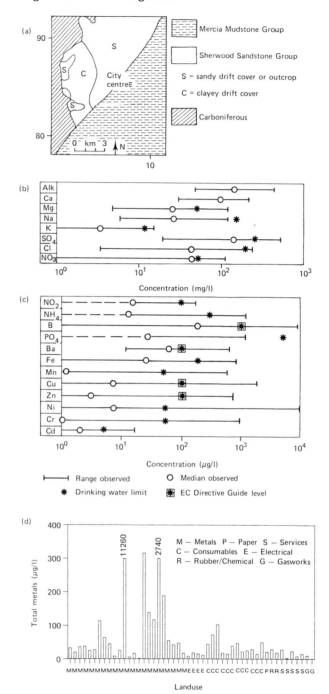

Fig. 4. Birmingham groundwater survey: (a) geology; (b) concentrations of major ions;
(c) concentrations of minor/trace determinands; (d) total concentrations of all
metals listed in (c).

The sewerage system is overloaded, and many of the hut dwellers have no access to sewers for disposal. The city rivers are often little better than open sewers. In addition there is a substantial population of animals. The many industries, ranging from large tanneries to roadside vehicle repairers, sometimes dispose of effluents to sewers but often only have access to soakaway drainage.

Surveys of groundwater and river quality were carried out in 1986-7 and 1989 at sites throughout the city, with a zone of more detailed investigation around the River Cooum (Fig. 3 (a)) [4]. Analyses were carried out for major inorganic ions (NO_3, Cl, SO_4, HCO_3, Ca, Mg, Na, K), fluoride and, for the river area, bacteria and metals. Inorganic quality was poor. Of particular note were nitrate concentrations which ranged up to 1000 mg NO_3/l (Fig. 3(b)). For comparison the Indian limit for drinking water is 45 mg NO_3/l [28]. Seventy per cent of samples exceeded this limit, and significant microbiological pollution was found where analyses were carried out. The probable source of both the microbiological and nitrate pollution is sewage, whether from leaking sewers, soakaways, septic tanks, direct application, or river infiltration.

Birmingham, UK Midlands

Much of Birmingham is situated on an unconfined Triassic Sandstone aquifer covered by a variable thickness of Quaternary clays and sands (Fig. 4(a)). The city has been an industrial centre for longer than most cities in the world, and in the past has used groundwater for both public and industrial supplies. However, problems in quality and quantity led to the abandonment of public abstraction, and in recent years the decline in industrial water use from 55 Ml/d in 1940 to 11 Ml/d in 1989 has resulted in rising groundwater levels.

Two recent surveys of the inorganic quality of the Birmingham groundwaters have been carried out [23, 24, 7]. Most sampling points are production boreholes, usually deeper than 100 m. Major ion quality is surprisingly good (Fig. 4(b)), and in terms of median values it is only NO_3 which comes close to the EC maximum admissible concentration (MAC) [7]. A similar pattern emerges for the trace metals (Fig. 4(c)), with few of the median concentrations even approaching the EC MAC or guide levels. However, the following factors are important.

(1) Severe metal pollution is only expected where the sampling site is near metal-working establishments. Fig. 4(d) indicates a close correlation of land use and groundwater quality: of the two consumables' sites with high concentrations, one is situated adjacent to a metal-working site, and the second occupies an old metal-working site.

(2) Sampling boreholes are deep, and hence higher concentrations are being diluted by deep, less-polluted groundwater.

(3) The Triassic Sandstones have high sorption capacity (e.g. Spears [25]) and hence metal loading must be severe enough to cause overloading.

(4) As Ford [7] has shown, pH levels are falling with time, and this will cause increases in metal mobility.

(5) Acidification may be related to rising groundwater levels but, in any case, saturated zone groundwater pollution can only increase as rising water levels sweep up through the disproportionately polluted unsaturated zone.

The pattern of groundwater pollution is dominated by the localized nature of the pollution sources. The effect is exacerbated by the low transmissivity and high storage coefficient of the sandstones, and by the sampling points being abstraction sites.

It is clear that even aquifers of high sorption capacity cannot be assumed to be safe from metal pollution and, if the acidification process postulated by Ford [7] is observed in other aquifers, the situation will deteriorate with time.

SOURCES OF URBAN POLLUTION

The potential sources of pollution are the same in an urban area as elsewhere, and include: (i) waste disposal, both in sanitary landfill and unsupervised; (ii) leaking sewers, septic tanks and latrines; (iii) spillages in transportation, storage sites and places of use; and (iv) effluent soakaways, and treatment lagoons.

However, if the sources of pollutants are the same, urban groundwater is noticeably more polluted because of (a) the density of potential sources, and (b) pollutant loads. The density of sources seems to overwhelm man's ability to regulate and monitor them. The high pollutant loads in cities often overpower the attenuation and dilution capacity of an aquifer.

The case studies outlined above, and reviews of other studies (Table 1), suggest that two problems can be identified in cities, (a) lack of sewers, and (b) industry. There are many recorded instances of sewers causing groundwater pollution [26], but they are generally localized and correctable. The contrast between Madras (on one hand) and Birmingham and Coventry (on the other) suggests that adequate sewerage will eliminate much of the pollution, particularly microbiological pollution posing an acute threat to health.

It is the authors' experience that all industries foul their own land, often grossly, and there is a close correlation between land use and the types of pollutants found in the underlying groundwater [7, 12]. There is no evidence that it is all a legacy from the past; as new factories are established or new chemicals introduced, so new pollution is found. Some service industries may be excluded, but not all. Dry-cleaners and petrol stations are proven polluters, and many other service industries use and mis-use chemicals.

PROTECTING URBAN GROUNDWATER FROM INDUSTRY

What is needed?
Given that, in general, sewage pollution can be removed by the installation of sewers, the remainder of this chapter concentrates on industry and how industrial pollution can be minimized. Action in the following areas is needed: (a) minimizing spillages and other releases of chemicals; (b) protecting groundwater from any accidental releases; and (c) detecting and remedying any pollution that does occur.

Preventing spillages
Industries use a wide range of chemicals in large volumes, and Table 3 illustrates the range of potential pollutants used in a particular engine-manufacturing plant.

The main interest on the site is making the engines. The chemicals are seen as tools or raw materials and, as long as they work well and do not harm the workforce directly, are of little interest. Thus there is little control on their handling, storage or disposal, except as needed to comply with health and safety rules. No wonder that the soil and groundwater beneath the site are polluted.

Table 3. Chemicals used in an engine-manufacturing plant

Location	Chemicals/processes
Pickle shop	Hydrochloric acid (before 1981, phosphoric acid)
Boiler house	Hydrochloric acid
Nitrocarburizing plant	Molten salt consisting of calcium carbonate, cyanate, cyanide, various trace elements
Planting shop	Hydrochloric acid, Zn-cyanide, Cu-cyanide, Co-Zn-alkali, nitric acid with chromates, sulphuric acid with chromates, Zn-phosphate, alkali cleaner (cadmium plating prior to May 1990)
Hardening	Cyanide
Alu-alloys	Chromate conversion for aluminium alloys (pre 1987)
Various	Degreasing, TCE, TCA, TeCE
Various	Cutting oils, fuels, greases, engine and gearbox oils
Playing field, gardens	Fertilizers, pesticides

Based on Table 14 in Nazari [18].

Preventing spillages requires a new attitude by industry. Environmental auditing is needed, under which inventory control, good handling and disposal practices, and continual monitoring and evaluation would be introduced. It seems unlikely that such new attitudes will arise without external pressures, and the environmental lobby is helping to change attitudes, particularly in North America. But few people can discover what happens on any site – the occupiers rarely know – and therefore monitoring must be made compulsory. This could be by a state agency (Her Majesty's Inspectorate of Pollution, National Rivers Authority) or by the occupiers themselves. In the latter case an open register of the results would have to be compulsory, with very limited rights to claim commercial confidentiality.

Protection from accidental releases
The best way to protect groundwater is to confine industry to impermeable geologies. Even this is not the perfect solution, as there is evidence from Birmingham of pollution penetrating 45 m of Mercia Mudstone [12]. This approach does nothing to cure the problem of existing factories, but helps to establish the ideal policy before tailoring it to fit the complications of real life. In many cases there will be countervailing technical and economic pressure to site industries on vulnerable aquifers. Planning restrictions on the design and operation of plants can help to protect aquifers in such cases.

Any area where pollutants are handled, stored, or used should be lined and drained to collection sumps. Soil gas and groundwater monitors should be installed at strategic locations in chemical handling and other areas. These provisions would contain spills and detect any failure of the containment. There are clear parallels with landfills, for which 'dilute and disperse' designs without monitoring are becoming increasingly untenable. Containment is now the norm for landfills and should become the norm for industrial sites, as there is more evidence for groundwater pollution from industry than from landfills.

Cleaning up

All factories on aquifers should have boreholes and be required to use groundwater for on-site processes. They would then be directly affected by any pollution on their site; if it damaged processes then they would have to pretreat the water. Their trade effluent licence would also control any off-site discharge of pollutants, whether from factory effluents or polluted groundwater. Licences should be tightened to disallow the discharge of substances not explicitly mentioned. Factories may even save money by not buying expensive mains water. There would be little impact on groundwater resources, as the increased industrial abstraction would be matched by a decrease in use of public water supply. This is similar to the suggestion that river intakes for factories should be placed downstream of effluent discharges [27].

There may be cases where natural groundwater quality is inappropriate for use on site. Accordingly standby boreholes should be installed for monitoring and clean-up.

The law

It is illegal to pollute usable groundwater. This statement is of course a gross simplification of the law, but there are sufficient powers available to the National Rivers Authority, the Secretary of State for the Environment, and planning authorities to control all the important pollutants. These specifically include many categories of organics and most metals under EC Directive 80/68, and poisonous, noxious or polluting matter under the Control of Pollution Act, 1974. The case studies discussed above make clear that pollutants are not controlled in practice. The reasons are twofold: (a) the lack of political will to recognize the problems or to act upon them, and (b) the lack of resources for monitoring and enforcement.

Engineers and scientists need to point out the extent of environmental pollution to their political masters, to demonstrate that there are technical solutions available to the monitoring and enforcement problems, and argue the need to develop remediation technologies.

Politicians need more than technical arguments to persuade them but, without hard technical evidence, all lobbying will be an uphill struggle.

CONCLUSIONS

1. Urban groundwater is, or should be, a valuable resource. Unfortunately it is usually polluted. The main culprits are (a) unsewered sanitation, and (b) industry. The microbiological pollutants from sewage carry the most serious threat to health, and the provision of adequate sewerage systems must remain a

high priority for all developing nations to protect the health of their population.
2. In the UK it is time to attack the problem of industrial pollution of groundwater. Three fronts are suggested for this attack.
   (1) *Preventing spills.* By monitoring and publishing environmental quality on
       industrial sites, occupiers might be encouraged to undertake environmen-
       tal audits and introduce proper controls on chemical handling;
   (2) *Protecting against accidents.* All new factories should be on impermeable
       terrain. This can be natural, such as non-aquifers, or artificial site liners;
       and
   (3) *Cleaning up.* All factories on aquifers should be equipped with boreholes
       for monitoring and clean-up. Wherever possible they should have to pump
       groundwater to supply their own water needs.

ACKNOWLEDGEMENTS

Many colleagues have been involved in collecting the data for this chapter. They
are too numerous to mention individually, but the authors' thanks are extended
to all. The Birmingham study was partly sponsored by WRc and NERC; the
Coventry Groundwater Investigation is funded by the EC; The Madras study was
partly funded by the British Council and Indian University Grants Commission;
the authors are grateful for their support.

REFERENCES

[1] Lerner, D. N. (1986) Leaking pipes recharge groundwater. *Ground Water*, **24**,
    654–662
[2] Lerner, D. N. (1988) Unaccounted-for water – a groundwater resource? *Aqua*,
    (1), 33–42.
[3] Lerner, D. N. (1990) Groundwater recharge in urban areas. *Atmospheric En-
    vironment*, **24B**, pp. 29–33.
[4] Somasundaram, M. V., Ravindran, G. and Tellam, J. H. (July 1991) Ground-
    water pollution of the Madras urban aquifer, India. *Submitted to Ground
    Water.*
[5] Cruickshank, G., Aguirre, J., Kraemer, D. and Craviota, E. (1980) Bacterial
    contamination of the limestone aquifer beneath Merida, Mexico. In: *Aquifer
    Contamination and Protection.* (Jackson, R. E., ed.)Studies and Reports in
    Hydrology, 30, UNESCO, Paris, 341–345.
[6] Eisen, C. and Anderson, M. P. (1980) The effects of urbanisation on ground-
    water quality, Milwaukee, Wisconsin, USA. In: *Aquifer contamination and
    Protection.* (Jackson, R. E., ed.) Studies and Reports in Hydrology, 30, UN-
    ESCO, Paris, 378–390.
[7] Ford, M. (1990) Extent, type and sources of inorganic groundwater pollution
    below the Birmingham conurbation. Unpub. PhD Thesis, Earth Sciences,
    University of Birmingham.
[8] Gosk, E., Bishop, P. K., Lerner, D. N. and Burston, M. Field investigation of
    solvent pollution in the groundwaters of Coventry. In: *Subsurface Contamina-*

tion by Immiscible Fluids. (Weyer, K. U., ed.) Balkema Publ., Rotterdam (in press).

[9] Kimmel, G. E. (1984) Nonpoint contamination of groundwater on Long Island, New York. In: *Groundwater contamination.* National Academy Press, Washington DC, pp.120–126.

[10] Marton, J. and Mohler, I. The influence of urbanisation on the quality of groundwater. In: *Hydrological Processes and Water Management in Urban Areas.* Proc. Int. Symp., Duisburg, FRG, pp. 453–480.

[11] Razack, M., Drogue, C. and Baitelem, M. (1988) Impact of an urban area on hydrochemistry of a shallow groundwater (alluvial reservoir), town of Narbonne, France. In: *Hydrological Processes and Water Management in Urban Areas.* Proc. Int. Symp., Duisburg, FRG, pp. 487–494.

[12] Rivett, M. O., Lerner, D. N., Lloyd, J. W. and Clark, L. (1990) Organic contamination of the Birmingham aquifer, UK. *J. Hydrology*, **113**, 407–423.

[13] Roman, J., Nowicki, Zb. and Olszewska, I. (1980) Trace-element-pollution of groundwaters in industrial-urban areas in Poland. In: *Aquifer Contamination and Protection.* (Jackson, R. E., ed.) Studies and Reports in Hydrology, 30, UNESCO, Paris, pp. 296–304.

[14] Shahin, M. M. A. (1988) Impacts of urbanisation of the Greater Cairo area on the groundwater in the underlying aquifer. In: *Hydrological Processes and Water Management in Urban Areas.* Proc. Int. Symp., Duisburg, FRG, pp. 517–524.

[15] Sharma, V. P. (1988) Groundwater and surface water quality in and around Bhopal City in India. In: *Hydrological Processes and Water Management in Urban Areas.* Proc. Int. Symp., Duisburg, FRG, 525–532.

[16] Thomson, J. A. M. and Foster, S. S. D. (1986) Effects of urbanisation on groundwater of limestone islands: an analysis of the Bermuda case. *J. Instn Wat. Eng. Sci.*, **40**, 527–540.

[17] Lerner, D. N., Gosk, E. and Bourg, A. C. M. (1991) Coventry groundwater investigation. Final Report on Contract EV4V-0101-C(BA), *Commission of the European Communities.*

[18] Nazari, M. (1990) Inventory of groundwater pollution of Coventry (UK): a study of major and trace element and some chlorinated organic solvent pollution. Unpub. MSc thesis, Earth Sciences, University of Birmingham.

[19] *Statutory Instrument 1147*, Her Majesty's Stationery Office, 1989.

[20] Rivett, M. O., Lerner, D. N. and Lloyd, J. W. (1990) Chlorinated solvents in UK aquifers. *J. Instn. Wat. & Envir. Mangt.*, **4**, (3), 242–250.

[21] Cavellero, A., Corradi, C., De Felice, G. and Grassi, P. (1985) Underground water pollution in Milan by industrial chlorinated organic compounds. In: *Effects of Land Use upon Fresh Waters.* (Solbe, J. F. de L. G., ed.) Ellis Horwood, Chichester, pp. 68–84.

[22] Nelson, S., Kalifa, S. and Baumann, F. (1981) Purgeable organics in four groundwater basins. *Water Forum '81*, American Society of Civil Engineers, 411–418.

[23] Jackson, D. (1981) Hydrochemical aspects of the Triassic Sandstone aquifers of the West Midlands. Unpub. PhD Thesis, Earth Sciences, University of Birmingham.

[24] Jackson, D. and Lloyd, L. W. (1983) Groundwater chemistry of the Birmingham Triassic Sandstone and its relation to structure. *Quart. J. of Engng. Geol.*, **16**, 135–142.

[25] Spears, D. A. (1987) An investigation of metal enrichment in Triassic Sandstones and porewaters below an effluent spreading site, West Midlands, England. *Quart. J. of Engng. Geol.*, **20**, 117–129.

[26] Short, C. S. (1988) The Bramham incident, 1980 – an outbreak of waterborne disease. *J. Instn Wat. & Envir. Mangt*, **2** (4), 383–390.

[27] De Bono, E. (1980) Lateral Thinking. In: *The Schumacher Lectures.* Blond and Briggs, London.

[28] Indian Standards Institution. (1983) *Water Quality Standards.*

Part V Reservoirs

30

The impact of reservoir drawdown on water-based recreation

A. I. Hill, BA, MSc, MRTPI, MIEnvSc[†]

INTRODUCTION

Water resource development, particularly involving reservoirs, has been seen scientifically and politically as a major arena for environmental decisions. It is no accident that the development of decision-making techniques such as cost-benefit analysis and environmental impact assessment has been prominent in this area. The environmental impact of the operation and management of water resource projects has received less attention. An important aspect of impact is the relationship between reservoir operation, drawdown and water-based recreation.

Water-based recreation continues to be a major area of growth in the expanding field of leisure activities. In many regions, heavy use of limited coastal areas and shortages of inland lakes have added to pressures on reservoirs. In addition to supporting active pursuits, reservoirs are a focus for those seeking quiet enjoyment of the countryside. Yet reservoirs, by their nature, suffer from fluctuations in water level and drawdown which have adverse affects on the very activities that the reservoirs attract. The importance of water-based recreation, its needs, and the adverse impact of drawdown, have had some professional recognition in the water industry [1]. It is on the range and significance of these effects that this chapter concentrates.

---

† Senior Lecturer, School of Environmental Technology, Sunderland Polytechnic.

The significance of drawdown impact on recreation is important, for there may be scope for modifying reservoir operation to manipulate drawdown. Increasingly, both new and existing reservoirs are being called upon to support fully a range of functions, and the trend is towards multipurpose operation. As more water supply sources are interlinked, flexibility of operation increases; supplies can be maintained with a choice of releases and of resultant storage levels at the reservoirs in a system. In the USA, recreational objectives are sometimes included in the economic derivation of reservoir operating procedures. Already in two UK reservoir systems, the Tees reservoirs, Northumbria, and the Dee reservoirs, North Wales, recreation is explicitly taken account of in operation. In both cases, 'recreation amenity lines' are used on control rule curves to reduce summer drawdown on recreationally important reservoirs [2]. The inclusion of secondary purposes in operation has been facilitated by the use of the '10-component method' [3] in control rule derivation.

DRAWDOWN EFFECTS ON RECREATION

Influence of reservoir operation

The satisfaction of the requirements of water-based recreation depends on both the provision and usability of specific facilities, and on the state of certain environmental variables. The capacity supportable is determined mainly by the quantity of water in terms of surface water area, shoreline length and water depth. Access depends on water level in relation to launching facilities, and the nature of the shoreline exposed by drawdown. Usability is influenced by a range of characteristics, according to activity, including minimum areas, submerged and exposed obstacles, muddy shorelines, and catchability of fish. Reservoir ecology, with implications for angling and possible effects from weed and algal growth, is strongly influenced by water level fluctuation and drawdown. Aesthetic effects are felt from visual shoreline changes. These environmental variables, with impacts on recreation capacity, access, usability, ecology and aesthetics are closely related to fluctuations in water level and drawdown resulting from reservoir operation. The influence of reservoir operation on these variables will depend to an extent on the characteristics of a particular reservoir. For instance, average bank slope will determine surface area and shoreline length lost due to a certain depth of drawdown. Likewise, shoreline slope and material will influence the effects drawdown will have on access and shoreline usability (Figs 1 and 2); Jaakson [4] emphasizes the influence of reservoir shoreline morphology. Other usability aspects and ecological and aesthetic effects will also vary according to reservoir characteristics.

Analysis of drawdown impact

The relative influence of the variables, and hence of drawdown, will depend on the recreation activities involved, and the extent to which these activities are water-dependent or water-enhanced. Water-dependent pursuits (angling, sailing, boardsailing, water-skiing, canoeing, rowing, sub-aqua) tend to have more exacting requirements; they also tend to be active and organized. Water-enhanced pursuits (sight-seeing, picnicking, walking) have less water specific needs, and are more passive and casual. Thus, for a particular reservoir, the impact of drawdown will

Fig. 1. Selset Reservoir: straight launching ramp (note the extension to permit continuation of use for severe drawdown; island hazards are emerging).

Fig. 2. Selset Reservoir: difficult access (note also the exposed islands; sailing has had to cease, but canoeing is less affected by the drawdown).

depend on both the characteristics of the reservoir and the recreational activities present; this relationship is illustrated in Fig. 3.

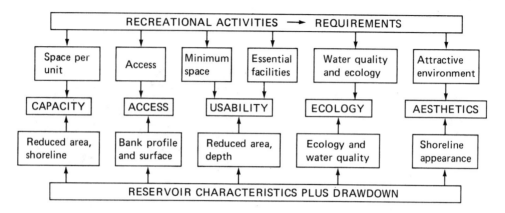

Fig. 3. Impact categories for drawdown impact analysis.

The impact categories: capacity, access, usability, ecology and aesthetics, have been used in the representation of impact in the matrix in Table 1. An earlier form of this matrix, including reservoir release impact on river recreation [5], has been adopted by Welsh Water [6]. The impact categories, recreational activities and impacts have been ranked in significance in Table 1 in the light of the three surveys discussed below.

The questionnaire surveys

It was decided to undertake a survey to find out more about the impact of drawdown on reservoir recreation, and more specifically to: (i) ascertain whether the effects considered in the original matrix are those that affect participants in the various activities; (ii) assess the significance of the impacts.

A direct approach to reservoir recreation participants was considered appropriate, with the use of questionnaires. Initially, the preferred approach was through personal interviews at reservoir sites. Unfortunately, it was not possible to obtain funding for a comprehensive survey, but limited pilot surveys (106 questionnaires completed) were carried out in the summers of 1985 and 1986. Postal surveys are relatively quick and cheap. A postal questionnaire survey of all known recreation clubs based on British reservoirs was undertaken, covering the summer of 1986. Most of the clubs were angling or sailing clubs; water-enhanced recreation was inevitably excluded because of its casual nature. The results of both surveys have been reported [7].

Conclusions from the site interview survey were limited because there was only slight to moderate drawdown at the survey sites; because of the numbers involved; and because almost all respondents were engaged in either sailing or water-enhanced activities. However, a pronounced distinction emerged between the views of the two groups, with 41% of those sailing aware of drawdown evidence compared with only 26% of casual users. Furthermore, only 3% of casual users had at some time experienced difficulties due to drawdown compared to 49% of those sailing; 19%

Table 1. Matrix of significance of reservoir drawdown effects

Variable / Activity	Access	Usability	Capacity	Ecology	Aesthetics
Sailing	Use of launching ramps† Awkward muddy approaches† Difficult moorings§	Reduced usable area† Muddy feet & dirty boats‡ Poorer wind conditions§ Disrupting modification of laid out courses§ Collisions with obstacles‡	Overcrowding through smaller area‡ Increased conflict with other activities§	Increased risk of algae§ (?) otherwise limited	Detrimental to scenery§ (?)
Other water sports	Awkward and muddy approaches† Problems with boat launching‡	Reduced usable area‡ Muddy feet & dirty craft‡ Problems with obstacles§ Poor visibility for sub-aqua§	Increased conflict with other activities§	Increased risk of algae§ (?) otherwise limited	Detrimental to scenery§ (?)
Angling	Awkward and muddy approaches† Use of launching ramps‡	Awkward stance‡ Lower angling success§ Increase catches, fish in smaller area (+) (?) Weed growth fouling lines§ Problems with boats on the water‡	Reduction in total fish available§ Increased conflict with other uses§	Long term productivity‡ Increased risk of algae§ (?)	Detrimental to scenery§ (?)
Casual water-enhanced	Muddy approaches§	Muddy shorelines§	Limited	Effects on wildlife§	Unnatural margins§ Exposed remains (+)

†Major and/or frequent impact; ‡Moderate and/or common impact; §Minor and/or last frequent impact; (?) Uncertain impact; (+) Positive impact

of casual users considered drawdown reduced their enjoyment of recreation, compared with 46% for sailing. The analysis of results from the 1986 postal survey has been combined with the results from the latest postal survey.

Table 2. Club survey questionnaire postings and returns

Questionnaires	1986 survey				1989 survey			
	Angling	Sailing	Other	Total	Angling	Sailing	Other	Total
Sent out (No.)	162	89	30	281	171	90	31	292
Usable returns (No.)	84	55	14	153	87	51	13	151
Usable response rate (%)	52%	62%	47%	54%	51%	57%	42%	52%

THE 1989 CLUB POSTAL SURVEY

The survey and response

The summer of 1986 was wettish, producing only limited reservoir drawdown, and it was decided to repeat the survey for a dry summer if the opportunity arose. This occurred with the drought year 1989. The club address list was updated, and the same structured questionnaire used. Throughout the analysis, and comparisons with 1986, the recreation categories used are angling, sailing and 'other'. The small numbers of 'other' (water-skiing, canoeing, rowing and sub-aqua) shown in Table 2 do not permit separation. The usable response rates of 54% and 52% are good by single probe postal survey standards [8]. Average club membership in 1989 was 145 for angling, 146 for sailing, and 84 for 'other'. Members' use of reservoirs was relatively intensive, with 58% of anglers and 62% of sailors using their reservoir at least weekly (although in the wetter 1986, 70% of sailors visited at this intensity).

Overall impact of drawdown

Clubs were asked to rate the overall impact of drawdown over the summer on their members' activities. An analysis of the results by activity is shown in Table 3.

A comparison of the results is shown in Fig. 4. Two key features emerge: (i) the greater significance of perceived effect on sailing and 'other' (where it is similar) than on angling; (ii) the greater significance of effect in 1989 than in 1986.

The relatively lower impact on angling is more marked in 1986 (20% moderate/significant for angling, 42% for sailing) than in the 1989 drought (48% moderate/significant for angling, 61% for sailing); on this basis, the impact on sailing was seen as double that for angling in 1986 but only 1.25 times as great in 1989. The difference between 1989 and 1986 is pronounced, with 53% of all respondents finding the effect moderate/significant in 1989 compared with 30% in 1986.

Table 3. Overall significance of drawdown effects

Effect of drawdown this summer on members activities	Angling %	Angling (No.)	Sailing %	Sailing (No.)	Other %	Other (No.)	Total %	Total (No.)
			1986		(wet summer)			
Significantly	12	(10)	22	(12)	29	(4)	17	(26)
Moderately	8	(7)	20	(11)	14	(2)	13	(20)
Slightly	24	(20)	20	(11)	36	(5)	24	(36)
Not at all	39	(33)	24	(13)	21	(3)	32	(49)
No noticeable drawdown	17	(14)	15	(8)	0	(0)	14	(22)
Total	100	(84)	101	(55)	100	(14)	100	(153)
			1989		(dry summer)			
Significantly	34	(30)	37	(19)	31	(4)	35	(53)
Moderately	14	(12)	24	(12)	23	(3)	18	(27)
Slightly	25	(22)	35	(18)	23	(3)	28	(43)
Not at all	22	(19)	4	(2)	23	(3)	16	(24)
No noticeable drawdown	5	(4)	0	(0)	0	(0)	3	(4)
Total	100	(87)	100	(51)	100	(13)	100	(151)

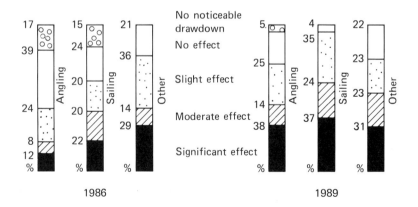

Fig. 4. Overall significance of drawdown effects.

Prevention of activities

Clubs were asked if any days of recreational activity had actually been lost through drawdown, and if so, the total number of days over the summer (May to September inclusive). The results are shown in Table 4.

Again, the distinction between angling and sailing/'other' is apparent. For both years, more than double the proportion of sailing/'other' clubs lost days. More than twice as many clubs lost days in 1989 than in 1986. The lost days for 1989 are not surprising, but even in 1986, a year of generally limited drawdown, 6% of angling clubs and 15% of sailing and other clubs lost days activities (although the drawdown in a few cases was exceptional, for repairs). In addition to those clubs losing days directly due to drawdown, some clubs lost; long periods of recreation due to problems with toxic blue-green algae; these have not been included in the figures, although it is possible that drawdown exacerbates the algae problem.

Specific problems

Clubs were asked to rate a series of particular problems, specific to their activities, as 'major', 'minor',or 'no problem'; the problems related to capacity, access and usability. The problems were largely those shown in the matrix in Table 1. Table 5 brings together the cumulative results for all problems. Thus the answers are for the aggregate of six questions for sailing and 'other', and five questions for angling (though only three questions for clubs without boats). As Fig. 5 highlights, the distinction between angling and sailing/'other' is even more apparent than that in Fig. 4. Major problems were twice as evident in 1989.

Table 4. Prevention of activities through drawdown

	Angling		Sailing		Other		Total	
	%	(No.)	%	(No.)	%	(No.)	%	(No.)
Were any day's recreation lost during summer 1986 due to drawdown?								
Yes	6.	(5)	15	(8)	14	(2)	10	(15)
No	94	(79)	85	(47)	86	(12)	90	(138)
Ave. No. of days where lost	—	(40)	—	(14)	—	(1)	—	(20)
TOTAL	100	(84)	100	(55)	100	(14)	100	(153)
Were any day's recreation lost during summer 1989 due to drawdown?								
Yes	13	(11)	33	(17)	38	(5)	22	(33)
No	87	(76)	67	(34)	62	(8)	78	(118)
Ave. No. of days where lost	—	(46)	—	(41)	—	(52)	—	(44)
TOTAL	100	(87)	100	(51)	100	(13)	100	(151)

The incidence of problems was also analysed according to category of impact (capacity, access, usability). The variations between activities shown in Table 5 were reflected fairly evenly across the three categories, although the level of angling problems relating to access were relatively closer to those for sailing and 'other' in both years. The difference between the two years emerged strongly, but was most marked for access related problems. This would suggest that for slight/moderate drawdown, a range of problems starts to appear, but that as drawdown increases, access constraints become rather more prominent. Considering the large reduction in surface area that even moderate drawdown can produce, lowering of capacity was not so highly rated. Membership numbers of most sailing clubs, usually with exclusive rights on a reservoir, are often limited in agreement with the water authority. It could be that limits are fixed to take account of reduction in capacity, with the result that, for much of the time, use is restricted to below capacity.

Table 5. Significance of specific problems overall

Significance of all listed problems	Angling % (No.)		Sailing % (No.)		Other % (No.)		Total % (No.)	
				1986	(wet summer)			
Major problem	7	(26)	21	(69)	23	(19)	15	(114)
Minor problem	22	(77)	37	(121)	31	(26)	29	(224)
No problem	71	(251)	42	(140)	46	(39)	56	(430)
TOTAL (1986)	100	(354)	100	(330)	100	(84)	100	(768)
				1989	(dry summer)			
Major problem	19	(72)	38	(115)	37	(29)	29	(216)
Minor problem	29	(108)	39	(117)	31	(24)	33	(249)
No problems	52	(195)	23	(71)	32	(25)	38	(291)
TOTAL (1989)	100	(375)	100	(303)	100	(78)	100	(756)

Although the survey did not specifically include interaction between recreation activities, several clubs commented that drawdown exacerbated conflict between activities, due to more interaction in confined space (in effect, a combined capacity

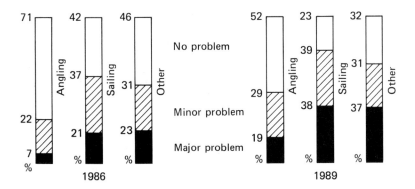

Fig. 5. Significance of specific problems overall.

limitation). Questions on aesthetic aspects were not included, due to individual subjectivity; however, several clubs mentioned that drawdown was detrimental to the scenery. It is not known how the significance of aesthetic aspects varies between the various water-dependent activities. (Figs 6 and 7.)

Fig. 6. Balderhead Reservoir: exposed margins. Wilder margins in the foreground, moderately steep in the middle distance; note the open, upland setting.

Fig. 7. Catcleugh Reservoir, Northumberland. Natural revegetation considerably softens the appearance of the margin; compare with Fig. 6.

For sailing, problems other than those listed included wind shadows or unpredictable wind patterns resulting from drawdown, and problems with mooring and foreshores. Danger from slackened moorings and the provision of new expensive moorings to avoid grounding were mentioned. Comments specifically relating to angling were more mixed. Some related to casting, such as anglers lines meeting from each side of reservoir, or the steep nature of bank causing major casting problems. Other comments included silting up of shallows forcing fish further out, and loss of caddis larvae and shrimp creating a major longer-term problem. Several angling clubs commented that the main effects of drawdown were not felt until the following season when the fish population declined significantly due to loss of margin flora and fauna. A few clubs noted that catches actually increased as the fish population was concentrated in smaller areas, although this might only be a short-term benefit. Conflicting views on the effects of drawdown on fish population and angling success reflect varying findings in research. However, Ploskey [9], in a major review of the research, concluded that deep drawdown and rapid fluctuations are detrimental. Angling clubs may not be in a position to interpret long-term drawdown effects, which may be more detrimental than the immediate effects. This may in part explain some of the differences between responses for angling, and those for sailing and 'other' where most detrimental effects are immediately obvious and generally shorter term.

EVALUATION OF IMPACT

Comparisons between surveys

The site interview surveys covered predominantly sailing and casual water-enhanced activities, whilst in the club surveys the emphasis was on angling and sailing with a smaller number of mixed active water-dependent pursuits. Hence sailing is the only activity common to both. This enables some limited comparisons. Fig. 8 is an attempt to compare the overall effect on the four activity groups, based on the question in each survey closest to an assessment of the overall impact of drawdown.

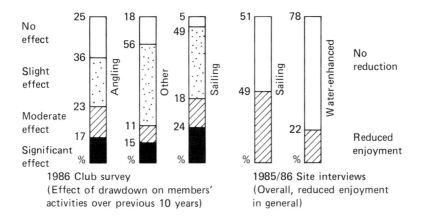

Fig. 8. Comparison of overall effect in the surveys.

In the postal surveys, clubs were asked to rate the overall effect of drawdown on their members' activities over the previous ten years, from 'significant effect' to 'no effect'; the responses in 1989 were very similar to those shown for 1986. In the site interview survey, respondents were asked 'And overall, would you say that such changes in water level at reservoirs in general reduce your enjoyment of recreation?'. The categories are not directly comparable having four and two subdivisions respectively.

However, looking at Figs 3, 4, and 5 together, a likely ranking of overall impact does emerge. It is reasonable to deduce that impact on angling is of a different order to that on water-enhanced activities, and no casual users in the site interviews had ever actually experienced difficulties. The conclusions on 'other' water-dependent and on water-enhanced activities are based on small samples. Thus, although it has not been possible from the surveys to quantify drawdown impact, a ranking, in descending order of impact on activities, is proposed:

(1) sailing (with less impact on boardsailing);
(2) other water-dependent sports in aggregate (though with variations between activities; canoeing and sub-aqua affected less);
(3) angling (though for some reservoirs, the long-term impact may be greater);
(4) casual, water-enhanced activities.

Further research is required, particularly on casual activities; a difference between impact on coarse and game fishing might also be established.

Evaluation of effects

No direct quantification of impact with regard to category of effects has been possible. The significance attached to problems did not show much variation between categories, the slight differences suggesting a ranking in order of priority of: access; usability; capacity. The relatively small number of comments made on aesthetic aspects suggests a lower order of priority for this category. Ecological effects are mainly important for angling, and are difficult to relate to short-term perceived impact.

The analysis of the surveys enabled a tentative three-point ranking of significance of drawdown impact for different recreational activities, categories of impact and specific problems. It would be inappropriate for the scale to be numeric, because the significance of an effect is related to several variables: (i) how severe it is when it occurs; (ii) how frequently it happens, in terms of numbers of drawdown reservoirs and numbers of participants affected; (iii) the scale of drawdown needed before it becomes a problem.

On point (iii) above, the morphological and other reservoir characteristics strongly influence the impact of drawdown at each reservoir. The ratings given in the matrix are a compromise between these interrelated variables.

CONCLUSIONS

The overall conclusion is that drawdown can have a significant detrimental impact on most forms of water-based recreation. The significance depends on the characteristics of individual reservoirs, the nature of the activity and the scale of drawdown. It has not so far been possible to quantify recreation benefit loss due to drawdown, but more has been learned on the effects of drawdown and the reactions of recreationists to these.

It has been shown that even the slight to moderate drawdown experienced in Britain in 1985 and 1986 did detract from enjoyment, though few days were actually lost. The effects were much more noticeable in the 1989 drought with many days' activities lost, many problems and much frustration. Sailing and other boat-based sports suffered most, but angling problems increased with more drawdown. The longer-term ecological effects on fish populations may still be a major factor. The impact of any but severe drawdown on casual water-enhanced activities may be more limited.

Sufficient evidence has emerged from the surveys to support the recommendation that more attention should be paid to the requirements of water-based recreation, and the impact of drawdown on these, in the operation of reservoirs. With increased flexibility from interlinked sources, it should be possible to manipulate drawdown regimes to the great benefit of recreation, and possibly at small cost to other reservoir functions.

In moving towards the derivation of multipurpose operation, both the mix of recreational activities and individual reservoir characteristics should be taken into account. Thus, water levels for each reservoir significant to key recreation interests could be established, and used either with economic loss functions, or as constraint lines in operation. Ameliorative measures, such as special launching facilities, beach replenishment and margin revegetation can help to some extent, but cannot be really effective alone without recreation-sympathetic reservoir operation.

ACKNOWLEDGEMENTS

The author is grateful to John Mawdsley of the University of Newcastle upon Tyne for his advice, to the water authorities and companies for providing club addresses, and to all the club secretaries and others who completed and returned questionnaires.

REFERENCES

[1] Dangerfield, B. J. (ed.), *(1981) Recreation: water and land*, Inst. Wat. Eng. & Sc., London, England.

[2] Hill, A. I. and Mawdsley, J. A. (1987) The application of recreation amenity lines in UK reservoir operation: case studies of the Dee and Tees schemes, *Proc. British Hydrological Society National Symposium*, Hull, England.

[3] Lambert, A. (1988) *An introduction to operational control rules using the 10-component method*, BHS Occasional Paper No. 1, British Hydrological Society/Institute of Hydrology, England.

[4] Jakson, R. (1970) *Wat. Resour Research*, **6**, 421.

[5] Hill, A. I. (1987). Allowance for water-based recreation in multipurpose river basin management. In: *Water for the Future* (Wunderlich, W. D., Prins, J. E. and Balkema A. A., eds.), Rotterdam, Netherlands.

[6] Stokes, R. K. (1987) *Regulated rivers*, **1**, 354.

[7] Hill, A. I. (1988). Reservoir operation and impact of drawdown on water-

based recreation: the assessment of user reactions. *Proc. VIth IWRA World Congress on Water Resources*, IV, Ottawa, Canada.

[8] Davidson, J. (1970). *Outdoor recreation surveys*: the design and use of questionnaires for site surveys, Countryside Commission, Cheltenham, England.

[9] Ploskey, G. R. (1983). *A review of the effects of water-level changes on reservoir fisheries and recommendations for improved management*, National Technical Information Service, Springfield, Va., U.S.A.

31

Management of Derwent Valley reservoirs

N. H. Groocock, LRSC, MIWEM, and M. S. Farrimond, BSc, PhD, FIWEM[†]

INTRODUCTION

Background

The series of three reservoirs in the Upper Derwent Valley (Ladybower, Derwent and Howden) are the source waters for treatment to provide the water supply, either wholly or in part, to over one million people in the East Midlands. The catchment is mainly open moorland with some afforestation around the reservoirs. However, a small number of sheep and cattle farms are situated on the lower slopes and in the valleys, and about 23 km of roads run alongside the reservoirs and feeder streams. Over half of this length is the A57 Snake Pass road which is one of the major thoroughfares between Sheffield and Manchester (Fig. 1).

Additionally, water which is brought by tunnel from the River Noe in the adjacent Edale Valley into Ladybower drains a further 6 km of road, and carries the effluent from four small community water reclamation works, including the tourist centre at Edale as well as the drainage from several more farms.

Water Use

About 80 Ml/day are released into the Rivers Derwent and Noe as compensation water. The release may be varied at time of drought or to allow more water to be abstracted at points downstream. About 40 Ml/day of raw water are supplied

[†] Process Development Officer, and Technology Development Manager, respectively, North Derbyshire Disrict, Severn Trent Water Limited.

through a tunnel to Riveline Reservoirs near Sheffield. About 200 Ml/day are available for treatment at Severn Trent's Bamford Works where treatment to drinking water standard is provided, comprising coagulation, upflow clarification, filtration, pH adjustment and disinfection.

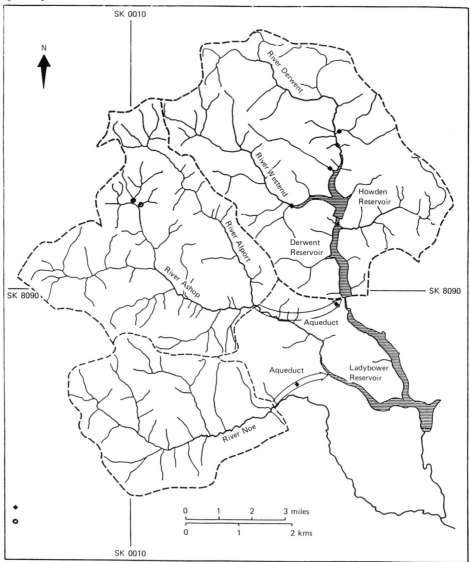

Fig. 1. The study area.

Problems

Being a sparsely populated upland catchment little faecal pollution of the water occurs, apart from the contribution from the Edale resource. However, potential runoff from sheep dip and other farm waste remains, and with the sensitive methods now available for detecting pesticides and herbicides in water even the massive dilution available for any release of material may not in future avoid the presence of the contaminant being detected. Algae are rarely found in significant numbers in the three reservoirs, although in 1976 and 1990 low numbers of blue-green algae (Oscillatoria) were noticeable in Ladybower despite the very low nutrient concentrations present. The main problem associated with this water source concerns the presence (and subsequent requirement for removal) at certain times of the year of highly coloured humic substances extracted from the peat, and the associated iron, manganese and aluminium compounds. The rate of leaching, and hence the concentration of these materials in the water passing to treatment, is variable from season to season and from year to year, and the treatability and costs associated with removal varies similarly. Fig. 2 shows the colour trends in the three reservoirs, and Fig. 3 shows the trend for Howden Reservoir in more detail. Fig. 4 indicates the relationship between the colour in the raw water and the dose of coagulant required to remove it.

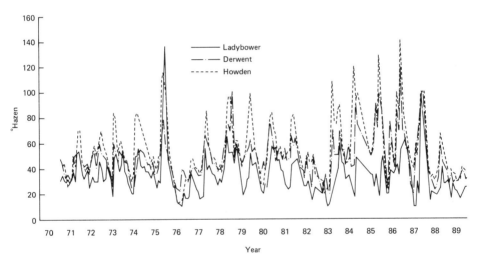

Fig. 2. Colour trends, Derwent Valley Reservoirs.

To cope with this variable raw water quality and to improve the quality of the water treated, in 1988 a new plant costing in excess of £5 million was commissioned. The improvement in colour, iron and manganese removal is shown in Fig. 5. However, should the quality of the water flowing from the catchment deteriorate significantly from that previously experienced, further treatment provisions might be necessary in the future.

Objectives of the work carried out

In 1987 in order to try to obtain some basic information about the catchment, and in particular what factors might affect the future treatability of the water, the Water Research Centre were commissioned to investigate, with the following objectives.

- To identify the processes occurring in the catchment which are involved in the formulation of coloured water, and the extent to which this problem will persist in the longer term.
- To identify the processes which affect the nature and presence of colour within reservoirs.
- To recommend means of management of the catchment and reservoirs to minimize colour in the raw water abstracted for treatment, and potential pollution from other catchment activities.

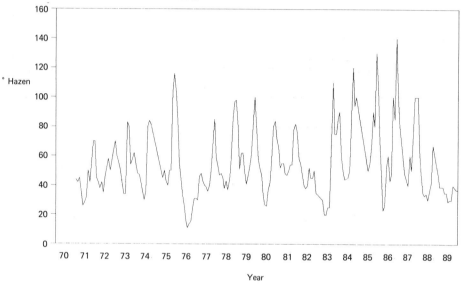

Fig. 3. Colour trends, Howden Reservoir.

The Soil Survey and Land Management Centre was subcontracted to carry out a study of soils and erosion, and the School of Geography at Leeds University was engaged to study catchment land use, vegetation, morphology, lysimetry studies and snapshot sampling.

This chapter summarizes the above work, and is based on the subsequent report produced by R. L. Norton and C. P. Mainstone of the Water Research Centre [1]. Other related works [2,3] were also carried out at that time.

GEOLOGY, PHYSIOGRAPHY AND CLIMATE

Geology of the Area

The study area was the catchments of the Rivers Noe, Ashop, Alport, West End and Derwent, these being the rivers which provide most of the water for the three reservoirs (Fig. 1). Peat is the main superficial deposit and occurs principally on the high moorland parts of the catchments. The only other significant deposits are landslips which occur mostly in the Ashop and Alport catchments.

The underlying rocks in the area are Carboniferous strata forming the Millstone Grit series. In the River Noe area the underlying shales of the series rest on Carboniferous limestone and although the latter does not appear at the surface, thin

bands of limestone within the surface shales result in the water passing to Ladybower from the River Noe being marginally harder than the remaining catchment
drainage, and the pH being closer to 7. The sandstones of the Grit series which
predominate in the North part of the catchment are very hard and of low permeability, and because the main constituent is quartz, which is of low solubility, the
water is very soft (Fig. 6). The dissolved silica content is significant, however, and
in a more nutrient rich water might give rise to blooms of diatoms which rely on
silica to provide their skeletal framework.

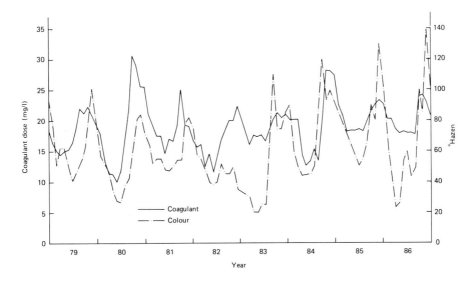

Fig. 4. Colour and coagulant relationship.

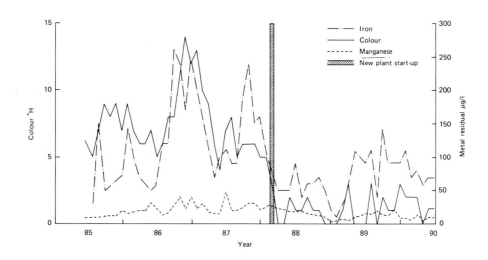

Fig. 5. Bamford final water quality.

Little water is held in storage within the rock and this is evidenced by there being only a few springs in the area, particularly at the shale/gritstone interface where they might logically be expected to rise.

Physiography

In order to study the catchment in detail it was divided into 32 subcatchments. Randomly-selected points over the area were examined for slope elevation and aspect (degrees deviation from North). Additional points were added where sub-catchments were not considered to be adequately represented. This work was used to determine the steep slope areas of the incised drainage network, and the areas of shallow slope or plateau where subsurface water would be flowing slowly through the peat, picking up soluble materials including colour. A further characteristic occurs where a higher plateau provides flows to a lower plateau, waterlogging the peat thereon and forming a 'flush peat zone' [4].

Climate

The rainfall data across the study zone were obtained from the Meteorological Office figures quoted in the Moorland Erosion Study [5] (Fig. 7). It indicates that 1975, then 1971, 1976 and 1964 were the driest years. The years 1986 and 1965 were particularly wet.

CATCHMENT SOILS
Methods

Soil survey and analysis
A soil survey of the study area was conducted by recording soil properties at 125 m inervals along transects crossing each of the main catchments. At each point the soil was augered to a depth of 1 metre or to the base of the peat if this was deeper. The area of each soil type was measured using a digitizer. At every fifth observation point along the transects a small pit was dug, and samples obtained for analysis. The latter comprised loss on ignition, pyrophosphate extract [6], colour and pH. The pH values were lower than expected according to the soil survey and were all within the range 2.9 to 3.9.

Peat Erosion

A review of earlier work
A number of studies have been carried out over the years into peatland erosion. The most significant in respect of this exercise is the Moorland Erosion Study carried out by Phillips *et al.* [5] on behalf of the Peak Park Board, and to which Severn Trent contributed. This indicated areas which have been subject to erosion from changes in land management practices, public access and atmospheric pollution.

In their synopsis the authors concluded that 8% of the Peak moorlands are now bare ground. The extent of the erosion increases with altitude, showing it to be a

Fig. 6. Simplified geology of the study area.

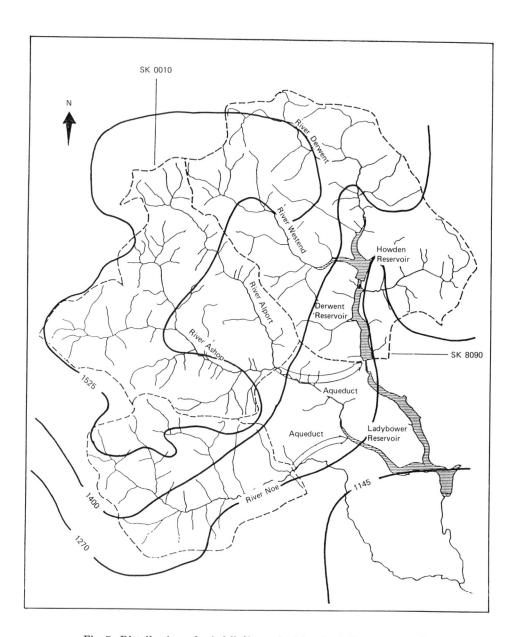

Fig. 7. Distribution of rainfall (from the Moorland Erosion Study).

key factor. The worst affected areas as far as this study was concerned were in the upper parts of the Noe, Alport and Ashop catchments, where many areas appeared with the surface bare of peat, and the mineral soil exposed (Fig. 8).

Although there was some evidence that atmospheric pollution has played a part in the erosion, this was by no means conclusive. There was also evidence [7] from pollen studies that erosion had been a feature of the peat blanket for over a thousand years. Other factors affecting erosion include uncontrolled burning and overgrazing [8]. The improvement of soil drainage by ploughing furrows and drainage grips also has an effect as does the furrowing of wet moorland when planting trees, and if this is aligned downhill it can create a particular risk [9].

One study on revegetation [10] trials at Doctors Gate, which is within the study area, concluded that revegetation of extremely eroded peat soils was very difficult and that therefore prevention was better than cure.

A further study into the effect of sheep grazing intensity on erosion [11] in Hey Clough, suggested that a fall in sheep numbers was linked to revegetation of scars on mainly mineral soils, but Phillips *et al.* [1] were somewhat inconclusive about the role of grazing on the initiation of erosion.

Air photographs

Two sets of air photographs, one taken in 1972 and one in 1984 were chosen to assess the occurrence and extent of the various forms of erosion, and the changes between the two dates. Although the quality of the latter survey is insufficient to allow finer details of erosion to be assessed, it was apparent that although there had been some progression of already existing erosion areas, the only major new area was that related to the Pennine Way.

Fig. 8. Eroded peat on Bleaklow Moor.

Soil associations

The soils of the catchment were grouped for convenience into associations and mapped according to vegetation patterns on air photographs.

STREAM WATER QUALITY

Routine stream monitoring

Sampling of six major streams feeding the three reservoirs was undertaken either manually, on a two-weekly basis, or automatically, on a daily basis, as follows:

River Noe   manual
Ashop Aqueduct   manual + automatic   from 17.09.87
                     —   to 28.11.88
River West End   manual + automatic   from 17.09.87
                     —   to 22.04.88
Linch Clough   manual + automatic   from 17.09.87
                     —   to 17.01.88
River Derwent   manual + automatic   from 17.09.87
                     —   to 18.03.89
Cranberry Clough   manual

Linch Clough flows from west to east into the northern arm of Howden Reservoir
and Cranberry Clough stream flows from east to west into the River Derwent to
the north of Howden Reservoir.

Samples were filtered through a 0.45 micron membrane prior to colour and metals determination. Samples for total organic carbon (TOC) were not filtered but
allowed to settle. Electrical conductivity (EC), pH and temperature were measured
on-site.

The general conclusions were as follows:

(a) The overall patterns of colour behaviour are generally similar for all streams,
    with the possible exception of Cranberry Clough. Differences exist in the
    magnitude of the flushes and this may be reflected in the localized rainfall
    as well as differences in catchment characteristics.

(b) A strong relationship between colour and TOC exists for all streams.

(c) A weak negative relationship exists between colour and pH for all streams
    except Cranberry Clough and the River Noe.

(d) There is a correlation between colour and iron for all streams except the
    River Noe.

(e) A relationship exists between colour and aluminium but is weak for Cran-
    berry Clough and the River Noe.

(f) There is no relationship between colour and calcium although a weak neg-
    ative relationship is evident for colour and magnesium. An attempt was
    made to fit an explanatory statistical model to predict colour runoff in the
    Upper Derwent. The best fit using daily data was found for the following
    parameters:

	Cumulative percentage variance accounted for
Effective rainfall	14.6%
+ soil moisture deficit (with no lag)	23.4%
+ flow	28.7%

Information over at least a decade would probably be required to furnish a better
explanation. Several factors contributed to this poor accountability. Firstly, the

daily colours were a single value measured at noon, and did not necessarily reflect the colour associated with peak runoff. Secondly, there is a time lag between colour and flow during a hydrograph, which can be up to eight hours. This lag cannot be modelled using daily single point data.

Snapshot sampling exercises

The purpose of snapshot sampling was to obtain water quality data under high and low flow conditions for a large number of subcatchments in order to correlate colour levels and metal concentrations with catchment characteristics. The results indicated that the Edale catchments appeared to be lower yielding than any of the other catchments, and that the Alport and Ashop catchments generated the highest colour yields. Some subcatchments such as Howden Clough had very low yields whereas others such as North Grain, Doctors Gate and Birchin Clough had consistently high yields. Similar conclusions were drawn from the metal results. Especially high iron levels were found in the Upper Ashop subcatchments.

RESERVOIR WATER QUALITY

The reservoirs were sampled during November 1987 and May 1988. Each reservoir was sampled 1 m below the surface along its length at 500 m intervals for Howden and Derwent and at 1000 m intervals for Ladybower. Several depth samples were taken at selected sites. Samples were treated in a similar way to those for the streams. Some sites were not available in the May exercise due to the lower reservoir levels (Fig. 9).

The November samples indicated, as expected, that the reservoirs were well mixed. Also, as expected, the colour and TOC decreased in the order of Howden, Derwent and Ladybower while the pH increased. Additionally, a trend along Ladybower was accounted for by the overspill from Derwent Reservoir, while the influx of the River Noe water was evident adjacent to the tunnel entrance. One set of data in Derwent Reservoir was anomalous but thought to reflect the ingress of water from one of the few springs on the catchment. Colour levels measured in May were approximately half those obtained in November but remained higher than those current in the steams.

STORM EVENT SAMPLING

At Slippery Stones on the River Derwent, north of Howden Reservoir, an automatic sampler was installed which was triggered by a float switch when the stream rose to a predetermined level. A number of storm sequences were sampled during the latter half of 1988. In the six events which were used for analysis, the pattern of colour variation followed flow closely, but with a peak lagtime of up to eight hours. The declining limb of the colour graph was often confounded by further rainfall events and by the lack of data due to the limited capacity of the sampling machine to cope with more than 24 hourly samples. However, it was evident that the colour appeared to take longer to return to the pre-event level than the flow.

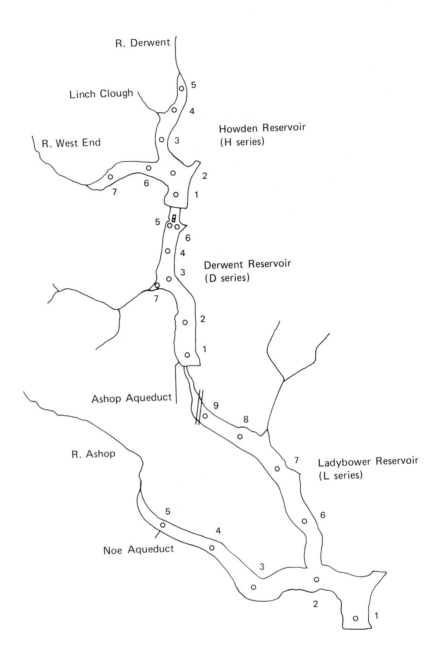

Fig. 9. Location of the reservoir sampling points.

In terms of colour contribution to the reservoir system, in all cases pre-event baseflow was low in colour, typically 1 to 2 Au/m[†], and returned to relatively low levels soon after the main body of the event had passed. Enhanced colour was therefore only being delivered to the reservoirs during periods of runoff. Based on crude estimations of runoff volumes, over the period July to December 1988, if an average of 23% of effective rainfall was manifest as stream flow, at a conservative estimate approximately 2.3×10^6 m^3 of surface runoff flowed down the Upper Derwent, compared to approximately 5.7×10^6 m^3 of baseflow over the same period. These figures highlight the importance of the surface runoff in supplying the reservoir system during the 'Autumn flush' period.

The low colour of the baseflow is perhaps surprising. As there is only a small potential for groundwater flow within the catchment, it may be assumed that the baseflow originates as slow release from the peat blanket. Comparison of rainfall with baseflow indicated that contact and hence leaching time within the peat was considerable. The lack of colour may be due to resorption by deeply humified lower peat layers prior to release.

COLOUR LEACHING AND GENERATION PROCESSES

To achieve a standard test of the leaching effects on peat materials found in various parts of the catchment, cores were obtained and the work carried out under laboratory conditions. The cores were allowed to dry under normal atmospheric conditions and the rewetting was carried out using a needle-based rain droplet system.

The results indicated that in all cases the colour intensity of leachates from the eroded samples was greater than that from vegetated samples.

The colour levels derived from the longest dry storage (150 days) were more than double those from the 50 and 100 day storage periods. Indications are, however, that subcatchments in the Derwent area would rarely be subject to drying for 150 days and that indeed most would be fully saturated for 250 to 300 days per year. Colour did not appear to relate to pH, but the pH differences found in this exercise were small.

USE OF STUDY RESULTS

Colour generation and release

Observations have confirmed that stream colour levels decrease from north to south within the catchment, with the River Noe being the least coloured of the four major streams. The seasonal pattern of high colour in late summer and autumn was also confirmed, although in this study the high colours persisted into the winter months, possibly due to the milder than normal conditions.

Up to 73% of colour variance between subcatchments could be explained in terms of catchment characteristics, of which the most significant was the total area of peat. High colours also appeared to be related to height, plateau area, stream length and drainage density, although this was less certain due to cross-correlation

---

[†] Absorbance units/metre at 400 mm wavelength.

between parameters. Nevertheless, this confirmed the observation that the most highly coloured streams were to be found in high altitude peat catchments with a high degree of dissection.

The release of colour to streams was directly related to flow, and controlled by rainfall intensity and antecedent soil moisture deficit. Stream water colour was successfully modelled using a simple mixing model based upon flow and runoff. It indicated that colour in runoff during a hydrograph remains fairly constant but decreased with successive storm events throughout the late summer and autumn. The lysimeter studies indicated that dry periods of greater than 100 to 150 days prior to rainfall were required for generation of higher than average colours.

Fig. 10. Ashopton aqueduct inlet to Derwent Reservoir.

The study period was likely to be one of average colour production, as there were only 32 days of soil moisture deficit in 1987 and 60 days in 1988.

Metals release
With the exception of the River Noe data there was a strong relationship between iron and colour, indicated that iron release could be predicted from colour levels. A similar relationship existed for colour and aluminium, except for Cranberry Clough. There appeared to be no relationship between colour and manganese.

Implications for reservoir management
The seemingly obvious choice of using as much of the better quality water within the catchment, namely that flowing into Ladybower, is not valid in practice. Firstly this water has to be pumped to treatment, which is not the case for water from Derwent and Howden Reservoirs. There is therefore a cost penalty. Secondly, failure to draw water from the other two reservoirs would lead to their overflowing into Ladybower with consequent acidification and colouring of the water there. This might also cause a problem with the fishery on Ladybower. However, the current inflow of highly coloured Ashop water into Derwent Reservoir adjacent to the dam and the short-circuiting of this to treatment needs to be addressed, and this flow

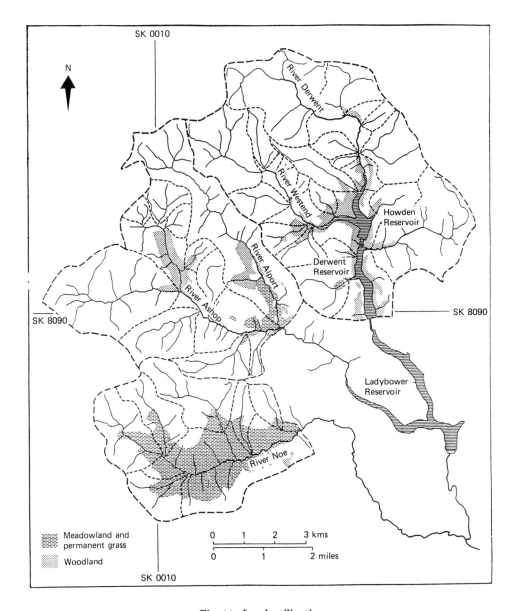

Fig. 11. Land utilization.

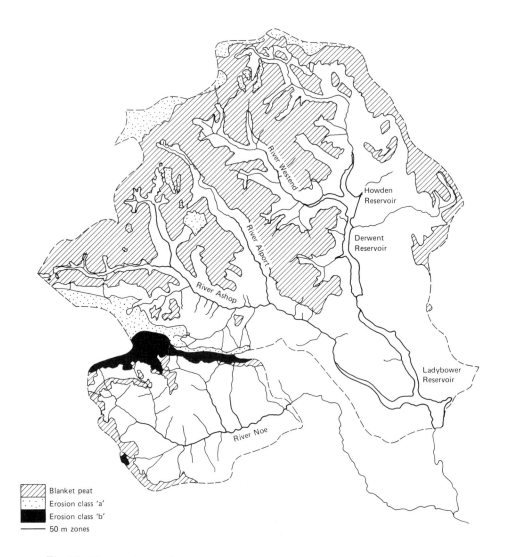

Blanket peat
Erosion class 'a'
Erosion class 'b'
50 m zones

Fig. 12. The catchment boundary showing the peat blanket and critical zones.

either diverted to a point further along the reservoir bank, where better mixing will take place, or an artificial air induced mixing system installed (Fig. 10).

Implications for catchment management

The utilization of the main areas of land in cultivation is shown on Fig. 11.

Forestry

Most of the forestry land is situated in the valleys, and the plantations are in most cases fairly mature, with extensive canopy closure.

The potential for extending forestry to the higher slopes has not been fully evaluated, although the altitude and the severity of the weather are likely to reduce economic return, and would probably preclude planting on the higher moors of Kinder and Bleaklow. While the establishment of mixed woodland on steep slopes may reduce erosion and subsequent sediment transport, forestry on deep peat soils is not an attractive proposition because it is normally preceded by drainage through deep ploughing. Even minimal soil disturbance methods result in enhanced soil moisture deficits, leading to increases in colour runoff. This may continue for several years, possibly until canopy closure. Also, such afforestation would result in increased levels of sediments, enhanced acidity and aluminium in streams, and in loss of yield. Strategies for controlling acidification and aluminium would be vitally important.

Grazing

The majority of the higher altitude area is classified as poorer than grade 4, and suitable only for rough grazing. The inclusion of almost the whole area in the North Peak Environmentally Sensitive Area will bring pressure to reduce grazing density and to fence off areas of land sensitive to erosion. Lower altitude areas have good grazing value which might be improved by liming and fertilization. Nutrient runoff following this course of action could result in reservoir eutrophication.

Recreation

Management of the moors for grouse shooting is extensive. This involves regular burning of heather to promote the new growth which forms the diet of the birds. Even good burning practice will lead to drying of the soil and subsequent increases in colour runoff. The burning of senescent heather should be especially discouraged however. Such heather burns at a much higher temperature than normal, and may lead to temporary devegetation or damage to the peat soil with subsequent erosion. Cutting or scarifying such heather should be considered as an alternative.

The moors are also extensively used for rambling, and erosion often takes place close to well used pathways. This is unlikely to be of sufficient extent to cause significant water quality deterioration, but management schemes to minimize the effect should be undertaken. Moorland fires started by careless walkers are also a potential contributor to peat erosion and colour production.

APPROACH TO CATCHMENT MANAGEMENT GUIDELINES

The results of this work and a review of best practices in other areas are being used to draw up management guidelines and a code of practice for planning authorities, land owners and tenants on how land in the area of the catchment should be managed, so that the quality of the drinking water supply and the aquatic environment can be protected, while preserving the benefits of the catchments for all users. The following aspects need to be addressed.

Control on land use activities

The Peak Park Planning Board controls activities within the area, and meetings between the Board, major land owners and Local Authority representatives are held regularly. The entire area, with the exception of the south west corner of the river Noe catchment, falls within the area designated as the North Peak Environmentally

Sensitive area, and much of the moorland coincides with the Kinder-Bleaklow Site of Special Scientific Interest.

Critical zones

In this type of catchment it is desirable to identify areas of particular risk in respect of water quality determinates. The critical zones shown in Fig. 10 are areas in the Derwent reservoir catchment where it is desirable to restrict activities to protect water quality. These areas fall into two categories.

(a) Zones within 50 m horizontally of the top water level of reservoirs and main feeder streams, in which there is a very strong likelihood that accidents or poorly managed activities will result in damage to the aquatic environment, or pollution of the water supply by such things as farm wastes, fertilizers and pesticides. In steeply sloping areas, or where wind drift may be a problem, it may be desirable to increase the width of these zones.

(b) Moorland areas where devegetation has lead to exposed, eroding peat and where as a result, the water draining from the area has suffered quality deterioration (higher colour, iron and aluminium levels). These are also shown on Fig. 12 but it should be noted that the entire area of peat blanket moorland is potentially at risk from devegetation and erosion, with high altitude, exposed areas being at highest risk.

Livestock
Animals
There are currently no intensive livestock units located in the area, and this position should be maintained. Farm animals, with the exception of sheep and goats, should be excluded from critical zones. Dead animals should be buried outside these zones.

Sheep dipping
All sheep dip should be disposed of under an official 'Consent to Discharge' and must not be discharged to any watercourse or soakaway. Spreading on soil well outside a critical zone may be an option. Mobile dips or new dips should be located on level ground outside the critical zones. Existing dipping arrangements within the critical zones or on steep slopes should be notified to the NRA and examined for suitability.

Silage
Silage production should be discouraged, but where it is necessary every precaution should be taken to ensure that watercourses are not polluted. Clamps should on no account be located within critical zones [12].

Grazing
Landowners and tenants should maintain sheep grazing densities in line with ESA agreements. In high moorlands, however, grazing densities should be regulated to ensure that overgrazing and subsequent devegetation is prevented. Particular attention needs to be paid to the sensitive areas and the erection of temporary fencing and active shepherding may be necessary to prevent overgrazing.

Where encroachment of bracken is a serious problem, control using *asulam* may be permitted provided that prior approval is sought, and the appropriate bodies notified. Spraying should not take place under windy conditions [13].

Grassland improvements should not be permitted within the critical zones, as there is a strong likelihood that ploughing, drainage, re-seeding and fertilization will result in the pollution of watercourses.

Arable crops
Except for small scale activities, intensive arable crops should be discouraged.

Organic fertilizers and wastes
No manure, slurry or sewage sludge should be brought on to the catchment. Manures and slurries produced there should be kept in secure, bunded storage outside the critical zones, and spreading of these is to be well outside these zones. Manures and slurries should not be spread on snow or frozen ground [14,15].

Chemical fertilizers
Chemical fertilizers should not be stored in critical zones nor where they can leak into watercourses accidentally. They should not be spread within critical zones without prior consultation, nor should application take place in windy conditions. Application or bulk storage of NPK fertilizers in excess of 5 tonnes anywhere on the catchment should also be subject to prior consultation. Again, spreading should not take place on snow or frozen ground.

Chemicals, oils and toxic materials
There is no restriction on the use of lime within the catchment, but only *asulam* and *glyphosate* herbicides may be used within the critical zones. On no account should 2,4D or other pesticides be used. Spraying should not take place on windy days.

No agrochemicals, fuel oils, wood preservatives or other potentially noxious pollutants may be stored or handled in bulk within the critical zones or discharged at any place where they may pollute a watercourse. All storage areas should be covered and bunded [16,17].

Spillages should be contained or absorbed using sand or sawdust and the NRA contacted for advice. On no account should they be flushed away [18].

Forestry
All forestry activities should be undertaken according to the 'Forests and water guidelines' [19]. Afforestation of the peat blanket moorland should be avoided. The planting of mixed shelterwoods for landscape or conservation reasons, in other than blanket peat moorland, and the planting of broadleaves such as alder and rowan in steeply incised valleys with eroding mineral soils is acceptable providing that the requirements of the SSSI are met. Wherever practicable a broadleaf/conifer mix should be used rather than a conifer monoculture. The re-establishment of ground cover following felling should be a priority, and felling of large areas at one time should be avoided.

Moorland management
Heather
Heather management, whether for conservation or for grouse rearing, must be undertaken according to 'The heather and grass burning code'. Hot burns have been shown to result in exceptionally high colour in drainage waters. Devegetated peat also produces more colour and may erode causing sediment transport in streams.

Particular care must be taken under dry conditions such as on south facing slopes, or following dry weather, to avoid hot burns which may cause devegetation and possible damage to the underlying peat. Senescent heather should be cut and not burnt as it tends to burn at high temperature.

Erosion control
Landowners and tenants should be advised of the benefits of revegetation, particularly in areas of severe erosion, and persuaded to embark on revegetation programmes. These programmes should include the exclusion of sheep, through the erection of temporary fencing.

Drainage
Drainage of the peat blanket (moorland gripping) should be avoided unless absolutely necessary. Ploughing of the peat blanket should be generally avoided and should not be permitted in the critical zones. Where it is unavoidable, furrows should not cut completely through the peat horizon.

Recreation and tourism
Any increase or change in the following activities should be subject to prior consultation: boating; canoeing; boardsailing; angling; camping; caravaning; carparking; autocross; rallying; motocross; and trials riding.

Petrol and diesel powered boats should not be permitted except in special circumstances. Large gatherings of people are generally to be discouraged, but where access to the public is permitted, toilet facilities should be provided. Informal camping should not be permitted within critical zones, and permission to camp in other areas should be given only to properly organized groups. Camping and caravan sites for more than six units should be provided with toilet and waste disposal systems connected to sewerage or septic tank.

Where footpath erosion can be demonstrated to be triggering more widespread erosion problems, the provision and signposting of alternative routes should be considered, and in severe cases a programme of footpath stabilization could be required.

Vehicular access to grouse butts should be permitted only along suitable tracks and driving over soft, especially peaty, areas should be avoided.

Conservation
Burning of bilberry, grass or bracken for the conservation of plant communities or animal habitats should be properly controlled. Burning of bracken is counterproductive and *asulam* should be used, as above.

Roads, tracks, and vehicular access

The use of vehicles within critical zones except on properly made roads should be minimized. Minor roads and tracks should be constructed with limestone chatter, rather than gravel and tar and pitch as the runoff may carry tarry residues which can lead to taint problems in drinking water.

Miscellaneous

Fly tipping should be discouraged and any landowner or tenant who finds rubbish tipped on his land should take steps to have it removed.

General summary

Severn Trent Water has commissioned the Water Research Centre to investigate both the actual and potential causes of deterioration of the quality of water passing to treatment from the Upper Derwent Valley catchment and suggested an approach to the management of land within the catchment which will minimize such deterioration.

REFERENCES

[1] Norton, R. L. and Mainstone, C. P. (1989) *Colour in upland supplies – Derwent catchment study.* Water Research Centre (1989) plc Report UC 482 (Report to Severn Trent Water).

[2] Simpson, P. J., Watts, C. D., Winnard, D. A., Gunn, A. M. and Welch, D. I. (1989) *Characterisation of colour in upland waters.* Water Research Centre (1989) plc Report PRS 2152-M.

[3] Jackson, P. J. (1989) *Treatability of coloured water by coagulation – studies of seasonal and physiochemical effects.* Water Research Centre plc Report 899- S.

[4] Boon, R., Crowther, J. and Kay, D. (1988) *Land use and water quality within the Elan Valley.* St. Davids University College, Lampeter. (Report to Severn Trent and Welsh Water.)

[5] Phillips, J., Yalden, D. W. and Tallis, J. H. (1981) *Peak District moorland erosion study. Phase 1 Report.* Peak Park Joint Planning Board, Bakewell.

[6] Stanek, W. and Silc, T. (1977) Comparisons of four methods for determination of degree of peat humification (decomposition with emphasis on the Van Post method). *Can. J. Soil Sci.*, **57**, 109–117.

[7]. Tallis, J. H. (1985) Erosion of blanket peat in the Southern Pennines: new light on an old problem. In: *The geomorphology or Northwest England* (Johnson, R. H., ed.) Manchester University Press, Manchester, 313–336.

[8] Kinako, P. D. S. and Gimingham, C. H. (1980) Heather burning and soil erosion on upland heaths in Scotland. *J. Environ. Management*, **10**, 277–284.

[9] Burt, T. B., Donohoe, M. A. and Vann, R. A. (1984) Changes in the yield of sediment from a small upland catchment following open ditching for forestry drainage. *Catena*, Supplement 5, pp. 63–74.

[10] Tallis, J. H. and Yalden, D. W. (1983) *Peak District Moorland restoration project. Phase 2 Report: revegetation trials.* Peak Park Joint Planning Board, Bakewell.

[11] Evans R. (1977) Overgrazing and soil erosion on hill pastures with particular reference to the Peak District. *J. Brit. Grassland Soc.*, **32**, 65–67.

[12] Department of the Environment Welsh Office. *Public consultation on proposed regulations to control silage slurry and agricultural fuel oil installations.*

[13] Pesticides: Guide to new controls. *Control of Pesticide Regulations.* Ministry of Agriculture, Fisheries and Food, (1986).

[14]. *Advice on avoiding pollution from manures and other slurry wastes.* ADAS Booklet 2200 (1985).

[15] *The storage of farm manures and slurries.* ADAS Booklet 2273.

[16] *Avoidance of pollution by oil,* Oil and Water Industries Working Group (1967).

[17] *Farm chemical stores.* ADAS Leaflet 767 (1984).

[18] *Guidelines for the disposal of unwanted pesticides and containers on farms and holdings.* ADAS Booklet 2198.

[19] *Use of herbicides in the forest.* Forestry Commission Booklet 51 (1986).

32

The Roadford Scheme: planning, reservoir construction and the environment

P. W. Gilkes, BSc, CEng, MICE, MIWEM[†] J. P. Millmore, BSc, CEng, FICE, MIWEM[‡], and J. E. Bell, BSc, CEng, MICE, MIWEM[§]

INTRODUCTION

The Roadford Reservoir scheme has been awarded six prizes from various organizations for different aspects of the works. Two of these prizes are for high environmental standards achieved on the scheme. This chapter outlines how the development of infrastructure on a major water engineering project can be achieved whilst safeguarding the environment.

Since the early years of South West Water, a regional strategy based on the use of three key reservoirs has been central to resource development within the area. Colliford, Roadford, and Wimbleball Reservoirs have quadrupled the resources available. The operation of the three reservoir strategy is outlined in Fig. 1.

Construction of Roadford Dam was completed in October 1989, with a usable storage of 34 500 Ml. Roadford is by far the largest of the three reservoirs in the strategy; it ensures that customers of South West Water will benefit from secure supplies for the foreseeable future. The basis of the scheme's design is to regulate the River Tamar, permitting the supply of water to the key areas of Plymouth,

[†] Clean Water Programme Manager, South West Water Services Ltd.
[‡] Partner, Babtie Shaw and Morton.
[§] Technical Director, Babtie Environmental Sciences.

South Hams and Torbay and to allow direct abstraction at Roadford to supply water to north Devon and north-east Cornwall.

SCHEME STRATEGY

The original concept was to convey water to where it was needed by intercatchment transfers, but the strategy has envolved fundamentally to direct supply by pipeline. The revisions to the scheme strategy are shown in Figs 2(a) and 2(b). Environmental factors were an important consideration in the change of strategy, although there were other contributory factors.

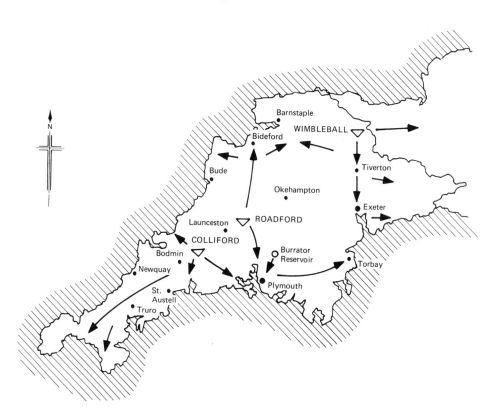

Fig. 1. Central reservoir strategy in south-west England.

Southern strategy

Rivers Tamar and Tavy

In March 1986 the Board of the then South West Water Authority decided to revise the southern strategy as a consequence of environmental implications. They were concerned about the different water quality in the donor River Tamar relative to the recipient Rivers Lumburn and Tavy. As this was seen as untenable, a direct pipeline from the Tamar to Crownhill water-treatment works (WTW) was provided.

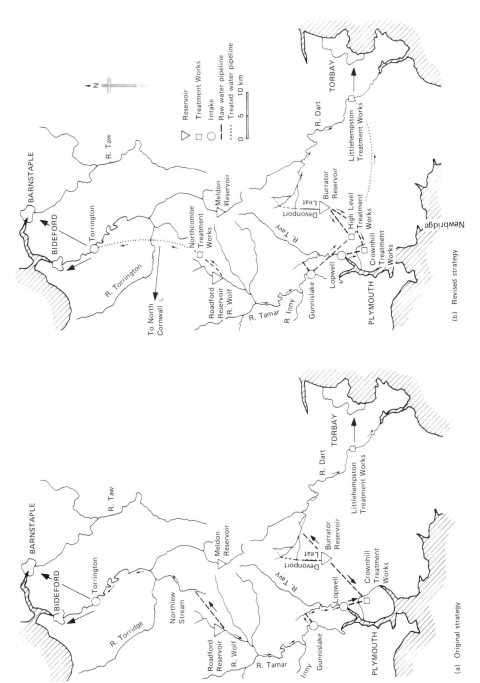

Fig. 2. Strategy for the Roadford Scheme.

Further advantages from this direct transfer of water to Crownhill are:

(1) slight increase in yield of the scheme by the reduction of regulating river losses at times of naturally low river flows;
(2) elimination of bankside storage at Milton Brook to protect the supply against major pollution incidents;
(3) reduction of the risk of migratory salmonids homing to the 'wrong' river; and
(4) no risk of changing the quality of the River Tavy.

River Dart

At the original public inquiry in 1976 it was proposed that increased abstration at the Littlehempston works of the River Dart would be supported by reinstatement of the Devonport Leats and pumped redeployment from Burrator Reservoir.

The strategy was reviewed following an environmental assessment in January 1988. The review confirmed that redeployment of Burrator and Devonport was insufficient to satisfy the resource demands and the environmental integrity of the Dart. An alternative strategy was developed to supply the Torbay area as set out in Fig. 2(b).

The key benefits of the revised southern strategy are:

(a) it allows for the supply of potable water to meet strategic growth on the highway corridor between Plymouth and Exeter into the foreseeable future;
(b) it will prevent the risk of environmental damage to the River Dart with some enhancement being probable in the upper reaches; and
(c) it will allow a gravity supply from the High Level WTW at Plymouth by a trunk main to the upper supply area of Plymouth and into the Torbay distribution area.

Northern strategy

The original northern strategy was based on a river transfer system as shown on Fig. 2(a). Objections to this strategy were:

(1) the scheme consisted of double pumping, that is from Roadford Reservoir to the catchment watersheds and again from Torrington WTW into supply;
(2) a massive increase in flows would be required to a very small stream in the upper reaches (92 Ml/d);
(3) differing water quality could be expected (reservoir to river);
(4) a likely transfer of different forms of aquatic life;
(5) fish movements could be changed. For instance they could be confused by increased flows in some streams when moving to their original spawning grounds. Any loss in fishing to riparian owners could be the subject of substantial claims with adverse publicity; and
(6) safety concerns would complicate operational management.

Because of these objections a new strategy was adopted as shown in Fig. 2(b). It consists of the following elements: (a) a high level water-treatment works at Northcombe; (b) trunk mains from Northcombe to Torrington linking to strategic potable reservoirs serving Barnstaple, Bideford and Bude; and (c) provision of water resources to Torrington area into the next century, with eventual provision of a treatment works to serve winter demand.

This revised northern strategy has the major benefits of: (i) protecting the natural environment of the River Torridge; (ii) increasing the yield because there would no longer be river losses; and (iii) providing substantial savings in energy.

EVOLUTION OF ROADFORD ENVIRONMENTAL STRATEGY

Public inquiry undertakings

The promotion of the Roadford Reservoir was both difficult and lengthy as the objectors to the scheme were well organized and articulate.

As a result of consistent lobbying, many undertakings with substantial ramifications were given to allay the concerns of interested bodies. The environmental strategy which has evolved from these commitments has three key points, namely:

(1) The comprehensive monitoring of the affected catchments.
(2) The compensation of all affected parties for damage and loss of earnings during reservoir construction.
(3) The establishment of a 'Roadford Fisheries Liaison Committee' to ensure adequate consultation with riparian owners.

The liaison committee's work covers two areas: (a) the catchment at and above the dam site; and (b) the rivers and reservoirs that would be utilized in different circumstances once Roadford is impounded.

This committee is central to South West Water's policy of full and open consultation with riparian owners. It enables representatives of riparian owners, the National Rivers Authority (NRA) and the Company to debate all the issues related to the development of the Roadford scheme.

Roadford environmental team

A clear policy decision was taken by South West Water to: (i) set up a strong environmental team to assess the detailed investigation work required; (ii) build a substantial data base of historical information on all aspects in the areas likely to be affected by the scheme; and (iii) utilize local knowledge and expertise wherever possible.

By comparison with today's environmental assessments those before the Roadford Inquiry were unsophisticated. Policies were based on judgements rather than being soundly based on scientific evidence. The development of the Roadford Environmental Team was a milestone in environmental work and the change to the Tamar/Tavy strategy in 1985–86 demonstrated a growing awareness of environmental issues.

Public interest/management

The Roadford Reservoir scheme has attracted extensive coverage in the media and customer interest. Throughout the period of construction the site has attracted a large number of visitors, with over 2000 on some weekends. To avoid interference with site activities and local residents the visitors' car park was twice enlarged to cope with additional vehicles.

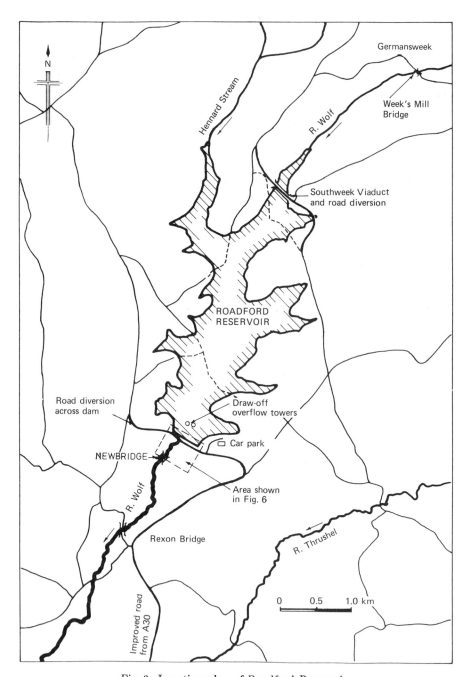

Fig. 3. Location plan of Roadford Reservoir.

PREPARATIONS FOR RESERVOIR

Road routes

At the Public Inquiry South West Water undertook to improve the minor county roads to be used by the construction traffic (Fig. 3). It was also accepted that the design should retain the characteristics of the area. Typically the original roads were narrow winding lanes with high hedgebanks on each side.

In consultation with interested parties the consultants developed a design which retained as far as possible the sinuous nature of existing routes, the existing hedge-banks, trees, and vegetation adjacent to the road corridor. This design task was complex as it was necessary to balance these criteria with the Devon County Council's requirement for Department of Transport standards for highway design.

The first Roadford contract, which was awarded in December 1984 to Macbar Construction, was for an improved access from the A30 trunk road to the dam site. In a few locations it was possible to thread the new road between existing boundaries, but in most instances a new boundary was necessary while retaining the old boundary on one side. It was decided to construct the new boundaries to the ancient Devon tradition of stone-faced earth banks, and some 5.2 km were built in this way at a cost of £250 000 out of the total roadworks cost of £2 million.

The road contractor maintained accesses to farms, and a new bridge was built over the River Thrushel as called for in South West Water's undertakings at the Public Inquiry. As the existing stone bridge was a Category B listed structure it was preserved *in situ*.

Other environmental features of the road contract included the sowing of mixes of traditional grasses and herbs along the verges, new blocks of indigenous trees and shrubs to complement existing species, the erection of traditional-style finger posts for directions, and the erection of safety boundaries of turf banks rather than steel crash barriers.

As shown in Fig. 3, the reservoirs severed some cross valley roads, and a new road was built across the head of the valley. This second contract valued at £2 million was let in 1985 to M. J. Gleeson Ltd. It included a temporary road for all construction traffic along the floor of the reservoir to avoid damage to the surrounding county highways. The new highway across the valley was partly carried on an embankment composed of locally quarried rock, and partly on a 15 m high viaduct over the reservoir at Southweek. The viaduct was built on a horizontal curve of 220 m length and supported on concrete piers. The design received recognition in the 1986 Concrete Society Awards for its appearance and integration into its setting (Fig. 4).

The final link in the network was the 1.5 km road across the 40 m high embankment of the dam to connect with country roads on either side of the valley.

Environmental management of basin clearance

The reservoir covers an area of 3 km² in what was an undisturbed rural valley about which little was known in detail before the advent of the scheme. In 1986 the Devon Trust for Nature Conservation was commissioned by South West Water to survey the wildlife of the area. The survey report made a number of recommendations on how the impact of the reservoir could be mitigated and how important species could be rescued or protected.

The advice was carefully studied and most aspects were incorporated into a reservoir clearance contract which was awarded to E. Thomas Construction in 1987 at a value of £400 000. The environmental aspects of basin clearance activities were supervised by a pollution control officer whose responsibilities also covered the main constructional works for the dam, which are described below.

Fig. 4. Viaduct across head of reservoir.

Flora

To ensure satisfactory quality of reservoir water and to prevent interference with water supplies it was necessary to fell 30 000 trees in areas to be flooded. The conservation survey identified particular areas where indigenous species such as oak, ash, hazel and birch could be planted to best effect and where coniferous trees around the reservoir should be felled and replaced by broadleaved plantings. Landscape consultants arranged for planting some 120 000 indigenous trees and shrubs as part of the basin management plan.

The conservation survey identified areas of 'Culm Measure Grasslands' in the valley as being of special botanical interest. As part of the clearance contract 400 m² of grassland containing such typical herbs as devil's bit scabious, greater bird's-foot trefoil and ragged robin were cut out in blocks and transplanted to specially prepared reception sites above the water line.

Fauna

To avoid interference with birds, clearance activities were programmed to avoid the nesting season. Where building demolition was necessary this was done at times which minimized the impact on the breeding of bats, and alternative homes were provided for them in the form of boxes on trees.

A number of other specific conservation measures were undertaken as part of the clearance programme. These included the trapping of dormice for release in a nature reserve; the provision of stick pile holts and culvert passageways for otters; and gateways for badgers through the perimeter fence.

A full-time ranger was appointed by South West Water and part of his duties were to co-ordinate wildlife conservation measures. There is a particular interest in developing the new reservoir to encourage bird life as far as this is compatible with water supply and other statutory requirements. The British Trust for Ornithology has reported on the potential of the new waterbody, and on its advice three islands have been constructed to provide refuges and nesting sites for birds in the sheltered north-west arm of the new lake. In the 1989 winter about 650 wildfowl were counted and in January 1991 the number had increased to 1750. In the spring of 1990 a number of interesting aquatic breeding species were recorded, the most significant being the great-crested grebe.

Archaeology

Archaeological work was carried out in the period 1987–89 by the Exeter Museum Archaeology Field Unit and Devon County Council with the aid of grants from South West Water and English Heritage.

Particular interest was centred on the five farmsteads in the valley which had been medieval longhouses. These had not been extensively modernized and were found to contain important relics of traditional rural life in the form of granaries, threshing barns, root stores, poundhouses (for making cider), pigsties, linhays, shippons, horse mills and dung pits. Where possible historic architectural features, farm implements and building materials were rescued prior to demolition for future use in public displays. A large Victorian water-wheel was loaned to Okehampton Museum in 1989 and re-erected in working order.

An archaeological excavation at the site of a mill abandoned in the last century revealed a remarkable echo of the present when a masonry dam was discovered. This previously forgotten structure dating from the 13th century had been used to impound a tributary of the River Wolf to provide the power for both a corn mill and a fulling mill (for wool manufacture), which together served this remote community in ages past.

Devon County Council's archaeologist has described the Wolf Valley as being 'the most closely researched piece of landscape in the county, off Dartmoor' [1]. The Roadford project has thus made an important contribution to the understanding of the past.

CONSTRUCTION OF DAM

Use of local materials

One of the advantages of the Roadford site was that the local rock quarried from the area to be flooded could be used to form an embankment dam. This had the

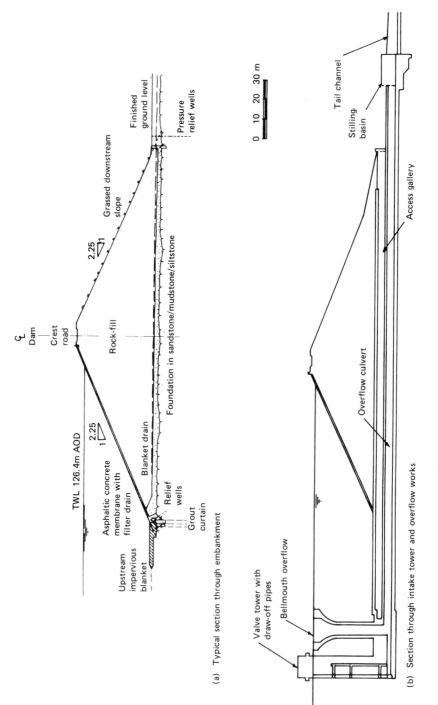

TWL 126.4m AOD

Upstream
impervious
blanket

Asphaltic concrete
membrane with
filter drain

2.25
1

Relief
wells

Grout
curtain

Blanket drain

Rock-fill

Foundation in sandstone/mudstone/siltstone

C
Dam

Crest
road

2.25
1

Grassed downstream
slope

Finished
ground level

Pressure
relief wells

(a) Typical section through embankment

Valve tower with
draw-off pipes

Bellmouth overflow

Overflow culvert

Access gallery

Stilling
basin

Tail channel

0 10 20 30 m

(b) Section through intake tower and overflow works

Fig. 5. Cross-section through Roadford Dam.

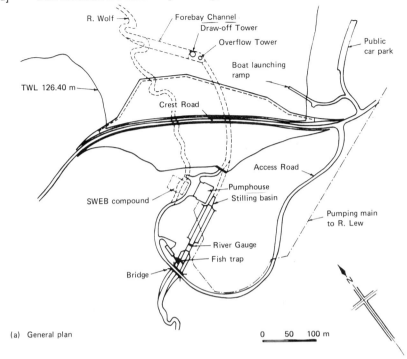

(a) General plan

0   50   100 m

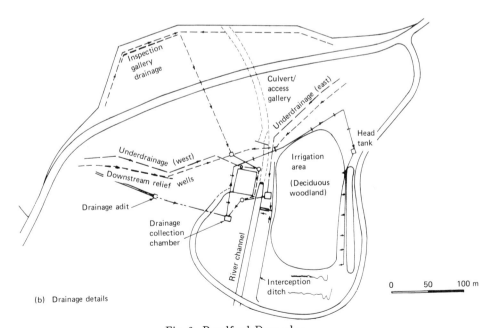

(b) Drainage details

0   50   100 m

Fig. 6. Roadford Dam plan.

environmental advantage of minimizing haulage disturbance through the valley and the economic advantage of using naturally occurring material. The local rock consists of closely folded strata of mudstones, siltstones and sandstones of the Upper Carboniferous period. Extensive geotechnical tests were carried out on the structural and chemical properties of these rocks.

It was found that the chemical constituents of the formations varied considerably and that there was an average of one per cent by weight of the potentially degradable mineral, iron pyrites. Leaching tests were carried out in the laboratory to simulate the effects of weathering, and trial embankments were made on site to monitor the structural aspects. The road embankment at Southweek was used as a trial for the dam, and information was gathered on settlement, pore pressures and the quality of drainage water. It was concluded that some mineral degradation would occur at full scale but that if suitable engineering measures were adopted to minimize water entry into the embankment fill the effects of weathering could be satisfactorily controlled.

The characteristics of the dam, with its massive structure containing $1\,000\,000$ m^3 of rock, are illustrated in Fig. 5. The main features controlling the weathering effects are the grout curtain cut-off in the foundation, drainage blanket, the waterproof upstream membrane, and the crest roadway. A plan of the dam is shown in Fig. 6(a) and the complex arrangements for drainage of the membrane area and relief wells are illustrated in a more simplified form in Fig. 6(b). Fig. 7 shows the completed dam.

The rapid construction of the embankment in the summer of 1988 also minimized weathering effects.

Measures for pollution prevention

Measures for the prevention of water pollution had two main aspects. The first concerned the specification and the second was the monitoring programme.

The contract for the dam specified that the contractor should provide storage lagoons, sufficient for the retention for 24 h of 50 mm of rainfall falling on disturbed areas, and for water pumped from excavations. Two criteria were also laid down to govern the release to the river of water stored in lagoons. These were, firstly, that the quality of water discharged should have:

Turbidity — less than 20 FTU

Suspended solids — less than 45 mg/l

pH — between 6 and 9

Total aluminium — less than 0.5 mg/l

The second comparative criterion was that the rate of dicharge should be limited so that the quality of water in the River Wolf downstream of the site would not suffer material deterioration. This was defined as anything which (a) would affect the natural river life, (b) would raise the suspended solids by more than 10% above the naturally occurring level at the time, or (c) would depress the dissolved oxygen level of the river by more than 10%.

In accordance with the terms of the specification, the contractor built a chain of temporary lagoons in the valley on both sides of the river totalling $27\,000$ m^3. Runoff water from the quarry, batching plant excavations, channels and ditches was collected in the lagoons. These were linked by pipelines to convey flow towards the final lagoon, which had a continuously monitored outlet to the river.

The system was operated by the contractor, but it was supervised by a full-time pollution control officer (PCO) who was part of the resident engineer's staff. An elaborate monitoring system was set up and managed by the PCO on a day-to-day basis. The monitoring programme is summarized in Table 1.

Fig. 7. Completed Roadford Dam.

Review of anti-pollution operations

Discharges to the river were made when conditions met the criteria quoted above. It was found that the quality criteria such as pH and suspended solids could be met fairly readily but that in terms of appearance the lagoon water was more turbid than the river in dry weather. However, in natural spates the river becomes very turbid and the opportunity was taken to release stored lagoon water at these times.

The lagoon arrangements proved extremely effective during the construction period between 1987 and 1989. In all, 16 pollution incidents were investigated by the PCO. All but two of the incidents over a two-year period concerned discoloured water downstream of the site. The PCO investigated the reports, took samples and reported the conclusion of the investigation to the complainant. On five occasions, the source of the incident was found to be vehicles working close to the river upstream in the basin clearance activities. Four incidents were associated with bursts during the pumping of drainage water around the site. There was only one report of an oil leak (minor), and three dead fish were reported after river diversion work (these were thought to have been missed in previous electrofishing).

The appointment of the PCO at the planning stage was seen as being of fundamental importance to the success of the anti-pollution measures, and this judgement was confirmed in the event.

Table 1. Monitoring programme during construction.

Monitoring station	Freqency		Sampled by[†]
River Wolf			
Week's Mill Bridge	Monthly	SWW	manual
Southweek	Weekly	PCO	manual
Roadford Bridge WQMS	Continuous	auto	print-out from SWW
Roadford Bridge (fish trap)	3 per week	SWW	manual check WQMS
Roadford Bridge, auto-sampler	Hourly	PCO	
Downstream of dam, auto-sampler	Hourly	Cont.	samples of previous week available
Rexon Bridge WQMS	Continuous	auto	print-out from SWW
Rexon Bridge	3 per week	SWW	manual check WQMS
Catchment streams			
Southweek	Weekly	PCO	manual
Hennard	Monthly	SWW	manual
Wortha	Weekly	PCO	manual
Roadford	Weekly	PCO	manual
Construction areas			
Southweek Viaduct	Weekly	PCO	manual
Combepark Quarry	Weekly	PCO	manual
Dam site, final lagoon WQMS	Continuous	auto	print-out from Cont.
Reservoir area			
Observation boreholes	3-monthly	SWW	manual

[†]Sampled by: PCO = Pollution Control Officer; Cont. = Contractor; SWW = South West Water; WQMS = Water Quality Monitoring Station. Analysis carried out as follows: WQMS (SWW station), pH, SS, turbidity, DO and conductivity. WQMS (Contractor's station), pH, SS, and conductivity. Manual samples, pH, SS turbidity, colour, sulphate, aluminium, iron and manganese. Borehole waters, a 'full analysis' to establish ionic balance.

MAINTENANCE OF WATER QUALITY AFTER IMPOUNDMENT

Development of monitoring programme

The programme which began in the construction period has continued into the impoundment period and there is close liaison with the NRA on quality aspects. To prevent duplication of effort and to ensure the efficient use of resources, routine monitoring information on rivers and reservoir water is supplied to the NRA. The present monitoring programme is summarized in Table 2.

Dam drainage

As would be expected, there has been a build-up in the quantity of drainage water at the dam since construction began in 1988. By the time of impoundment in October 1989 the dam drainage flow was 300 m³/d. This quantity rose after impoundment to a peak of 1300 m³/d in the wet winter months but declined to 900 m³/d in the dry summer and autumn of 1990 (Fig. 8). The depth of water throughout this period was approximately 80% of the final level (which is commensurate with 55% of the volume).

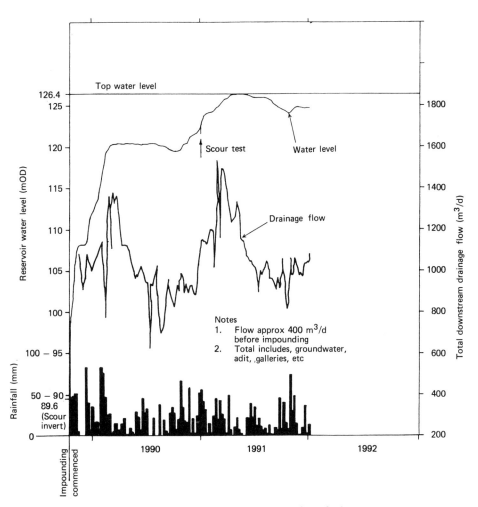

Fig. 8. Rainfall, reservoir level and dam drainage.

As predicted in the geotechnical studies, the dam drainage contained products of primary rock degradation such as iron, manganese and sulphate: secondary reactions produced carbonates, especially those of calcium and magnesium. The chief concern voiced by NRA officials has been the elevated level of manganese in the untreated dam drainage, which reached 5 mg/l in the period immediately after impoundment. The low oxygen at around 50% of saturation in the untreated drainage water also makes it unsuitable for direct river discharge.

Before impoundment, two approaches were investigated on pilot scale for treatment of drainage waters: namely a package pressure-filtration plant and land irrigation. Pressure filtration in itself was found to have little effect on manganese, but land irrigation was proved to be reasonably effective.

Table 2. Monitoring programme after impoundment

Location	Frequency
Reservoir draw-offs top, middle, bottom and scour	Twice weekly
River Wolf sampled from: Germansweek, Newbridge and Rexon Bridge	Weekly
Untreated dam drainage	Weekly
Treated dam drainage, woodland irrigation interceptor ditch	Weekly
Inspection gallery drainage	Monthly
Inspection gallery pressure relief wells, pressure relief well	Fortnightly
Membrane drains	Fortnightly
Downstream toe pressure relief wells	Fortnightly
Drainage adit	Fortnightly
Embankment drainage	Weekly
Boreholes	3-monthly

Notes on chemical analysis

1. Normally the following tests are carried out: pH, conductivity, turbidity, colour, dissolved oxygen, ammonia, nitrate, suspended solids ($105\,^{\circ}$C and $500\,^{\circ}$C), alkalinity (to pH 8.3 and 4.5), calcium, magnesium, chloride, orthophosphate, silicate, dissolved sulphate, sodium, potassium, magnesium, aluminium, manganese (total and dissolved), iron (total and dissolved). Other parameters such as hardness are calculated from these basic analyses.

2. On selected samples the following additional analyses are carried out: BOD, TOC, algal counts, pesticides, lead, cadmium, mercury, zinc, copper and nickel.

In 1990 the land irrigation system was developed and the treatment arrangement now employed is illustrated in Fig. 6(b). In Table 3 the quality of the dam drainage is compared with that of drainage collected in the ditch at the foot of the slope after land irrigation. The chief points to note are the 90% reduction in manganese and the elevation of the dissolved oxygen from about 50% to 100% saturation level. In the first year the total quantity of dissolved solids in untreated drainage waters declined by about 50%, but continued monitoring is needed to establish the pattern over subsequent years.

The flora and microfauna of the woodland irrigation area are also being monitored at three-monthly intervals. First indications have shown that the number of species is becoming more diverse due to the development of marshy areas at the foot of the slope.

Monitoring of reservoir water

There is an automatic probe in the valve tower to measure dissolved oxygen, pH and temperature at various depths below the water surface. There is also a system of pipework which allows water samples to be drawn from the reservoir at each of the draw-off levels and the scour. Continuous measurement of temperature, dissolved oxygen and pH is carried out and a complete chemical analysis is performed on samples twice a week.

The variation of certain of the key parameters in the first year is shown in Fig. 9. It will be seen that the dissolved-oxygen saturation was near 100% until May 1990 and that soluble iron and manganese were low. However, when stratification began to develop in the reservoir, manganese and iron increased and there was a corresponding fall in dissolved oxygen. At this point the aeration mixing system was put into operation to reverse the tendency for stratification. However, it was not possible to operate this on a continuous basis until August 1990, and the graph illustrates the tendency for stratification to develop very quickly. The destratification system was switched off in November 1990 to monitor the resultant effect in winter conditions. There was no immediate evidence of significant algal activity, and nutrient levels are very low: phosphate level being about 0.01 mg/l and nitrate around 0.9 mg/l.

River water monitoring

Monitoring of the river as part of the reservoir scheme was started in 1986 and has continued throughout the period of impoundment. Comparative data on the water quality over the year after impoundment, upstream of Germansweek and at the two downstream stations at Newbridge and Rexon Bridge, are summarized in Table 4.

The main changes in quality are slight rises in manganese, iron and sulphate. However, there is no indication that these parameters would have any significant effect in terms of the EC Directive on 'Quality of freshwaters needing protection or improvement in order to support fish life' or in terms of the National Water Council's classification of water quality, Class 1A.

The agreement for the discharge of compensation waters contains an allowance for freshets; therefore part of the reservoir storage is reserved for the protection of fisheries.

Scour operation

The commissioning programme included a test operation of the scour valve in the draw-off tower. Following consultations with the NRA a procedure was agreed to mimic a natural hydrograph of river flow. The whole operation began at 06.00 hours one morning in November 1990 and continued for 70 hours with a maximum flow of 600 Ml/d occuring over a peak of 15 minutes. Throughout the operation of the scour valve and shortly before, the NRA monitored water quality and river flows between the reservoir and the confluence of the Tamar. Results were successful in demonstrating a beneficial rather than detrimental effect; for example the operation lifted and disposed of rotting matter, thereby improving the habitat available for egglaying by migratory fish in the upper reaches of the River Wolf.

Table 3. Summarized data for dam drainage before and after treatment

Parameter†	Untreated dam drainage		Treated drainage (pilot scheme)		Treated drainage (improved scheme)	
	(Oct 1989) Range	Oct 1990 Mean	(Oct 1989) Range	June 1990 Mean	(July Oct 1990) Range	Oct 1990 Mean
Dissolved oxygen	3.3–9.0	6.0	9.5–11.8	10.6	9.3–10.5	9.
pH	7.1–7.5		7.4–8.4		7.0–7.7	
Conductivity	424–666	529	290–484	420	277–589	338
Suspended solids (105°C)	1–17	5	1–34	5	3–11	5
Suspended solids (500°C)	1–11	4	1–31	11	2–9	4
Colour	1–16	3	1–9	4	3–8	5
Turbidity	2–15	7	1–38	7	1–8	3
Soluble sulphate	51–77	61	48–68	62	61–110	75
Manganese (total)	0.83–1.5	1.2	0.13–0.59	0.52	0.02–0.08	0.04
Manganese (soluble)	0.78–1.49	1.13	0.13–0.51	0.42	0.03–0.06	0.04
Iron (total)	0.17–1.58	0.88	0.12–1.26	0.45	0.03–0.05	0.06
Iron (soluble)	0.14–1.00	0.59	0.08–0.17	0.12	0.03–0.09	0.06

†All parameters expressed as mg/l except pH, turbidity (FTU), colour (Hazen) and conductivity (μS/cm).

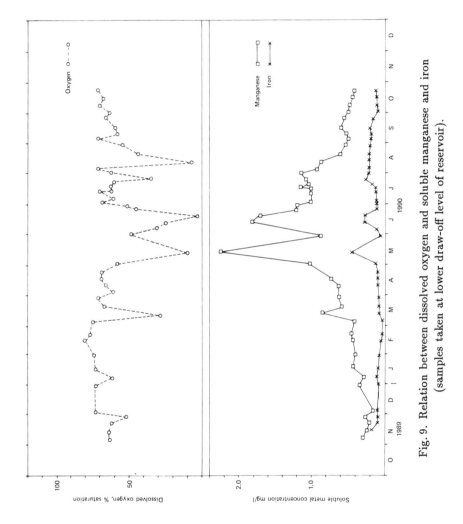

Fig. 9. Relation between dissolved oxygen and soluble manganese and iron (samples taken at lower draw-off level of reservoir).

COSTS

Costs associated with environmental work are listed below:

At reservoir		£000s
Destratification equipment	175	
Provision of settlement lagoons	68	
Water quality monitoring station downstream	10	
Fish trap and broodstock facility	153	
Employment of pollution control officer	36	
Dam draining	250	
		692
Other environmental work		
Devon Trust for Nature Conservation	40	
Archaeological study	162	
Landscaping	533	
Hydrometry, fisheries and water quality monitoring	3788	
		4523
Total		**5215**

The overall costs associated with environmental work are, therefore, about 6% of the total scheme budget of £85 million.

Table 4. Summarized data for river water quality
(October 1989 – October 1990)

Parameter[†]	Weeks Mill Bridge		Newbridge		Rexon Bridge	
	Range	Mean	Range	Mean	Range	Mean
Dissolved oxygen	7.7–10.9	9.7	7.5–11.6	9.2	8.1–10.5	8.9
pH	6.0–7.7		6.2–7.6		6.5–9.0	
Conductivity	122–156	137	153–211	176	149–207	174
Susp. solids (105°C)	1.0–18	6.2	1–25	4.7	1.2–14	5.9
Susp. solids (500°C)	0.4–16	5.1	0.4–20	3.8	0.4–10	5.1
Colour	9–28	19.2	1–82	30.0	19–74	31
Turbidity	1.3–31	5.1	1.2–17	4.4	1.5–9.0	4.8
Soluble sulphate	8.7–14.9	11.0	12.1–27.4	16.6	11.7–19.3	15.7
Manganese (total)	0.01–0.11	0.03	0.10–1.96	0.40	0.06–0.73	0.22
Aluminium (total)	0.05–1.00	0.16	0.02–0.13	0.05	0.02–0.14	0.06
Iron (total)	0.19–0.91	0.37	0.52–6.88	1.8	0.45–4.5	1.5

[†]All parameters expressed as mg/l except for pH, turbidity (FTU), colour (Hazen) and conductivity (μS/cm).

CONCLUSIONS

1. The promotion of any major civil-engineering project within the south west of England is challenging because of the high profile of environmental issues associated with one of the most beautiful areas of the UK.

2. Throughout the period of promotion, design and construction of the Roadford Reservoir scheme, South West Water and its consultants (Babtie Shaw & Morton) have regarded environmental protection as a fundamental part of their overall strategy. The main environmental aspects have been:

   (1) modification of the transfer strategy from the reservoir, so minimizing the environmental impact on river systems and fisheries;

   (2) design of works to minimize disturbance;

   (3) control of potential pollution during construction, in which the role of the pollution control officer was very important;

   (4) Rescue and protection of wildlife and archaeological features, treatment of drainage waters; and

   (5) the continued monitoring of reservoir water, drainage waters and the rivers throughout the impoundment period.

3. The application of thorough procedures in investigation, planning, design and supervision have ensured the successful completion of the scheme. The community has benefited by an essential service, while the environment has been safeguarded and in some respects enhanced.

REFERENCE

[1] Timms, S. (1990) Making of Roadford Reservoir. *Transactions of Devon Association for the Advancement of Science*, 122. December.

33

The Roadford Scheme: minimizing environmental impact on affected catchments

J. D. Lawson, MA, CEng, FICE, MIWEM[†], H. T. Sambrook, BSc, MSc,[‡], D. J. Solomon, BSc, PhD, MIBiol, FIFM[§], and G. Weilding, BSc, MSc[¶]

INTRODUCTION

Roadford Lake follows Wimbleball and Colliford as the third reservoir to be built under South West Water's three-reservoir strategy, which will meet the region's water demand into the 21st century. Construction of the lake started in 1986 and the dam was completed in October 1989, when filling of the lake started.

Roadford Lake lies in the centre of a complex water resource development which is shown schematically in Fig. 1. The reservoir is designed to operate in conjunction with abstractions from the Rivers Tamar, Tavy, Dart, Torridge and Taw, as well as with Burrator reservoir in the Plym catchment and Meldon reservoir in the Torridge catchment. The concept of the scheme is that abstraction is substantially increased from present levels at times when there is sufficient water in the rivers; during dry periods, the river abstraction is reduced and instead the increased supplies are obtained from Roadford. The scheme has an impact on six river catchments. The operating rules needed to control the amount of abstraction from the rivers and the

[†] Sir William Halcrow and Partners.
[‡] National Rivers Authority, South West Region.
[§] Fisheries Consultant.
[¶] South West Water.

use of the reservoirs will have a profound influence on the environmental impact of the scheme as well as its drought reliable yield and operating costs.

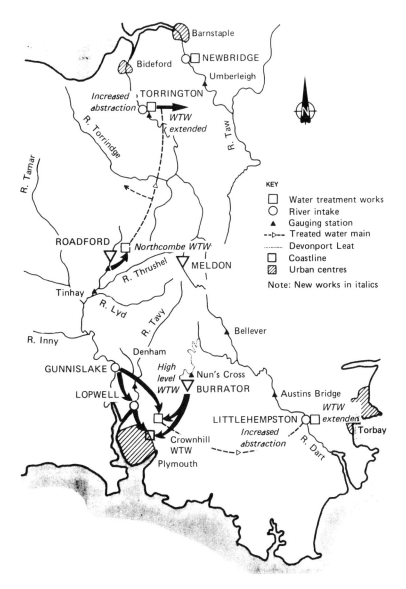

Fig. 1. Roadford supply system.

All six rivers support important sport and commercial fisheries for salmon, sea trout and brown trout. As a result, the fisheries interest is an influential and well-informed lobby. Water quality is generally good throughout the catchments, except in specific areas of the Tamar and Torridge where problems exist due to agricultural practices. In recent years, the Torridge has suffered a considerable decline in stocks of migratory salmonids.

In recognition of the potential environmental impact of the scheme, South West

Water undertook an extensive programme of environmental investigations. These studies were aimed at assessing the environmental impact, establishing a comprehensive data base, derivation of operating rules for the scheme (including the design of measures to alleviate adverse effects) and, where possible, to promote benefits. The work started in 1985 and was undertaken by in-house staff and consultants.

In August 1989, Sir William Halcrow and Partners Ltd were appointed to undertake the Roadford Operational and Environmental Study. The study was controlled by a steering group comprising members of South West Water Services Ltd and the National Rivers Authority (South West Region). The objectives of the study were: (a) to assess the environmental impact of the scheme; (b) to make recommendations for the priorities for abstraction and the operating rules of the scheme; (c) to obtain the optimum balance between environmental impact and operating costs; (d) to propose the terms of a reservoir operating agreement; and (e) to make detailed recommendations for future environmental monitoring.

This chapter describes the general approach taken and derivation of some of the operating rules. Implementation of the recommendations followed formal negotiations between the regulatory body, the National Rivers Authority (NRA), and the water company.

APPROACH TO REDUCING RIVER IMPACT

It must be recognized that any water resource development involving impoundment, regulation, and abstraction will have an impact on the environment. One of the main objectives of the study was to minimize the impact of the scheme operation on each river and, where possible, to maximize the benefits. Inherent in the study is the recognition of the catchment management philosophy and the integration of multi-objective river planning.

The approach taken to reducing river impact can be addressed via an iterative process involving the following five stages:

(1) during the design of the scheme and for its promotion through public inquiries, derivation of simple and conservative operating rules;
(2) a programme of data collection designed to determine the quality of the existing environment and the extent of its dependency on river flows and water quality;
(3) consultation with fisheries and conservation interests;
(4) determination of detailed operating rules based upon the programme of data collection and consultations; and
(5) further refinement of operating rules based on monitoring the impact of the scheme during its early years of operation.

The stages of development of the scheme operating rules are described below.

OPERATING RULES ORIGINALLY PROPOSED

The scheme promoted at the public inquiry in 1978 included increased or new abstractions for the Rivers Tamar, Torridge, Tavy and Dart, in conjunction with the use of the new reservoir. At the time, the concept of the scheme was that water would be abstracted from the rivers and local sources before drawing upon

Roadford. The original operating rules were based on a combination of prescribed flow and percentage take rules. New licences were proposed for the Rivers Tavy, Torridge, Dart and Tamar.

For the abstraction from the River Tamar at Gunnislake, it was agreed at the public inquiry that for the first ten years of operation (commencing in 1992), the prescribed flow would be 477 Ml/d, and 50% of flows above this level could be abstracted up to a maximum of 148 Ml/d. After ten years of operation, the prescribed flow would be reviewed and possibly reduced to 245 Ml/d, depending upon the outcome of the continuing investigations. At the time of the original preparation of the scheme it was envisaged that there would be no additional abstraction from the river Taw and that the North Devon area would be supplied by the increased abstraction from the River Torridge in conjunction with supplies from Roadford.

The compensation flow from Roadford was to be a fixed amount of 9 Ml/d released at all times in addition to any regulation releases being made from the reservoir.

During the promotion of the scheme an undertaking was given that volume would be reserved for fisheries purposes. This fisheries bank of 2270 Ml represents about 7% of the net available storage and is to be held in reserve to be released to the River Tamar for the benefit of its fisheries over any two-year period. The precise means of using the reserve was not specified during the original promotion of the scheme.

There were no new rules put forward governing abstraction from Burrator Reservoir. However, with the increase in reservoir storage because of Roadford, Burrator could be drawn upon more frequently, which could result in increased reservoir drawdown. This could cause a delay in the reservoir refilling, and spilling from Burrator would take place later in the year than at present.

In addition to these operating rules, there were several other measures put forward to reduce the impact of the scheme on the rivers. These measures included introduction of prescribed flows on existing abstractions from the Upper Dart into the Devonport Leat which feeds Burrator reservoir, and reduction in some existing hydro-electric abstractions on the River Tavy.

DATA COLLECTION

Since 1985, South West Water have initiated and funded an extensive programme of data collection, modelling, analysis and interpretation (Fig. 2). The major study areas have been related to fisheries, water quality and hydrology but also included ecological and river corridor studies.

This work has concentrated on the Tamar catchment, where an extensive data base has been established, and its principal components have been:

   (a) juvenile and adult salmonid surveys on the Lyd subcatchment;

   (b) trapping of adult salmon and sea trout at Gunnislake;

   (c) radio tracking of adult salmon and monitoring behaviour from the estuary;

   (d) analysis of records of historic rod (private and statutory returns) and net catches plus an anglers' census;

   (e) extensive water quality sampling (continuous and spot) and modelling;

   (f) development of an adequate hydrometric network, development of a flow model, and time of travel studies;

(g) macroinvertebrate surveys on the rivers of the Lyd sub-catchment and main Tamar;

(h) sedimentation study of affected river reaches; and

(i) wetlands and river-corridor surveys.

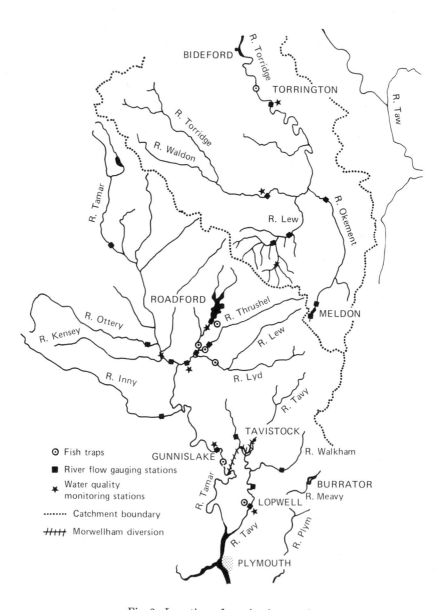

Fig. 2. Location of monitoring stations.

The majority of the studies have had a distinctive fisheries bias, but even so it must be recognized that the Tamar data base has been extensively used to formulate the basis for determining the operating rules for Roadford Reservoir and abstraction at Gunnislake. However, the concepts developed on the Tamar have been applied to the other rivers where fewer data were available and where further investigations were thus needed.

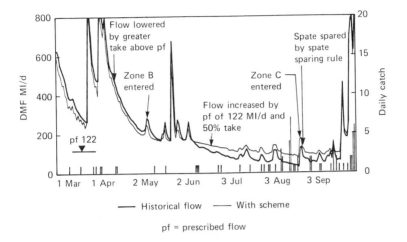

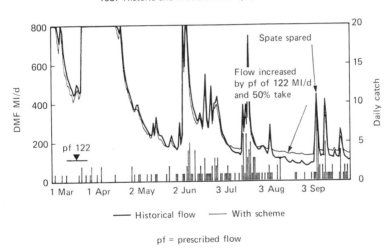

Fig. 3. River Dart salmon catches and flows.

An important source of fisheries data has been angling catch statistics. These have been obtained through individual fishery owners and from the statutory rod returns. These data have been particularly useful for the rivers where no research has been carried out into movements of salmonid fish (i.e. all rivers except the Tamar), making use of the correlation which is known to exist between salmon migration and catchability.

CONSULTATION

It was recognized at an early stage that consultation with fishery, riparian and conservation interests would be a vital aspect of the successful promotion of the scheme. Following the public inquiry, the Roadford Fisheries Liaison Committee was established. This committee was the main forum for consultation of fisheries matters related to the Tamar, Tavy and Torridge. In addition to this original forum, the study expanded the circle of consultees to include the other rivers and a total of 10 national and regional conservation groups. The level of consultation increased during the period of refinement of the scheme operating rules.

REFINEMENT OF OPERATING RULES

Preliminary assessment of impact

The effect of the scheme on river flows was assessed using a computer simulation model developed by South West Water. The model simulated the operation of the scheme on a daily basis for the period of 1957 to 1991. The model, although simple in principle, was complex in detail and took account of all the abstraction licence conditions relating to potable supply, reservoir control rules for Roadford, Burrator and Meldon Reservoirs and conjunctive use with a number of minor sources in the region. The model had many outputs including daily flow hydrographs for various locations. These compared natural flows with the estimated flows when the scheme was in operation. Examples are shown in Fig. 3. These were essential tools when assessing the environmental impact of the scheme.

The impact of the scheme with its originally proposed operating rules can be summarized as:

(1) loss of salmonid spawning areas in the Wolf catchment. This could be compensated for by adoption of an adequate salmonid enhancement programme;

(2) the impact of the scheme elsewhere in the Tamar catchment was expected to be very slight due to the conservative abstraction conditions proposed and the lower priority given to abstraction from the Tamar in relation to the other sources of Burrator and the Tavy;

(3) the increased abstraction from the Torridge, Tavy and Dart would have an impact on the flow regimes of these rivers, particularly at times of fairly low flows though not drought conditions. These changes could possibly have adverse effects on salmon migration, angling and water quality;

(4) delayed spillage of Burrator Reservoir in the autumn could have an effect on salmon runs and angling prospects in the late winter fishery of the River Plym; and

(5) the abstraction of water from all the rivers affected at times of small summer spates could have a particular impact on migration of salmon.

The sensitivity of salmon migration to the availability of water during small spates is generally well known, but was confirmed by the programme of investigations on the River Tamar and by examination of angling catch statistics (Fig. 4).

From this assessment it was concluded that although the measures proposed at the time of the public inquiry would go a long way towards minimizing the environmental impact of the scheme, there were still a number of areas where substantial improvements could be achieved by adoption of more sophisticated operating rules.

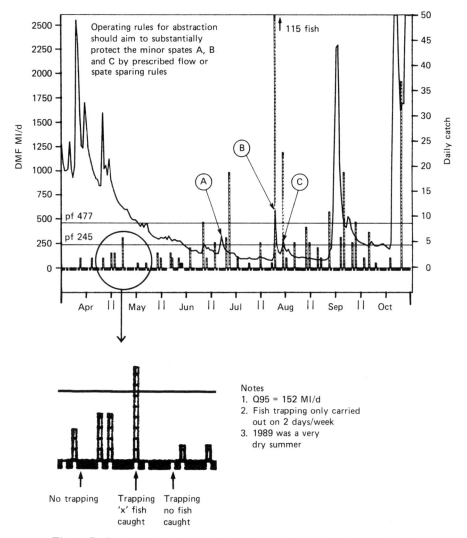

Fig. 4. Daily catch of salmon and daily mean flow at Gunnislake, 1989.

Choice of operating rules

The principal adverse impact of the scheme with its originally proposed operating rules would arise not in the Tamar catchment, which would be well-protected, but in the five other catchments, where increased abstraction would be supported by Roadford only during droughts. In non-drought years, under the originally proposed operating rules, only a small part of Roadford Lake capacity would be used. Therefore, the first refinement of the operating rules was to introduce a system which would allow more use to be made of water stored in Roadford Lake during non-drought years to reduce the amount of abstraction from the other rivers. This could be achieved by linking the abstraction conditions with control rules for Roadford Lake. The amount of storage remaining in Roadford Lake could be divided into three zones, and abstraction conditions such as prescribed flows could vary depending on the storage. An example of this is given in Table 1.

Table 1. Variation of prescribed flows (Ml/d)

| River | Zone A | Reservoir storage | | |
		Zone B	Zone C	Q95
Tamar	304	220	90	152
Tavy	73	73	41	93
Dart	122	122	80	122
Torridge	200	150	80	80
Taw	208	150	104	104

Fig. 5. The River Tamar below the Gunnislake abstraction.
Flow approximately 150 Ml/d (Q95).

The choice of the prescribed flows in Zones A, B and C is based upon the assessment of the impact of low flows on the migration and catches of salmon and sea trout on each river, and other uses. In most of the rivers, salmon migration is minimal at the level of flows being considered for prescribed flows, and thus is not a major consideration. However, on the Tamar, for example, decreasing salmon migration does take place at flows down to the Q95[†] (Fig. 5) and below – hence the case for greater protection of such conditions. On the Dart and Tavy, analysis of angling catch records suggests that sea trout migration past the abstraction takes place even at very low flows, and that this activity is not sensitive to discharge within the range being considered. On the Torridge and Taw, there is evidence of a sharp falling-off of sea trout migration at flows below Q95 – hence the protection of such flows even in Zone C. In the case of all the rivers other than the Tamar these conclusions are based upon inadequate data and any operating rules based upon them are considered interim pending further investigation (Fig. 6).

Fig. 6. River Torridge at Torrington. Flow approximately 200 Ml/d
(more than double Q95).

Small summer spates, with a peak flow above the prescribed flow, may be very important for salmon migration in dry years. Although the relatively large abstractions in this scheme can greatly reduce the size of such spates, they can be protected by adopting a spate sparing rule.

On the Tamar, trap and tracking observations were used in conjunction with private catch statistics to characterize flow requirements needed for salmon migration. As a result the following rules were formulated:

(a) as spate sparing is only needed to protect 'small spates' at times of low flow, an upper threshold of 500 Ml/d has been applied above which abstraction is not affected;

(b) the falling water following a spate is most important to salmon migration and protection is given for a duration of five days following the peak. Aug-

---

[†] Q95 = Flow which is equalled or exceeded for 95% of the time

mentation releases are required to support any abstraction in this period; and

(c) following a period of low flows, the first spate is most important to salmon migration, and as a result the first spate is spared. No spate is spared for a period of 14 days after commencement of sparing an earlier spate.

Similar rules proposed for the other rivers were based upon daily catch statistics and hydrological information for each specific catchment.

Consideration was given to varying the compensation flow releases from Roadford seasonally in order to optimize the benefits in terms of migration, spawning and juvenile rearing conditions for salmonids. The possibility of making extra releases of water during the pre-spawning migration period when reservoir storage allows was also examined.

It appeared that successful angling and optimal dispersion of spawning salmon in the Plym system was heavily dependent upon timely spilling of Burrator Reservoir in the autumn. A control rule to ensure spill by 1 December was investigated, although the possibility of making releases from the reservoir in the event of failure to spill was also considered.

Use of fisheries bank

The possible uses of the 2270 Ml allocation of storage for fisheries purposes was carefully considered. The initial proposal was for block releases to be made at specific times of the tidal cycle in the summer to attempt to stimulate salmon migration from the estuary at times when water quality there can be inimical to the wellbeing of the fish. This would have advantages both for the health of the stock of salmon and for angling purposes.

Impact of the improved operating rules

As already discussed, the 'improved' operating rules were based upon a specific assessment of the flow requirements for migratory fish and fisheries for that river. Thus very low flows would be protected, and above that abstraction would be allowed when it would cause minimal disruption of fish migration. This meant that abstraction would be permitted at certain low flow levels, whereas the somewhat higher flows associated with minor spates (and significant movements of fish) would be protected. Sea trout movements take place at low flow levels and appear to be affected by very low flows to a different extent in the different rivers. This presumed water quality factor was reflected in the derivation of the variable prescribed flow rules.

Although based upon the best available data, in all cases except the Tamar this was considered to be a preliminary assessment that had to be backed up or modified by the acquisition of much more direct observation of fish migration.

Operational conditions

In choosing the improved scheme operating rules, a fundamental assumption was that the drought reliable yield of the scheme would remain as assumed during the original promotion of the scheme and the public inquiries. The design drought for this work had been taken to be the period 1975/76, with a 10% reduction in demand due to drought measures during the critical period, and allowing 20% of total reservoir storage to be held in reserve. These assumptions are rather more severe than assumed at the time of original scheme promotion and reflect the increased frequency of droughts since 1976, that is 1984, 1989 and 1990.

The proposed operating rules may involve an increase in pumping and more use of the less easily treated waters, which would result in increased operating costs. This increase has been considered by the Company, taking into account its environmental policy [1].

MONITORING AND REFINEMENT OF OPERATING RULES

The proposals for operating rules to minimize the environmental impact of the scheme were based in some cases upon sophisticated computer models combined with extensive environmental data collection, but in other cases by making informed judgements using tenuous data such as angling catch statistics. The early years of the scheme operation present the opportunity to monitor its impact and refine its operating rules. The areas where further investigations are needed are:

(1) monitoring of salmon migration in the Rivers Taw, Torridge, Tavy and Dart, particularly from the estuary into freshwater, and in the vicinity of the abstraction points;

(2) ecological and water quality monitoring, particularly in the lower reaches of the rivers and upper estuaries;

(3) surveys of salmon spawning and juvenile fish in the River Wolf and the Lyd subcatchment; and

(4) monitoring salmon and sea trout migration and water quality in the rivers Taw and Torridge from the joint estuary.

These monitoring studies will allow further refinement of the operating rules, particularly those for spate sparing, and a selection of prescribed flows for abstractions.

CONCLUSIONS

1. The impact of the Roadford scheme on the six affected catchments can be substantially reduced by the adoption of complex operating rules which reduce river abstraction and make more use of reservoir storage during periods which are critical for the environment.

2. The design of such sophisticated operating rules depends upon a detailed knowledge of the interrelationship of such factors as river flows, water quality, behaviour of fish and aquatic life on a daily or even hourly basis. Such data are not available from programmes of routine environmental monitoring as normally adopted for river catchment management, and specific programmes of investigation need to be designed for any particular scheme to cater for the particular environmental impact envisaged.

3. In the case of Roadford Lake, the main means of reducing the impact on the rivers could be:

   (a) variable prescribed flows which can be maintained at higher levels except during droughts;

(b) reduction of abstraction during summer spates; and

(c) making use of 'fisheries' bank storage in the reservoir to encourage salmon migration from the upper Tamar estuary, at times when water quality is insufficient to sustain salmonid life.

4. Consultation with fisheries and nature conservation interests is as important as scientific data collection and can be an equally important source of data and other relevant information.

5. The use of sophisticated scheme operating rules, developed through a combination of data collection, consultation, computer modelling and expert judgement, can substantially reduce the impact of water resource developments without adding significantly to their capital or operating costs.

REFERENCE

[1] South West Water (1990) *Environmental Policy*.

34

The environmental impacts of Mahaweli River engineering and reservoir construction project

A. D. Moonasingha, BSc, MEng

INTRODUCTION

The Mahaweli Ganga (river) rises at an altitude of over 1200 m in the Hatton plateau of the westerly central hills of Sri Lanka. It first flows in northerly and easterly directions, winding its way around Kandy, the historic hill capital, before flowing through the low lying plains of the eastern dry zone and entering the Indian Ocean at Trincomalee (Fig. 1). The Mahaweli Ganga, at 206 miles, is the longest river in Sri Lanka, and discharges an annual flow of some 7900×10^6 m^3 [1].

Construction work of the Mahaweli Project commenced in 1970 with the interbasin transfer diversion headworks at Polgolla near Kandy [1, 2]. The Polgolla complex, comprising a barrage, tunnel and a hydropower station at Ukuwela, was completed in 1976. This initial programme, in phase 1, stage 1, also included a hydropower station at Bowatenna in the Amban Ganga basin – a tributary of Mahaweli Ganga – which was commissioned in 1981. This complex also diverts water for irrigation in the North Central Province.

The master plan of the Mahaweli Project envisaged completion in 30 years, but in 1977 metaplanning by the key figures of the government led to the accelerated implementation of the remaining major phase of the project to meet the growing socio-economic and power supply needs of the country, using bilateral and multilateral aid. This programme comprised the construction of four major reservoirs and three hydropower stations which were completed and commissioned from 1981 to 1990. The accelerated programme also encompassed most of the major irrigation settlement zones in which settlement and development were nearing completion [4]. The Mahaweli Development Project irrigates about 160 000 ha of land in the dry zone, including enhancement of irrigation supplies to over 26 000 ha of existing irrigated land, and has the capacity to generate 2×10^6 MW-hr of hydroelectric energy per annum.

ACCELERATED MAHAWELI PROGRAMME

The largest phase of the Mahaweli Project comprised the construction of the major headworks (Fig. 1). The Kotmale Project, the most upstream of the Mahaweli cascade, is located on the Kotmale Oya, a tributary of the Mahaweli Ganga. The Kotmale Project posed both serious geological and economic problems [1, 5] and consists of a rockfill dam incorporating a reinforced concrete face [6]. The Victoria Project is located on the Mahaweli Ganga approximately 22 km downstream of the Polgolla barrage, and consists of a double curvature concrete arch dam. Below Victoria is the Mahaweli's biggest reservoir, Randenigala and downstream of this is the medium sized Rantembe Project, the last of the reservoirs on the Mahaweli cascade, which was completed in 1990 [3]. The Maduru Oya Project is another major reservoir in the Mahaweli Accelerated Programme, located on the Maduru Oya basin in the Eastern Province adjacent to the lower Mahaweli catchment [1, 4]. The completion of the Accelerated Programme has allowed settlement 'systems' B, C and G to be developed.

DAM GEOLOGY

The geology of the Mahaweli Ganga dam sites comprise primarily metamorphic rocks of pre-Cambrian age [7, 8]. Site investigations have revealed that rock strata underlying the Kotmale project area comprise cavernous limestone beds with widened joints dipping downstream, and weak charnokites with an orientation of foliation shear. These geological weaknesses led to the relocation of Kotmale dam to a position 200 m downstream of the original location, which contributed to increased cost and subsequent reduction in dam height, with a reduced capacity compared to the original design [5, 7].

Sri Lanka is considered to be an area of low seismicity. However, in other parts of the world cases of seismicity attributed to too rapid filling behind dams of 90 m or more in height have been cited [5, 8]. Because of this potential risk, and the large size of the dams of Kotmale, Victoria and Randenigala, as well as the presence of well developed major lineaments in and around these projects, a microseismic network was installed in 1982 to measure any increase in seismic activity during and after impounding of the reservoirs. No such increase in seismicity along the lineaments has been observed [8].

During the period of monitoring, clusters of seismic events of less than Richter scale 2 were recorded in the major reservoir region in the hills of the Mahaweli upper catchment, and in the Madura Oya region in the east [9].

DOWNSTREAM SETTLEMENTS

The downstream land settlements in the dry zone are part of the Mahaweli Project, and are demarcated into areas which are called Systems [1, 4]. These systems occupy large areas of the Eastern Region and North Central Province.

Over 256 000 ha, or almost two thirds of the total proposed land area of Mahaweli Project, were covered with forest. Chena (slash and burn, or shifting) cultivation extended over 39 660 ha or 10% of the total land area under the project. About 6070 ha of privately-owned paddy land and highland in System H were acquired under the Mahaweli Project for resettlement [1].

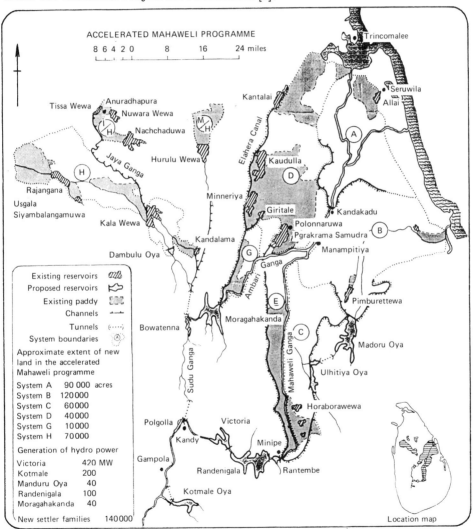

Fig. 1. Plan of the Accelerated Mahaweli Programme area.

System H, lying in the Kalawewa catchment in the North Central Province, has a total area of 30 769 ha, and is fed by the waters diverted at Polgolla. The irrigation infrastructure development in the System H area included improvements to the major Kalawewa and Kandalama reservoirs, construction of new reservoirs, the improvement of existing village reservoirs, and the construction of numerous distributor canals [1].

System C, in the Eastern Region, was an underdeveloped area of 23 886 ha, which required many roads, about 1600 km of irrigation canals, and a vast number of social infrastructure buildings. The conditions in System B, with 47 773 ha fed from Maduru Oya, were similar to those of System C.

Apart from increased rice production, alternative subsidiary crop production during the Yala (south west monsoon, May–July; September) season has proved to be a successful programme in the Mahaweli Development [1, 4].

IRRIGATION

Before the availability of Mahaweli augmentation, most irrigation schemes in the dry zone were operated in isolation. Since the first century A.D. there has been augmentation of a few major reservoirs from diversion works (anicuts) of the Amban Ganga and the Mahaweli Ganga [1]. Most of the diversions feeding the reservoirs in the dry zone received their yield from the north east monsoon, or Maha, from October to February, with an annual rainfall of about 1650 mm. In most schemes this north east monsoon yield was inadequate for the successful irrigation of two cropping seasons a year. A significant advantage of the Mahaweli Ganga diversion interbasin transfer is that the upper catchment receives rainfall from both monsoons, and with an annual rainfall of 4000 mm to 5000 mm provides runoff in most months of the year [1, 10]. The increased reliability of irrigation water for both cultivation seasons, Maha (October–February, April) and Yala (May–September), made available by Mahaweli development, provides a contrast to the isolated local irrigation schemes that do not benefit from the Mahaweli waters.

Seepage losses are an important factor in irrigation schemes [1, 16, 19]. Therefore the few canals that are in well drained soils in the Eastern region, such as the Madura Oya main and branch canals, are concrete lined [1].

Few cases of salinity have been reported in the System H irrigation area. Various soil and irrigation water tests have been carried out, with one report deducing that the water was of a suitable quality for irrigation, albeit with adequate provision for leaching [21]. Another survey, however, suggests that significant increases in total salt concentration, sodium (SAR) and residual carbonates (RSC) will occur in irrigation waters in the Kalawewa area of System H in April (end of Maha harvesting period), and to a certain extent in November (end of late Yala harvesting period) [22]. Some well waters sampled in the System C area in the Eastern region have been found to be unsuitable for irrigation because of high salinity indicators [12].

PUBLIC HEALTH

Malaria has been an endemic tropical disease in Sri Lanka from time immemorial. But with modern preventative and curative campaigns, malaria was virtually eradicated from the island in the early 1960s. Stagnant pools of water are favorable sites for the malaria vector Anopheles mosquito to breed [23]. However, since the late 1970s, when the Mahaweli settlements commenced, malaria has been an endemic disease in the North Central Province [25], and is regarded as a threat to community health in many regions of Sri Lanka [13, 23, 24]. Vigorous preventative and curative campaigns have therefore been relaunched since settlement began in the Mahaweli Project areas [26]. Schistosomiasis (bilharzia), a disease commonly associated with large irrigation projects in the tropics, is not found in Sri Lanka [13, 27]. Water-borne gastro-enteritis and other diarrhoeal infections are also common in the Mahaweli Projects areas, although the problem is not yet acute [26, 28]. There was an epidemic of cholera in 1974 in the Mahaweli settlement area, which receded in 1976 [13] and mortality of infants and children from diarrhoea is almost absent [28]. Japanese encephalitis is reported to be a major health problem in Sri Lanka, as the main vector breeds in rice fields, and outbreaks have been reported in the Mahaweli settlement areas [23]. The health situation in the Mahaweli Development area is not, overall, markedly different from the rest of the country, except for the danger of malaria.

The main sources of drinking water are shallow wells fitted with hand pumps, with a small percentage of households having their own wells. After a few years such wells fall into disrepair with covers removed and handpumps broken, and apart from the wells near the reservoirs many dry out during the dry season. Drinking water supply from shallow wells in the region is therefore inadequate, and settlers travel long distances in the dry season to wells with water, to main canals and to reservoirs, in order to fetch water [13, 28]. To improve conditions deep wells are being constructed in favorable locations [28]. Deep wells do not dry out and can serve about 50 families, and although the capital cost is high (five times the cost of a shallow well) these are more economically viable. Standpipe water supplies are available in many townships in the Mahaweli Development area [29]. The construction and use of lavatories is poor, but improving. Of the permanent latrines built, about 35% are the water sealed type and the remainder are pit latrines [28].

SOCIAL IMPACT

The major social impacts of the scheme were the displacement of the local population from the reservoir construction areas, and their resettlement in the new irrigation system areas.

Many families were evacuated from the Victoria and Kotmale reservoir sites, and although the majority were resettled in the downstream irrigation settlements, others have opted to resettle in areas made available in Kandy District and in Kotmale [2, 8]. Also private land and chena cultivation areas that came under the new development area were acquired, and their owners were resettled [1]. As the Victoria Reservoir inundated old shanty-towns, new towns were built to replace those lost [8]. Fishing is practised in all large reservoirs, and the availability of fish and the growing demand have encouraged people to take up fishing as a new livelihood [8, 29].

New settlers in the downstream developments suffer from both physiological and psychological stress. Apart from the small proportion who come from the dry zone, most settlers come from a milder climate than the harsh dry climate of the downstream settlements. Physical tiredness, and sickness associated with the change in climate and poor drinking water quality tend to hinder the progress of the settlers' initial work. *Inter alia*, the fear of wild animals, the danger of snakes, and personal conflicts between the new arrivals causes a general feeling of insecurity [8].

A well-formulated socio-economic welfare system is provided for the benefit of the women and children. The Mahaweli irrigation settlement system areas are managed by a team including professional advisers of several disciplines.

Training programmes for the benefit of junior managers, and for community participation, have been successful and popular [28]. Farmers in Sri Lanka have in general adapted to semi-mechanical and scientific cultivation practices, and although the overall mechanization of rice growing is still in its infancy, it is developing.

A significant impact outside the Mahaweli development area itself is the extensive rural electrification, as electricity generation has increased.

ENVIRONMENT

The Mahaweli Project area can be considered as two separate regions; the upper catchment, and the downstream development.

The extent of the upper catchment, which comprises the major reservoir catchments, is more than 316 000 ha [11]. About 49% of this area was intensively managed land, of which two thirds was mainly tea, and 14% was covered by forest. The rest of the land supported dry land agriculture and chena cultivation, or was abandoned tea and grassland. The mean annual temperature in the region varies little from 27°C [10].

The previous practice of tobacco and chena cultivation has been banned from the immediate reservoir catchments in order to reduce soil erosion, and farmers are encouraged to cultivate more permanent crops. Where tobacco is grown, farmers are encouraged to implement soil conservation techniques. All estates in the catchment are required to provide sufficient ground cover in all ravines and gullies [29]. The Kotmale valley has a record of mass movement, but there is no conclusive scientific evidence to show that the reservoir and its operations have caused cracks or serious damage to houses in the area [27].

It is cited that soil erosion can range from 308 to 913 tonnes/ha annum in badly-managed tobacco lands [8]. Neglected tea plantations are also a source of heavy soil erosion [27], although in comparison a well maintained tea cover provides the best protection to soil [8].

With an estimated average annual sediment yield from the catchment of $300 \, m^3/km^2$, and with allowance for bed load, the total sediment load is about $10^6 m^3$/annum. It is considered that this sedimentation rate would not impair the effective operating capacity of the reservoirs for at least 50 years [10].

The Water Resources Board found that groundwater was heavily polluted by agrochemicals in some of the tubewells drilled in the area around Nuwara Eliya, making the water unsuitable for human consumption [8]. The heavy use of fertilizers and agrochemicals in the tea estates and the vegetable plots poses a serious problem to the Mahaweli reservoirs. The fact that Mahaweli waters are conveyed a considerable distance through the interbasin transfers highlights the need for vigilance on monitoring water quality in the Mahaweli areas [8].

Flood peaks in the Mahaweli Ganga will be considerably reduced by the storage in the new reservoirs and controls in the headworks, although Kotmale reservoir is likely to spill annually because it has been built below its optimum capacity. Victoria reservoir is built for maximum capacity without flooding Polgolla. The new hydrology of Mahaweli Ganga, particularly below the dam sites, could increase the risk of stagnant pools, which are a favourite breeding ground for mosquitoes which spread malaria[23, 24, 27].

High flow river conditions of the Amban Ganga attributable to overloading by the Mahaweli diversion operations have been cited. The replenishment and spilling of many reservoirs in the dry zone that are augmented by the Mahaweli waters are now influenced more by the diversion operations than the mass curve criteria based on local catchment inflows. In the downstream areas the clearance of forest will increase the proportion of rainfall runoff, although the paddy fields, with water retaining bunds, will control the rate of runoff, and increase the base flow. In the same context evapotranspiration will be reduced, but the evaporation will increase from the imported waters covering large surface areas. The reduction in the downstream flows of the Mahaweli Ganga may influence the saline intrusion to the lower flood plain, which could be aggravated by return flows from irrigation[10].

All reservoirs will show some thermal and chemical stratification with reduced oxygen at the bottom layers, although apart from the very deep major reservoirs on the Mahaweli Ganga, the reservoirs will probably mix one or more times a year[10, 12]. In the Kotmale reservoir the establishment of aquatic and semiaquatic vegetation has been slow [27]. The reservoirs on the Mahaweli Ganga will adversely affect the river fish 'marhsier' and several other species. Many other species, such as tilapia and carp, with demand as a food fish, will proliferate[10, 29]. The diminution of the species 'marhsier' is not considered a loss, because it occurs in large numbers in other rivers [10]. The flood plain 'villus' (river marshlands) are a landscape feature liable to be adversely affected . Reduction in seasonal inflow of water may cause gradual changes in these semiaquatic ecosystems which favour the survival of several wildlife species, particularly wading birds[10, 11].

Several thousand acres have already been reforested under the reforestation programme, as a means of protecting the upper catchment of Mahaweli [29]. Fuel wood plantations are also included in the reforestation, and those in the downstream development Systems have already been implemented. In Systems B and C, however, all the available supplies will be exhausted before the plantations are ready for use, and this will result in degradation of the fuelwood development [29].

As the downstream settlement projects are implemented, mammals are displaced and move into the remaining forest habitats in the surrounding areas, thus causing overcrowding. Imbalances in the natural ecosystems can then result in a decrease in wildlife populations. Damage to paddy and other crops by displaced elephants has increased since the accelerated downstream development [3, 10], although water holes in the Wilpattu National Park have been improved to increase the carrying capacity [2, 9]. Additionally, seven protected wildlife areas have been declared covering a land area of over 230 000 ha. These reserves have varying degrees of protection, from strict nature reserves to wildlife corridors. Encroachment of forest land in the settlement System areas and in the Flood Plain National Park is, however, a persistent problem [29].

There is potential for tourism development in the Victoria, Randenigala and Madura Oya reservoirs, with wildlife attractions at the latter.

Archaeological remains in the project areas have benefited by the development as action has been taken to preserve and exhibit them.

Apart from a few new roads constructed to re-route the stretches of inundated main roads, and a link road facilitated by the Victoria Dam, the old highway infrastructure in the upper catchment had been neglected. Modern road links have now been constructed to remote settlement regions.

OPERATIONS

A key issue in the water use of the Mahaweli Project is the flow of water diverted at Polgolla, compared to the remaining flow routed along the Mahaweli Ganga to Victoria and Randenigala Reservoirs. Victoria Reservoir tends to be given priority with respect to the availability of Mahaweli water, because it has the better economic criteria of water use [20]. The Mahaweli Project has more than doubled the power generation capacity of the country with an increased capacity of 600 MW. The augmentation facility has changed the mean water level of the large reservoirs fed by the Mahaweli, attributable to the increased operational possibilities now available. Fig. 2 depicts the Mahaweli operations network.

CONCLUSIONS

As a result of the metaplanning impetus of the political authority, the major headworks and other infrastructure of the Mahaweli Development Project were implemented during the period 1981–1990 under an Accelerated Programme, with bilateral and multilateral aid.

The geology of the upper catchment of the Mahaweli includes several weak strata, and their adverse effects have aggravated the construction problems of the Kotmale headworks. Microseismic activity has been monitored, with concentrations of less than 2 on the Richter scale recorded adjacent to Victoria Reservoir, and in the Madura Oya region. An increase of microseismic activity due to impoundment of these large dams has not been recorded.

The river ecosystem in the upper catchment has been replaced with a deep reservoir ecosystem. A programme of soil conservation, planned vegetation, and reforestation to protect the Mahaweli upper catchment has been implemented.

In the Mahaweli distribution system more water is routed through Victoria along the Mahaweli cascade, which is considered to be the most favorable economic operational policy. Downstream flood prevention may gradually affect the flood plain villus.

Mahaweli Project has provided irrigation with increased reliability to a large area of existing and new land in the dry zone of Sri Lanka. The Mahaweli Project land exhibits a sustained green environment in contrast to the irrigated land in the dry zone which is not fed by Mahaweli waters. Mild effects of salinity have been reported in one area. Overall the health conditions in the Mahaweli settlements are not greatly different to the rest of the country. The agricultural development

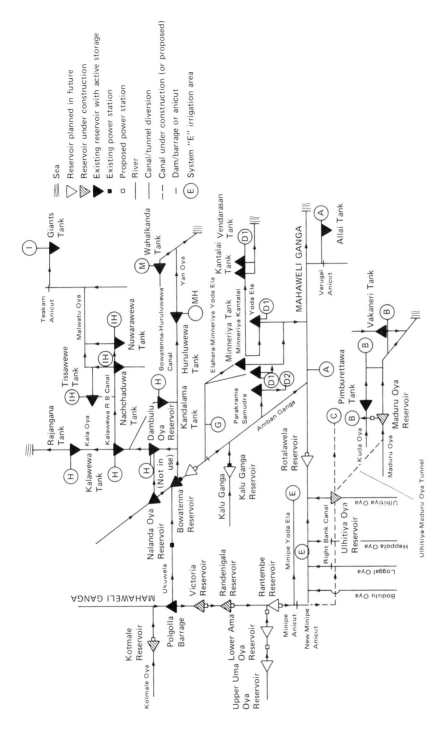

Fig. 2. Schematic layout of the Mahaweli water conveyance system.

of the Mahaweli Project has cleared a vast area of forest land in the dry zone and destroyed wildlife habitats. In consequence damage by elephants to cultivated areas has increased. The downstream natural forest environment has been replaced with an irrigated agricultural environment beneficial to humans, rich in aquatic resources and lush vegetation. New wildlife parks have been designated. The Mahaweli Project may not have caused the extinction of any species of fauna or flora, but has caused environmental disturbances which contribute to a reduction in the populations of already endangered species.

ACKNOWLEDGMENTS

The author is grateful to the Minister of Lands, Irrigation and Mahaweli Development, Mr P. Dayaratne, and the Mahaweli Authority of Sri Lanka, for their kind permission to carry out research for this chapter. Grateful acknowledgements are also due to other institutions and officials for assisting with the site visits and enquiries. The author gratefully appreciates the publications and other works listed in the references, on which this chapter is based.

REFERENCES

[1] Karunatilake, H. N. S. (1988), *The accelerated Mahaweli Programme and its impact*, Centre for Demographic and Socio-economic Studies, Sri Lanka.

[2] Vistas of Mahaweli (1990), *Newsletter of the Mahaweli Centre of the Mahaweli Authority of Sri Lanka*, **5**, No. 3, 6.

[3] *Ibid*, (1990), **5**, No. 2.

[4] *Accelerated Mahaweli Programme*, A summary report, Mahaweli Authority of Sri Lanka, undated.

[5] Brown, K. (1982), Kotmale – the dam that moved, *Construction News Magazine*, **8**, No. 8, 18–25.

[6] Gosschalk, E. M. and Kulasinghe, Dr., A. N. S. (1985), Kotmale Dam and observations on CFRD, *Proceedings of a symposium sponsored by the ASCE, Concrete Face Rockfill Dams*, Detroit, pp. 379–395.

[7] Sir William Halcrow and Partners, *Kotmale Hydropower Project*, Technical brochure, undated.

[8] *Victoria Dam and its impact* (1986), A Seminar held at Hotel Hantana, Kandy, Sri Lanka.

[9] *Final report on Kotmale microseismic studies*, Sri Lanka (1984), Department of Earthquake Engineering, University of Roorkee, U.P. India.

[10] TAMS, USAID, (1980), *Environmental Assessment, Mahaweli Development Program*, Ministry of Mahaweli Development, Colombo, I.

[11] *Ibid*, Terrestrial Environment, II.

[12] *Ibid*, Aquatic Environment, III.

[13] *Ibid*, Human Environment, IV.

[14] Acres International Ltd, (1984) *Organizational aspects*, Mahaweli Water Resources Management Project, Water Management Secretariat, Colombo, Canadian International Development Agency, Canada.

[15] Ibid, (1985), *Studies of operating policy options*, Main Report.

[16] *Ibid, Policy studies briefing document*, Appendix I.

[17] *Ibid, Irrigation water requirements*, Appendix II.

[18] *Ibid, Basic data*, Appendix III.

[19] *Ibid, Acres Reservoir simulation program*, Appendix IV.

[20] *Ibid, Economic aspects of water use*, Appendix V.

[21] Handawela, H. *A study on inland salinity in Mahaweli H Area*, Mahaweli Authority of Sri Lanka, undated.

[22] Gunawardhana, H. D., Kumudini, A. M. and Adikari, R. (1981), Studies on the quality of irrigation waters in Kalawewa Area, University of Colombo, *J. Natn. Sci. Coun. Sri Lanka*, 9(2): 121–148.

[23] Joint WHO/FAO/UNEP, Panel of Experts on Environmental Management for Vector Control, *Proceedings of the Workshop on Irrigation and vector-borne disease transmission*, International Irrigation Management Institute, Sri Lanka, 1986.

[24] Wijesundara, Dr., M. (1990), *Daily News*, Sri Lanka, Friday, April 6, p. 15.

[25] Patrick, W. K. (1983), *Primary Health Care in the Mahaweli*, UNICEF sponsored-Health Education Bureau Publication.

[26] Nichalson, Dr. A. et al. (1978), *Aid in confidence*, Report on and Overseas Development Ministry Health Mission to Sri Lanka.

[27] Johanson, D. (1989), *Kotmale environment*, Swedish International Development Authority Evaluation Report, SIDA, Stockholm.

[28] *Social development benefiting children and women in System H of Mahaweli Development Area*, Evaluation of the Project, National Institute of Business Studies Management, Colombo, 1989.

[29] Sobczak, Dr. M. T. *Mahaweli Environmental update* USAID, Colombo.

35

The natural development and water management of Lake Volkerak-Zoom, the Netherlands

F. L. G. de Bruijckere, MIN. T., P. W.& W. M.[†]

INTRODUCTION

In April 1987, with the closure of Philips Dam, a new lake was created in the Netherlands, Lake Volkerak-Zoom. Philips Dam was the final stage of the Delta Project, the specific aim of which was to protect south-west Netherlands against flooding and to improve freshwater supplies for agricultural purposes. After a desalination period of about a year, the previously estuarine section of the Eastern Scheldt had become a stagnant freshwater lake. Since then, this lake has been fed primarily by small tributary rivers and by supplementation water from the Rhine/Meuse system.

As a result of the closure, North Sea water no longer dilutes the incoming river water, and concentrations of nutrients, heavy metals and organic micropollutants have increased. Without preventative measures it is anticipated that an ecosystem dominated by algae will develop, and that lake sediments and organisms will become highly contaminated with micropollutants.

A third significant adverse effect of the new Lake Volkerak-Zoom is erosion of the shoreline. When the area was still tidal, alternate processes of erosion and sedimentation took place. Following the closure, only the process of erosion remains,

[†] Water Management Section, Zealand Division, Ministry of Transport, Public Works & Water Management, the Netherlands.

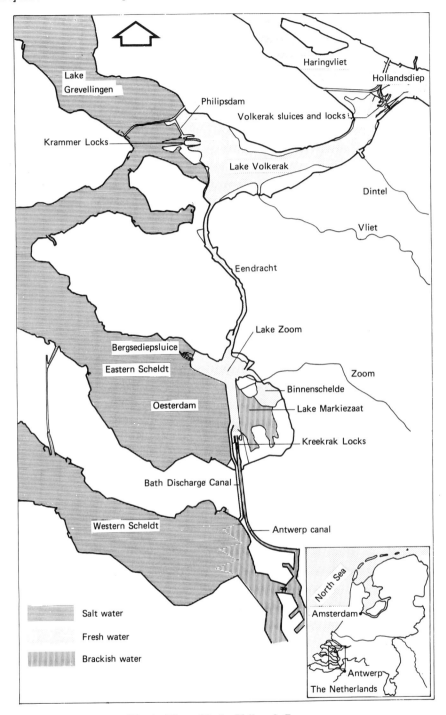

Fig. 1. Map of Lake Volkerak-Zoom.

due to concentrated wave activity at one level along the shoreline. This chapter examines the threats to the environment and the policy devised to tackle them.

MORPHOLOGY AND HYDROLOGY

Lake Volkerak-Zoom consists of two parts: Lake Volkerak in the north and Lake Zoom in the south (Fig. 1). The two lakes are connected by a channel, the Eendracht, and form a single water system. Table 1 shows the morphological data of both lakes.

Table 1. Morphology of Lake Volkerak and Lake Zoom

	Lake Volkerak	Lake Zoom
Surface area (km^2)	45.7	15.5
Volume (m^3)	238×10^6	80×10^6
Mean depth (m)	5	6
Max. depth (m)	24	20

An ingenious infrastructure has been created for the benefit of water management and shipping. Locks have been constructed to enable ships to pass between Lake Volkerak and the port of Rotterdam and the Rhine (the Volkerak locks), the Eastern Scheldt (the Krammer locks, Fig. 2) and the port of Antwerp (the Kreekrak locks). In addition to its function as a shipping route, however, Lake Volkerak-Zoom also provides a freshwater supply for agriculture. The Krammer and Kreekrak locks are therefore equipped with freshwater/saltwater separation systems to minimize the intrusion of salt or brackish water. In the north, the Volkerak locks are equipped with a sluice for flushing the lake to maintain the water level and to compensate for salt intrusion. To avoid damage to agriculture, chloride concentrations should be maintained below 400 mg/l. In the south excess water is carried to the Western Scheldt via the Bath Discharge Canal. Table 2 shows the water balance in the period 1988–1989.

Table 2. Water balance in Lake Volkerak-Zoom in 1988 and 1989 (m^3/sec)

Mean inflow	1988 m^3/sec	1989 m^3/sec	Mean outflow	1988 m^3/sec	1989 m^3/sec
Volkerak Sluice	9	9	Outflow Bath	22	7
Volkerak Locks	3	3	Locks	15	16
Tributary rivers	24	11	Agriculture	0.5	0.5
Precipitation	3	2	Evaporation	1.5	1.5
Total in	39	25	Total out	39	25

The tributary rivers (Dintel and the Vliet), which drain freely into the north west of Lake Volkerak, are an important source of freshwater. The size of the catchment is about 165 000 ha, and in a wet year such as 1988 Lake Volkerak-Zoom is filled largely with discharge from the tributary rivers. In a dry year such as 1989, however, the ratio between the inflow of Rhine/Meuse water and water from the tributary rivers is of a similar magnitude (Table 2).

Fig. 2. Krammer sluices and lock complex.

DESCRIPTION OF THE THREATS TO THE ENVIRONMENT

Eutrophication

Eutrophication is a widespread phenomenon in the Netherlands. Most freshwater lakes suffer from excessive algal growth due to excessive nutrient load. Light can no longer penetrate to the bottom and aquatic plants no longer grow. Pike, which depend on these plants for cover, then leave or do not breed. Species of white fish, particularly bream, benefit from the absence of this major predator, and quantities increase so rapidly that the waters become infested. This results in the growth of individual bream being stunted, as the high populations quickly deplete their main food resource, the bottom fauna. This, in turn, increases the pressure on the zooplankton population, and algae, the food of herbivorous zooplankton, are able to grow almost unhindered. In their search for food, the bream disturb large quantities of sediment; the water becomes turbid; nutrients are released from the bottom and aquatic plants are uprooted.

Two states of equilibrium are theoretically possible [1]. One is a eutrophic, relatively turbid system with high nutrient concentrations, and dominated by algae and whitefish and is indicated in Fig. 3. The other possibility is a system with clear water, aquatic plants, a predatory fish population that keeps the whitefish population under control, and sufficient zooplankton to be able to graze the algae effectively. This is also indicated in Fig. 3.

Fortunately, and contrary to all predictions, Lake Volkerak-Zoom is not yet dominated by algae. The lake contains elements of both equilibrium situations: many nutrients, clear water and high quantities of zooplankton; this too is indicated in Fig. 3. In the absence of a significant whitefish population, the grazing pressure of the zooplankton on the algae is still very high. For that reason, the algae population is low and the water of the lake is still transparent. Aquatic plant life has started to develop, but not yet in sufficient quantities to offer shelter and a good breeding environment for predatory fish and zooplankton. Although the light conditions seem favourable, the inflow of nutrients, especially phosphorus, to Lake Volkerak-Zoom is too high to maintain a stable, clear water equilibrium in the lake.

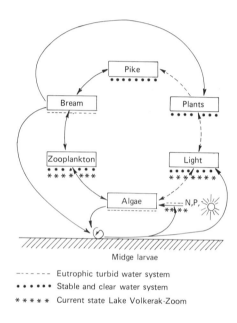

Fig. 3. Food chains in eutrophic shallow lakes.

Table 3 shows the average annual phosphorus load in Lake Volkerak in 1988 and 1989.

Table 3. The external phosphorus load of Lake Volkerak in 1989 and 1989

	1988	1989
Volkerak Lock and sluice	100	75
Tributary rivers	380	175
Inflow P† (10^3 kg/year)	480	250
External P load (g/m²/year)	10.5	4.9

†P is the chemical symbol for phosphorus

Computer simulations show that the most effective way to combat eutrophication is to reduce the phosphorus concentration. In order to limit phosphorus concentrations and thus algal growth, the average annual phosphorus-loading should not exceed 2–4 g P/m^2/year ($\approx$180 tonnes P/year for the lake). The exact rate to which the phosphorus load should be reduced also depends on the capacity of the developing aquatic system to compensate for the relatively high nutrient load.

Table 4 shows that 50% or more of the external phosphorous load remains in Lake Volkerak, the proportion depending on the discharge of the tributary rivers. Thus the process of phosphorus accumulation in the sediments of the lake has already commenced. As a result of the phosphorus accumulation, the release of phosphorus from the sediments of the lake will be increased. In turn, the lake water phosphorus concentration will increase, and with it the algal growth. In Lake Zoom there is not yet any significant accumulation of phosphorus.

Table 4. Phosphorus balance Lake Volkerak in 1988 and 1989

	1988	1989
Inflow P (10^3 kg/year)	480	250
Outflow P (10^3 kg/year)	230	80
Residue (10^3 kg/year)	250 (52%)	170 (68%)

Micropollutant contamination

The Netherlands lies on the delta of the Rhine, Meuse and Scheldt rivers. In delta areas rivers with braided channels discharge to the sea and the tidal cycle can affect water levels a considerable distance inland. Prior to 1965 the influence of the Rhine and Meuse extended as far as the Eastern Scheldt estuary. At that time, the Delta was in a state of near morphological equilibrium.

Until 1970 the amount of sand and silt brought down by the Rhine and Meuse rivers was roughly equal to that discharged through the estuaries to the North Sea. After 1970, successive closures in the Delta area produced changes in the flow distribution of the main rivers (Bijlsma and Kuipers). Micropollutants (heavy metals and organic compounds) are attached to the fines in the sediment carried by the rivers, with the result that the Delta region is also affected by these contaminants. The closure of the Haringvliet and the creation of Lake Volkerak-Zoom led to the formation of new sedimentation areas. Layers of contaminated material continue to settle in these sedimentation areas. A relatively large amount of silt transported by the Rhine and Meuse and the tributary rivers is deposited in these areas, and as a result, the bed has become contaminated with micropollutants. Table 5 shows the salient contaminants in the suspended sediments carried into Lake Volkerak-Zoom [3]. In order to categorize different sediments, four standard classes have been defined. Class 1 is ecotoxicologically safe. In class 2, ecotoxicologic effects are to be expected and in class 3 ecotoxicologic effects have been observed. Finally, class 4 shows direct toxic effects for aquatic organisms. Table 5 shows that the transported materials in the Volkerak inlet are more polluted than the suspended sediments carried by the tributary rivers. The suspended sediments in the lake are less polluted than the transported materials, and the suspended sediments in Lake Volkerak are more polluted than the suspended sediments in Lake Zoom, which implies that more micropollutants have accumulated on the bottom of Lake Volkerak than Lake Zoom. The bottom sediment of the lake was sampled both before and after the closure, and the results of this study showed that by 1989 the bed directly behind the Volkerak sluice and at the mouth of the tributary River Dintel was more contaminated with pollutants than prior to closure in 1986. The bottom sediment of the lake near the Volkerak inlet contained increased concentrations of PAH and PCB (class 3), and at the mouth of the Dintel, increased concentrations of PAH (class 3).

Table 5. The pollution grade of the suspended sediments transported into the Lake Volkerak-Zoom

Compounds[†]	Volkerak inlet	Tributary rivers	Lake Volkerak	Lake Zoom
Heavy metals	Cd, Hg, Cu, Pb, Ni, Zn	Cd, Cu, Ni, Zn	Cd, Hg, Cu, Ni	Hg, Cu, Ni
Organic pollutants	PAH and PCB HCB	PAH, PCB DDT, HCHs	PAH, PCB DDT, HCHs HCB	PAH, PCB DDT, HCB HCHs
Quality class	3–4	2–3	2–3	2

[†] Heavy metals: Cl = cadmium, Cu = copper, Hg = mercury, Pb = lead, Ni = nickel and Zn = zinc. Organic pollutants: PAH = polycyclic aromatic hydrocarbons, PCB = polychlorobiphenyls. DDT, HCB and HCHs are organochloro-pesticides.

Erosion of the shoreline

With tidal water-level variations no longer existing, wave activity causing erosion is concentrated in a narrow zone. Additionally this erosion is no longer compensated for by sediments deposited by tidal currents during high tides. The result is an increased net erosion of between 5 to 35 m/year in certain locations. In the 1987–1988 period about 20 ha of flats and former salt water marshes were disappearing annually from Lake Volkerak-Zoom, which is about 1% of the present area, although without protection measures, much more of the valuable remaining shore area will disappear.

In order to develop well-balanced aquatic and terrestrial communities, protection and conservation of the shoreline and the adjacent shallow parts of the lake are essential, and the designation of most of this area as a nature reserve will assist in this objective.

COMBATTING EUTROPHICATION

Emission-oriented measures

To reduce eutrophication in Dutch surface waters such as Lake Volkerak-Zoom, a dual policy is being pursued [4]. General emission-oriented measures are being emphasized as an essential first step. These measures, together with a substantial reduction of nutrient input via transboundary rivers, which has resulted from the Ministerial Declarations on the North Sea, the Rhine Action Plan, will provide a sound basis for the regional effect-oriented measures necessary for a sustainable development of the different water systems.

Municipal sewage treatment plants

At present the mean phosphorus removal efficiency is about 40%, but an agreement between the Government and the water authorities, signed in 1989, required a mean phosphorus removal efficiency of 75% by 1995 for municipal sewage treatment plants. Anticipating this policy, an agreement was signed last year between the Zealand Division and two adjacent water authorities, one in the Netherlands and one in Belgium.

Under this agreement thirteen sewage works discharging into Lake Volkerak-Zoom have been subject to this phosphorus removal requirement from 1992.

Agriculture

Leaching and runoff from agricultural land is a major source of nutrients for Dutch surface waters, and is closely related to the excessive application of manure.

In order to avoid adverse effects on soil and water quality, a comprehensive set of legal measures has been developed. The primary objective of the policy is to achieve a balance between the application of manure and chemical fertilizers, and their uptake by crops, by the year 2000. The transition period up to the year 2000 is needed for a phased implementation of far-reaching changes in Dutch agricultural practices. As a result of these stringent rules, the runoff of phosphorus will decrease. Unfortunately, there is a considerable time lag between the implementation of measures and the reduction of nutrient leaching. As a result of this, phosphorus leaching is anticipated to increase in the next few years, until the effect of the policy come to fruition.

Regional effect-oriented measures

To ensure adequate eutrophication reductions in inland and coastal waters, nutrient input should be further reduced to 70%–75% of the 1985 input, to be achieved by 1995. In the meantime regional effect-oriented measures are necessary to prevent Lake Volkerak-Zoom becoming a stable eutrophic system.

The following regional effect-oriented measures have been implemented, or are planned, to reduce the phosphorus load in Lake Volkerak-Zoom.

(a) Diversion of effluent from the largest sewage works in the catchment. In 1988, effluent from the Nieuwveer sewage works was diverted from the Dintel tributary to the Hollandsch Diep, a larger surface water, which reduced the phosphorus load by 200 tonnes P per year.

(b) Limitation of the water losses. In 1988, water losses from the locks were reduced by nearly 50% by the development of a sophisticated saltwater/freshwater separation system. This decrease in the inflow through the Volkerak sluices of about $15 \, m^3/sec$ equates to a reduction in the phosphorus load of about 100 tonnes P per year.

A third water control system (Bergsche Diep sluice), to prevent salt intrusion from the Eastern Scheldt to Lake Zoom, was constructed in 1991. An overall reduction of the phosphorus load of about 125 tonnes P per year can thus be achieved by optimizing the hydraulic operations of the lake.

(c) Treatment of intake and tributary water. To date semi-technical experiments have been carried out with biofilters consisting of freshwater mussels (zebra mussels), with plans to locate these at the Volkerak sluices. The principle is based on the filtering capacity of large quantities of zebra mussels suspended on nets, and the experiments proved that about 50% of suspended sediments could be filtered by the zebra mussels. As about 50% of total phosphorus concentration adheres to suspended sediments, the total phosphorus load could be reduced by about 25% with a biofilter. Application on a large scale is not yet possible, however.

(d) Biofilters are not suitable for use in the mouths of the tributary rivers because of their fluctuating discharges. A more appropriate measure to reduce the phosphorus load from these sources could be the creation of basins in the tributary river to encourage natural settlement processes. A model has been developed to choose the most suitable location for each basin, and to predict the reduction of the phosphorus load in different circumstances.

Areas with phosphorus-rich sediments are also being identified, and the most phosphorus-rich sediment areas will be dredged to prevent the phosphorus returning to suspension.

(e) Active biological management, or biomanipulation. Without appropriate measures, the aquatic system of Lake Volkerak-Zoom will develop into a turbid bream-dominated system within 5–10 years. Even with a reduction of the phosphorus load, it will be very hard to change such a stable eutrophic system into a clear pike-dominated system. Therefore, management of Lake Volkerak-Zoom is aimed at assisting the development of aquatic communities towards the establishment of a stable, clear system. This implies controlling the fish population and creating suitable habitats for waterplants and fish, and includes:

(1) an agreement on the fishing rights based on the management of the fish population;

(2) optimizing shallows and shorelines in order to maximize the breeding area for pike. This is achieved by the compartmentalization of shallows and the reconstruction of former creeks;

(3) creating large sheltered areas for waterplants in the shallows as a refuge for zooplankton, small pike and other fish. This can be achieved by the construction of offshore protection structures (see below);

(4) stimulating the growth of zebra mussels, which fulfil the same role as the zooplankton; filtering algae from the water and thereby contributing to the grazing pressure on the algae. Mussels require a hard substrate, which has been provided by dumping large amounts of shells in suitable places.

COMBATTING THE ACCUMULATION OF TOXIC COMPOUNDS

Emission-oriented measures

The first principle for restricting discharges is 'decreasing pollution at source'. An emission approach is foremost in nearly all forms of pollution control and, together with the substantial reduction of pollutant inputs via transboundary rivers resulting from the Ministerial Declarations on the North Sea, the Rhine Action Plan, will provide a sound basis for the regional effect-oriented measures necessary for sustainable development of the different water systems.

The most important emission-oriented measures to achieve a reduction in the micropollutants in Lake Volkerak-Zoom are: (a) the continued cleaning-up of industrial discharges; (b) research into clean technology and encouraging its application by industry; (c) stricter licensing; (d) the reduction of emissions from insecticides and herbicides; (e) a ban on the use of organotin as an antifouler on boats; (f) a reduction in the use of impregnated wood for shore protection; and (g) research to improve the quality of effluent from sewage works.

Fig. 4. Example of shoreline erosion.

Regional effect-oriented measures
Apart from tackling pollution at source, supplementary effect-oriented measures are required to protect Lake Volkerak-Zoom against accumulations of toxic compounds in sediments and organisms. The following regional effect-oriented measures are planned, or have already been implemented.

Dredging spoil
Spoil of a quality lower than the general environmental quality (i.e. class 2 or worse) must be stored under strictly isolated and controlled conditions, which effectively means controlled dumping on land or in deep underwater pits. For the Delta area, two large-scale storage depots are being developed (each with a capacity of 10–$20 \times 10^6 \, \mathrm{m}^3$) which should be completed in 1995, depending on the duration of the associated legal proceedings. Spoil of a quality equal to or better than the general environmental sediment quality (class 1) can be dispersed in the aquatic environment as long as there are no adverse effects.

Selective use of the Volkerak sluice
The flow of Rhine/Meuse water through the Volkerak sluice depends on:

(a) The quality of the river water. Flow is stopped in the event of calamities, emergencies and peak discharges, as peak discharges contain high concentrations of suspended sediments and micropollutants;

(b) The wind. At normal wind velocities most of the transported materials settle in the deep pit close to the Volkerak sluice. High wind velocities, at more than Force 5 on the Beafort scale, induce a flow path in the direction of the inlet sluice, and thus most of the transported materials are unable to settle in this pit, and are dispersed widely in the lake. To minimize dispersion, flushing should occur under low wind velocities whenever possible.

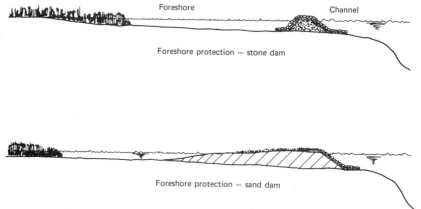

Fig. 5. Principle of foreshore protection structures.

COMBATTING EROSION

It was stated above that the shorelines and shallows are the areas with the highest potential for natural development, and they play a key role in active biological management. They should not therefore be lost by erosion (Fig. 4).

Erosion can be reduced by protection of the shoreline. The effect of this measure is, however, limited since erosion continues in the shallows and the natural development of the shoreline is inhibited. A more integrated objective is achieved with offshore protection structures (Fig. 5). A small dam or bank is built in the shallows at a certain distance from the shoreline and at a depth of about one metre. The waves are broken by this dam, erosion is minimized and sheltered lagoons are created. Gaps in the dam allow an exchange of water, plants and organisms between the lagoon and the shallows. These sheltered shallows are important for the development of waterplants, and as spawning beds for young fish.

About 30 km of the total shoreline (54 km) will be protected by offshore rubble dams and a further 10 km by protection of the shoreline directly. In 1990 the north flat, which had completely disappeared beneath the water surface as a result of erosion, was restored and protected by dams. About 30% of the shore protection plan had been completed by 1991, and the whole restoration and reconstruction plan is due for completion in 1995.

REFERENCES

[1] De Hoog, J. E. W. and Steenkamp, B. P. C. (1989). Eutrophication of fresh waters of the Delta. *Proceedings and information no. 41*, TNO Committee on Hydrological Research, The Hague, The Netherlands, pp. 43–44.

[2] Bijlsma, L. and Kuipers, J. W. M. (1989). River water and the quality of the Delta waters. *Proceedings and information no. 41*, TNO Committee on Hydrological Research, The Hague, The Netherlands, pp. 4–8.

[3] Ministry of Transport, Public Works & Water Management, (1990). *Watersysteem-onderzoek Volkerak-Zoommeer in 1989.* Dordrecht, The Netherlands (in Dutch).

[4] Ministry of Transport, Public Works & Water Management, (1989) *Water in The Netherlands: a time for action.* National policy document on water management. State Publishers, The Hague, The Netherlands.

Part VI Wastewater process engineering

36

Water treatment – an updated Hong Kong experience

J. C. K. Cheung, BSc(Eng), CEng, MICE, MIStrustE, MHKIE, ACIArb[†]

INTRODUCTION

Hong Kong saw a rapid growth in average daily water consumption from $1.41\,Mm^3$ to $2.34\,Mm^3$ in the past ten years. The lack of natural lakes or rivers or substantial underground water sources, and the scarcity of storage reservoir sites, have led to a shift in main source of raw water from rainfall from natural catchment to supply from China.

In 1964, an agreement was reached with the Chinese authorities whereby Hong Kong purchased $68 \times 10^6\,m^3$ of water each year. Subsequent agreements were made for progressive increases in water supply up to $660 \times 10^6\,m^3$/year by 1995. Recently, an agreement was signed for further increments beyond 1994 to an ultimate annual volume of $1100 \times 10^6\,m^3$. During 1989/90, the water supply from China amounted to $619 \times 10^6\,m^3$ which represents 72% of the total annual consumption.

To cope with the continued rise in demand for water, uprating or extension of existing treatment facilities was implemented in the last decade which brought an additional capacity of about $500\,Ml/d$. At the same time, a similar increase was effected by construction of a number of new treatment works.

[†] Engineer, Water Supplies Department, Hong Kong Government.

WATER TREATMENT

Raw water quality

In general terms, the raw water can be classified as being soft, low to moderate in mineral salt content, and very low in organic matter. The total dissolved solids concentration is comparatively low except that the initial runoff derived directly from local catchwaters could contain quite high levels of settleable solids. Some reservoirs contain, on occasions, moderately high levels of algae sufficient to cause minor operational problems at the water-treatment works. Some typical qualities of the major resources are shown in Table 1.

Drinking water standard

The standard for treated water conforms in all aspects, both chemically and bacteriologically, to international standards for drinking water recommended by the World Health Organization [1]. A fairly high pH value (8.2–8.8) is specified to suppress corrosion in the water supply system. Fluoridation, aimed at reducing dental decay particularly in children, has been practised since 1961. The current target fluoride level in the supply is 0.5 mg/l. Table 2 gives details of the standard.

Existing plants

There are 18 treatment plants in operation, the oldest dating back to 1928. A list of the plants in chronological order together with their characteristics is shown in Table 3. The geographical distribution of the treatment plants together with the raw water storage and transfer facilities is shown in Fig. 1.

Treatment process

Generally in Hong Kong, water-treatment works are either of single- or two-stage. Screens or settlement are not used.

Before 1963, all the treament works were of single stage. The process consists of the addition of a coagulant, rapid mixing, limited flocculation, followed by almost immediate filtration.

In 1963, a two-stage process was introduced in Hong Kong. The process consists of chemially-assisted sedimentation or clarification, which is designed to clear the water as much as possible prior to it passing to the filters. Since then, the two-stage process had been widely used until the adoption of a modern type of single-stage 'direct filtration' for Yau Kom Tau and Sheung Shui Treatment Works in 1985 and 1986, respectively. After a few years trial, the two-stage process has once again been found to be more suitable and is accepted as a preferred treatment method. Details will be discussed later.

Clarifier

At present, sludge-blanket or solid-contact clarifiers are commonly used in Hong Kong. These can either be circular or rectangular. The three major types of clarifier in use are the clariflocculator, the accelator, and the rectangular flat-bottomed, upward-flow, sludge-blanket clarifier. The rectangular flat-bottomed type has only been installed in recent years.

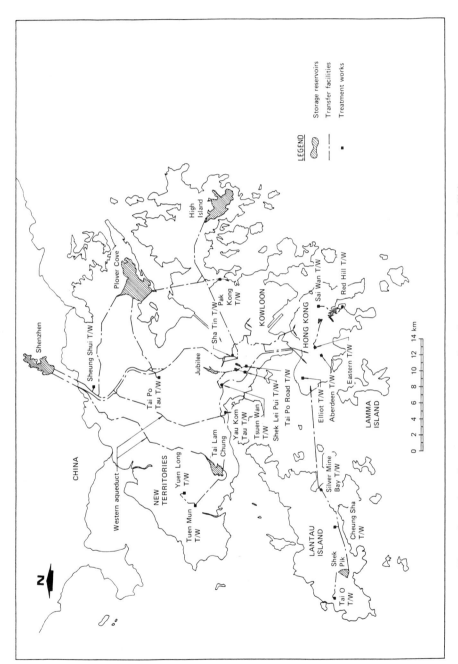

Fig. 1. Location of water-treatment work and raw water transfer facilities.

Table 1. Quality of resources (April 1989 – March 1990)

	From China Muk Wu A P/S		From China Muk Wu B P/S		Plover Cove Reservior		High Island Reservior	
	Ave.	Peak	Ave.	Peak	Ave.	Peak	Ave.	Peak
pH	7.1	7.7	7.2	7.6	7.3	7.8	7.3	7.6
Colour (Hazen)	22	80	26	80	5	5	5	5
Turbidity (FTU)	9.2	23	12.3	36.0	2.6	5.0	2.4	5.1
Ammoniacal Nitrogen (N)	0.06	0.25	0.07	0.33	0.04	0.07	0.01	0.03
Suspended solids	–	–	–	–	6	14	–	–
Dissolved solids	54	79	54	80	101	120	–	–
Total alkalinity ($CaCO_3$)	21	26	20	27	25	26	16	23
Total hardness ($CaCO_3$)	23	32	24	34	32	32	18	28
Calcium (Ca)	7.6	10.0	7.5	9.6	8.8	9.6	5.3	8.0
Magnesium (Mg)	1.0	2.2	1.3	2.7	2.3	2.4	1.2	1.9
Chlorides (Cl)	6	9	6	9	28	30	16	24
Sulphates (SO_4)	10	13	9	12	9	13	6	7
Fluoride (F)	0.21	0.26	0.21	0.26	0.16	0.18	0.19	0.26
Iron (Fe)	0.17	0.42	0.22	0.44	0.08	0.13	0.15	0.33
Manganese (Mn)	0.02	0.06	0.03	0.15	0.03	0.05	0.03	0.14
Temperature °C	24.8	32.0	23.9	30.5	23.6	28.0	25.2	28.9
Coliform (MPN/100 ml)	86	350	185	950	73	250	550	2500

Note: units in mg/l unless otherwise stated.

Filter

Because of the scarcity of land in Hong Kong, slow sand filters are not used. Pressure filters are used occasionally in minor treatment works where preservation of head is desirable. In general, the filters in use are of the rapid gravity type. The filter medium can be either graded sand or a dual medium of sand and anthracite. The total depth of the medium varies from 0.8 to 1.3 m depending on the water quality and treatment process.

Syphon wall and surface flush arrangement were once used in some of the filter installations to enhance backwash, but these constructions have ceased.

Recent construction

In the last decade, extension or uprating of some major treatment works were carried out to cope with the ever-increasing water demand. The extension or uprating generally adopted the same treatment process as the existing one so as to ensure uniformity and hence easier operation. The notable works were carried out at Shatin Treatment Works and Tuen Mun Treatment Works. After extension, the Shatin Treatment Works became the largest plant in Hong Kong and has now a capacity of 1227 Ml/d, which represents about half the total volume of treated water in Hong Kong. The total additional capacity brought in by the extension was 514 Ml/d.

Over the same period, five new treatment works were put into service, two of which are small pressure filters. They are at Tai O and Sai Kung. The latter was dismantled upon completion of Pak Kong Treatment Works. The other three are of medium to large size. They are Yau Kom Tau Treatment Works, Sheung Shui Treatment Works and Pak Kong Treatment Works. These three plants incorporate some new features which will be discussed in detail in the following paragraph. These new installations brought an additional capacity of 539 Ml/d.

At present, Pak Kong Treatment Works stage II and Au Tau Treatment Works stage I are under construction, while Au Tau Treatment Works stage II, Yau Kom Tau Treatment Works stage II, and Sheung Shui Treatment Works stage II are being designed. They will add an extra capacity of 1047 Ml/d after completion.

Table 2. Final Treated Water Standard and Quality

		Average quality (April 1989 – March 1990)			
Parameter	Standard	Shatin T/W	Yau Kom Tau T/W	Sheung Shui T/W	Pak Kong T/W
pH	8.2–8.8	8.5	8.5	8.4	8.1
Colour (Hazen)	< 5	< 5	< 5	< 5	< 5
Turbidity (FTU)	< 1.0 period to final pH conditioning	0.5	0.6	0.6	0.5
Iron (Fe)	< 0.1	0.01	< 0.01	< 0.01	< 0.01
Manganese (Mn)	< 0.05	0.01	< 0.01	< 0.01	< 0.01
Aluminium (Al)	< 0.01	0.04	0.05	0.06	0.04
Fluoride (F)	+ 10% of nominal level (current 0.5)	0.45	0.42	0.42	0.31[†]
Taste and odour	Unobjectionable	NIL	NIL	NIL	NIL
Coliform & E. Coli (MPN/100 ml)	Absent	NIL	NIL	NIL	NIL

Notes:

(1) For individual water quality parameters not referred to above, the parameters shall meet the values in the World Health Organization 1984 Guidelines.

(2) Units in mg/1 unless otherwise stated.

[†] Fluoride dosing plant under commissioning.

Yau Kom Tau

Yau Kom Tau Treatment Works was planned in 1981 to have a capacity of 160 Ml/d and to be capable of being extended, if required in future, to an ultimate capacity of 250 Ml/d.

Water to be treated is supplied from China through a western aqueduct except during two months of the year when the supply is shut down for maintenance, and water is then supplied from Plover Cove Reservoir.

Since the inception of supply from China, the raw water quality has remained remarkably good and fairly consistent. In view of this, a direct filtration process was considered to be a viable alternative to the conventional two-stage treatment process. In other words, there is no clarification prior to filtration.

Direct filtration has been in use for about 23 years mainly in the USA [2]. The advantage of employing this process is that it allows the sedimentation stage to be omitted thereby reducing plant, civil engineering, land and maintenance costs.

Table 3. Water treatment work in operation

Treatment works	Year first com-missioned	Present capacity (Ml/d)	Clarifier	Filter
Shek Li Pui	1928	91	—	Sand
Elliot	1931	6	—	Sand
Eastern	1949	50	—	Sand
Red Hill	1951	21	—	Anthracite/sand & sand
Tai Po Road	1956	32	—	Sand syphon wall
Tsuen Wan	1957	317	—	Anthracite/sand
Sai Wan	1957	14	—	Sand syphon wall
Aberdeen	1962	27	—	Anthracite/sand Syphon wall
Silvermine Bay	1963	159	Clariflocculator	Sand syphon wall
Yuen Long	1964	27	Clariflocculator	Sand syphon wall
Shatin	1965	1227	Clariflocculator & accelator	Anthracite/sand Syphon wall & flush
Tai Po Tau	1967	109	Accelator	Sand
Tuen Mun	1972	332	Accelator & reactivator	Anthracite/sand & sand
Tai O	1973	2	—	Sand pressure filter
Sai Kung[†]	1983	11	—	Sand pressure filter
Yau Kom Tau	1985	160	—	Anthracite/sand
Sheung Shui	1986	100	—	Anthracite/sand
Pak Kong	1989	273	Rectangular flat-bottomed	Anthracite/sand
Cheung Sha	1989	6	—	Sand pressure filter

[†]Dismantled upon completion of Pak Kong treatment works.

Pilot plant testing for the feasibility of direct filtration was carried out from November 1980 to mid-April 1981. Despite the short testing programme, the pilot testing concluded that direct filtration of Chinese raw water supply was feasible at a rate up to 14.7 m/h using dual medium filter and 6–12 mg/l alum in conjunction with 0.25 mg/l cationic polyelectrolyte.

Under certain conditions where the raw water contains a high level of turbidity and colour the filter runs are expected to be reduced to about eight hours. However, this occurs for only a small proportion of the available record. The design was considered acceptable in the knowledge that short reduction in plant output could be tolerated within the whole flexibility of the distribution system.

The special features of this direct filtration plant of modern concept (Fig. 2) include: (i) mixers with turbine agitation, which provides variable velocity gradients and optimum mixing conditions for varying quality and quantity of incoming water; (ii) variable time contact tanks to provide ranges of contact times from five minutes to 25 minutes, which is critical to the formation of pinhead floc before filtration and to allow part of the silt or algal load in poor raw water to settle out, thus reducing the loading on the filters. Additionally, polyelectrolyte is used to enhance the flocculation.

A filter rating of 8.3 m/h was chosen which should normally be readily achieved while maintaining filter runs of the order of 20 hours over most of the range of the water quality specified. Rapid-gravity, downward-flow, constant-level filters with sand and anthracite media are provided to deal with the high rates and range of filtration rates required.

Civil construction and plant erection commenced in November 1983 and were completed in May 1985 [3].

Fig. 2. Yau Kom Tau Treatment Works: special features of direct filtration
plant include rapid mixers and variable time contact tank.

Since Yau Kom Tau Treatment Works commenced operation in November 1985, the plant performance has been satisfactory under consistent quality of Chinese water. However, in March and April of 1987, there were instances of temporary suspension of plant operation after torrential rainfall in the course of reverse pumping from Plover Cove Reservoir because the Chinese supply was shut down for maintenance [4]. It was noted that reverse pumping had been interrupted and high turbidity local intake water dominated the raw water supply. The turbidity was reported to be in the range of 170–200 FTU.

Additionally, it was found that although the average turbidity of Chinese water is low, it sometimes fluctuates upwards to exceed the limit which has been taken as acceptable for direct filtration.

In view of the above and the inabilty of maintaining an uninterrupted supply should it be necessary to shut down the works for several, say ten, hours due to high turbidity water, it was decided that the works should be converted into a two-stage system at the time when the Stage II works is implemented.

Sheung Shui

Sheung Shui Treatment Works has a capacity of 100 Ml/d. It has many similarities to that of Yau Kom Tau Treatment Works: raw water is from China substituted by pumped water from Plover Cove Reservoir for two months in a year, and direct filtration is used. However, the rapid mixing tanks provide a shorter detention time of 32 seconds compared with the 60 seconds in Yau Kom Tau Treatment Works, while the flocculation tank provides a fixed detention time of only 1.7 minutes.

The rapid-gravity downward-flow filters consist of sand and anthracite media to a total thickness of 950 mm. A higher filter rating of 11.6 m/h is used. As a result, the plant is more sensitive to rapid changes in water quality. The filter runs are significantly reduced at the maximum designed throughput compared with that of Yau Kom Tau.

In view of the same arguments for Yau Kom Tau Treatment Works, the system will be converted into the conventional two-stage process when the stage II works is implemented.

Pak Kong

Pak Kong Treatment Works has been constructed in two stages. Design capacity of stage I is 273 Ml/d.

Raw water is mainly from Plover Cove Reservoir, supplemented by High Island Reservoir. At times when the upland catchments are discharging to the High Island tunnel system, water drawn off to Pak Kong Treatment Works will be a combination of surface runoff and stored reservoir water.

The raw water quality is generally good except occasionally moderately high levels of phytoplankton or high levels of settleable solids may be present.

The conventional treatment process with clarifiers and rapid-gravity filters is used. This is considered to be suitable in treating variable water occasionally containing relatively high levels of algae and suspended solids. In addition, considerable expertise and operational experience of this type of work are available.

Unlike those already in operation, which use circular clarifiers with an integral flocculation zone and solid recycling, Pak Kong Treatment Works introduced the rectangular flat-bottomed, sludge-blanket clarifier to Hong Kong (Fig. 3). The rectangular clarifier has the distinct advantage of less space being required. This is especially important in Hong Kong where restricted site area is the norm. The rectangular clarifier is also more economic to construct. An estimated saving of HK$9.1 million out of a total value of HK$202.1 million in the stage I construction costs was claimed.

The rectangular clarifiers have been in operation in the UK since the early 1970s. The clarifier uses suspended inlet distribution pipes (often referred to as 'chandeliers') with multiple downward-pointing jets of water which impinge on the tank floor, creating a toroidal zone of turbulence around each jet and keeping the flow of the tank generally free of settled material. In the upper part of the tank there is a quiescent layer in which nearly uniform upward flow of clarified water occurs. The clarified water is decanted at the surface into transverse collecting channels.

Sludge removal from the clarifier is effected by flexible cone sludge concentrators (Fig. 4). The equipment consists of a flexible sludge cone suspended in water and connected by a cable to a load cell. Control of desludging can be either by timer or by the load cell. The load cell is sufficiently sensitive that when the weight of the sludge reaches a preset value the load cell will initiate the opening of the desludging valve on the cone outlet. In theory this design would facilitate a stable suspended blanket while allowing excess floc to be bled off in a controlled manner.

For the filter, the dual anthracite/sand medium of a relatively fine grading was preferred to a mono-graded sand because of its satisfactory performance in treating algae-rich waters, and to continue operational similarity with other treatment works in Hong Kong.

Construction of stage I commenced in August 1985. Since commissioning of the plant in December 1989, certain problems have been encountered.

Fig. 3. The rectangular flat-bottomed sludge blanket clarifier at Pak Kong Treatment Works.

At the start-up of the plant, difficulties were met in building up the sludge blanket because of the relatively clean raw water. Attempts were made to adjust the dosage of aluminium sulphate and to regulate the pH value. The problem was solved later through better control over chemical dosing and the sludge blanket formation has been satisfactory since then.

The other problem is the sludge accumulation on the tank floor, although the turbidity and suspended solid level of the raw water are well within the designed limit. Accumulated sludge level up to 0.9 m in height was recorded. The problem seems to have resulted from low throughput and hence the 'toroidal circulation' failed to materialize by the reduced incoming water, leaving 'craters' of sludge on the tank floor. The problem is still under investigation. An attempt is being made to maintain the designed flow rate by limiting the number of clarifiers in operation under low throughput conditions.

Though rectangular flat-bottomed clarifiers have not yet demonstrated their long-term effectiveness, they are likely to be more widely used in Hong Kong. At present, the extension of Yau Kom Tau Treatment Works has adopted the rectangular clarifier due to space restriction of the site.

ENVIRONMENTAL ISSUES

Chlorine

Chlorine is used for disinfection. On most major treatment works in Hong Kong, chlorine is supplied in one-tonne drums and stored in the drum store; the duty and stand-by chlorine drums are transferred to a separate room. The only plant in Hong Kong using bulk steel tanks is Shatin Treatment Works. However, delivery of chlorine is also by one-tonne drums as bulk transport vehicles are not available. Therefore, chlorine is required to be transferred from the one-tonne drums to the bulk tanks after delivery to Shatin Treatment Works.

The chlorination installation is accommodated in a self-contained area, comprising a chlorine drum room, chlorinator room, pump room, switchgear room and store room. Special safety features such as air-tight glazed panels in internal walls, leak detection system, automatic closing of ventilation louvres and shutter doors are provided.

Fig. 4. The suspended inlet distribution pipes and flexible sludge cone of the rectangular clarifier at Pak Kong Treatment Works.

Chlorine stores having more than 10 tonnes in storage will require hazard assessments to be conducted with a view to assessing the risk levels associated with an accidental release and to identify the necessary measures in order to maintain such risks at an acceptable level. Such assessments to review the security of the chlorination installation of existing treatment works have been completed, and arrangements are in hand to upgrade the safety facilities as necessary.

Sludge Disposal

Currently, more than $12 \times 10^6\, m^3$ of water works sludge is generated each year from water-treatment works. The solid content varies from 0.15% to 3.5%. The estimated total solid amounts to more than 30 000 tonnes per annum.

Three methods of sludge disposal are being used: (a) direct disposal into a natural stream or sewer; (b) discharge into a sewer or stream following dilution to an acceptable concentration in sludge holding tanks; and (c) the sludge will be thickened in sludge concentration tanks and disposed of off-site by tankers, to nearby sewage-treatment plants and discharged to submarine outfall after primary treatment.

The first two methods process 86% of the sludge while the last method treats the remaining 14%.

As people in Hong Kong are becoming more environmentally conscious, the first two methods are no longer acceptable. The last method, principally employed at Pak Kong Treatment Works has caused a considerable amount of siltation around the submarine outfall and a further increase in the discharge rate is considered unacceptable.

Two alternatives for sludge disposal have been considered: (i) to thicken and transfer the sludge to a vessel-loading facility for onward disposal to sea, and (ii) to thicken and dewater the sludge prior to road transfer to landfill disposal.

The first option is being implemented for the marine disposal of sewage, and waterworks sludge arising from Shatin treatment works comprises 11% in volume of the total water works sludge. A 1500 dwt vessel has been built and will commence operation in early 1991. The sludge will be dumped at the extremity of Hong Kong waters some 15 km away. If proven to be environmentally acceptable, it may be extended to dispose of further quantities of sludges in later years. However, a decision on this is not expected to be made before the end of 1993 as the analysis of two full years of environmental monitoring will be needed.

The second option has been used in some sewage treatment plants in which plate-press filters and filter belts are being used. A recent trial of dewatering waterworks sludge using plate-press filters demonstrated that the solids content of the sludge cake could easily be produced well above 20%. Additionally, when immersed in water the sludge cake would not recombine to form a sludge even when mechanically rubbed. The sludge cake is considered suitable for landfill purposes. The drawback of this option is that potential landfill areas are limited and the cost of formation of a landfill site is increasing as a higher standard is required especially for controlling escapes of leachates and gases.

There is no clear conclusion of which alternative will provide a better and efficient solution to the sludge-disposal problem. Each case should be assessed on its own merits in the light of the evolving knowledge.

CONCLUSIONS

1. Water treatment in Hong Kong has undergone a steady development in the adoption of modern techniques, advanced plant design and new technologies.
2. This arises from the need to ensure a reliable and continuous supply of potable water in conformity with World Health Organization standards.
3. Further construction of water-treatment works will continue and more attention will be paid to the environmental impact associated with water treatment.

ACKNOWLEDGEMENT

The author would like to acknowledge the help from many colleagues in the preparation of this chapter and the permission of the Director of Water Supplies, Hong Kong government to publish this chapter.

REFERENCES

[1] World Health Organization (1984),*Guidelines for drinking water quality*, Vols 1–3. Geneva.
[2] Culp, R. L. (1980) The status of direct filtration. *J. Am. Wat. Wks. Assoc.* July, pp. 405–411.
[3] Chan, W. S., McMeekan, J. F. and Sinclair, J. (1985) Yau Kom Tau Treatment Works. *Hong Kong Engineer.* April, pp. 9–23.
[4] Chan, D. K. M. and Sinclair, J. (1991) Commissioning and operation of Yau Kom Tau water treatment works (Hong Kong) using direct filtration. *J. Instn. Wat. & Envir. Mangt.* April, 5, 2, 105–115.

37

The effect of the EC Urban Wastewater Treatment Directive on Marine Treatment Practice

R. Huntington, BSc, CEng, FICE, FIWEM[†]; B. Chambers, PhD, CEng, MIChE[‡]; and P. Dempsey, BA, CEng, MIWEM[§]

INTRODUCTION

In May 1991 the Council of the European Communities adopted a Directive [1] concerning Urban Wastewater Treatment (UWWTD). This Directive will have a considerable impact on the practice of wastewater treatment and disposal, an impact which, for the UK, will be most evident in coastal areas.

The dispersive characteristics and oxygen saturation of our coastal waters together with the large tidal range make them eminently suitable for marine treatment of wastewater. The assimilative capacity of such waters ensures that biodegradable contaminants are not normally critical design parameters.

Consequently, UK practice for the collection and treatment of wastewater from coastal communities has centred on the maximum use of the marine environment for the safe disposal of effluents.

Modern marine treatment schemes [2] include a long sea outfall for effluent dispersal, preceded by fine screening and grit removal for aesthetic and operational reasons. Such preliminary treatment facilities can usually be designed to fit into small compact headworks with minimal visual impact and low demand for land.

[†] Director of Engineering & Operations, Wessex Water.
[‡] Manager, Activated Sludge Group, WRc.
[§] Pollution Management Group, WRc.

Following the adoption of the UWWTD this approach to marine treatment will not be possible. It is expected that, for most UK coastal waters, only primary treatment will be required to satisfy the requirements of the Directive. However, where discharges to estuarial waters are concerned, or perhaps in areas of high recreational or amenity value, full secondary treatment will be required. In exceptional cases, perhaps in shallow estuaries or enclosed bays, additional treatment for the removal of nutrients will be necessary to protect the receiving waters from the possibility of eutrophic conditions developing.

IMPLICATIONS

This shift to a more demanding level of land-based treatment at coastal locations has considerable implications for sewerage undertakers. The more obvious ones are extra costs and longer lead-in times before overdue improvements are built and become operational. This chapter does not consider these aspects directly instead it concentrates on planning and design issues under the following four main headings: (i) treatment plant design; (ii) outfall design; (iii) stormwater management; (iv) catchment planning.

TREATMENT PLANT DESIGN

The topography of coastal towns and earlier marine treatment concepts mean that sewage usually drains through congested urban areas to the coast. The construction of sewage treatment facilities in such urban locations poses significant environmental and engineering problems.

The main problems facing the designer are those of: (a) process design in limited space; (b) sludge disposal; (c) visual impact, odours and noise; (d) maintaining flexibility.

These problems are all interconnected but it will be helpful to consider each in turn.

Process design in limited space

The limited space available in urban areas means that the traditional approach to process design needs rethinking. There is considerable scope for innovation in plant layout and for the installation of less conventional processes which can achieve the required degree of treatment in minimal area. Perhaps one of the first objectives should be to use alternatives to the traditional space-consuming primary sedimentation tanks; the use of Lamella separators, for example, can reduce space requirements by up to 90%.

Table 1 shows the minimum area requirements for different treatment processes. (No allowance is included for sludge disposal plant, odour control equipment or stormwater tanks.) The figures show that a secondary treatment plant will require at least twice the surface area of a primary treatment plant.

If secondary treatment is required the basis for a compact design using conventional processes would be Option 2. Activated sludge plants for the treatment of screened and degritted sewage are well understood and several process configurations are possible. Compared with conventional biological filters the activated sludge processes can produce equivalent effluent quality with about one-eighth of the land requirements.

Option 2 also illustrates the potential for saving space by eliminating primary sedimentation tanks. This option produces an effluent quality equivalent to Option 4 using only half the land requirement.

Table 1. Minimum area requirements for various treatment processes

	Process	Minimum area m²/1000 pe[†]
1.	Primary treatment only	20
2.	Compactly designed, conventional activated sludge plant treating finely screened, degritted sewage	35
3.	Primary treatment and carbonaceous oxidation by activated sludge	55
4.	Primary treatment and nitrifying activated sludge	70
5.	Primary treatment and nutrient removal activated sludge	85

† pe = population equivalent

Appropriately designed activated sludge plants can produce effluent of high quality. Several plants of this type are already in operation in the UK at coastal sites and many have been incorporated into enclosed structures which successfully disguise their true purpose.

Many new processes which are not based on conventional technology have been proposed for coastal sites. The UWWTD will no doubt give impetus to the development of such ideas. Those which show potential for achieving high rates of treatment with small space requirements merit detailed evaluation. The constraints imposed at some sites may result in processes being selected against different economic criteria from those normally applied at inland treatment works. It may be acceptable to trade capital investment against operational costs in some instances. An example is the elimination of phosphorus from the effluent by chemical precipitation. This process is expensive in chemical costs: biological systems of phosphorus removal are usually preferred at inland sites. However, the biological version of the process is capital intensive and requires a large land area.

For both conventional activated sludge and new process designs, thought needs to be given to underground and multifloor construction techniques to reduce land requirements still further in difficult urban sites.

Construction costs and subsequent pumping costs are likely to be greater for such layouts. However, these increases could be offset by savings in land purchase and by benefits from commercial or amenity use of surfaces created. For example, in Holland car parks have been established above utility installations, and in Japan sports facilities such as tennis courts have been provided. A large activated sludge plant in Rotterdam has also been constructed beneath urban parkland. Further examples of dual purpose sites exist in Switzerland and Southern France, where full secondary treatment and sludge thickening plants have been constructed under office blocks. In Marseilles a primary treatment plant for a population of 1.5 million is situated below a sports ground, and in Nice gardens have been established over the treatment plant.

Sludge disposal

Sewage treatment processes produce quantities of sludge which must be treated and disposed of in an acceptable manner. It is important to consider sludge disposal options at an early stage in the design procedure. Some coastal sites for sewage treatment may preclude the adoption of sludge treatment processes which may be acceptable elsewhere.

At small coastal sites it is obviously attractive to remove sludge to larger inland sites for treatment and disposal. This option will become increasingly expensive and unacceptable as the size of the works increases. For large schemes, on-site dewatering may be required to reduce the volume of sludge to be transported. Storage and dewatering problems may be overcome by pumping the sludge away, but installations would remain vulnerable to mechanical or electrical breakdown. Some buffer storage may need to be provided.

A final point to consider is that innovative, high-rate sewage treatment processes may give rise to unstable sludges which are difficult to dewater. It is therefore extremely important that sludge treatment and sewage treatment process design should be considered together.

Visual impact, odours and noise

Perhaps one of the biggest hurdles will be to produce urban treatment plant designs which gain public acceptance. There is a need to move from the existing negative image of dirty, smelly sewage treatment works to a positive image of clean 'purification centres'. To achieve this will require the adoption of high aesthetic standards for the buildings (and good PR!).

Treatment plants may need to be either partially or totally buried, and above-ground construction will have to blend with existing surroundings. Construction as an integral part of sea defence works or promenades may become more common.

Odour control will require careful attention. Forced ventilation with air filters or gas scrubbing installations will be needed, associated with the maintenance of negative pressures throughout the buildings. Where the plant is fully enclosed this will be relatively simple. Air locking systems for vehicle movements into and out of the plant will also be an essential part of odour control.

Maintaining flexibility

The UK water industry has prided itself in recent years on the flexibility of its designers to plan for the accommodation of future enhancements of treatment processes. This foresight will reap its rewards when plant improvements are able to proceed quickly and economically on an existing site.

In the future higher standards are likely for instance nutrient reduction may well become more common. It is essential, therefore, that the principles of flexibility are not lost in the attempt to fit treatment plants into congested sites, and provision for future expansion must always be considered.

In some instances this aim to maintain flexibility will dictate the need to pump sewage from confined urban sites to an alternative, perhaps inland, site where space constraints are less onerous. Many wider aspects come into the equation when transfer schemes are considered. Some of these are discussed in a later section.

STORMWATER MANAGEMENT

In coastal situations, good stormwater management is important because storm discharges from combined sewer overflows can compromise the achievement of amenity and recreational standards. The UWWTD does not change this situation.

However, the Directive does, perhaps, increase the need to intergrate the sewerage design with the treatment plant design, particularly with regard to maximum treatment rates (which can affect storage requirements) and the location of storage tanks. By suitable hydraulic design, maximum use should be made of any existing storage within the sewer system. Additional storage facilities should be designed at strategic locations so that storm discharges are limited to an acceptable volume and frequency.

OUTFALL DESIGN

Following the land-based treatment stage, in many cases a long sea outfall will still be the chosen option for effluent disposal. It is worth recalling the way an outfall is designed to achieve acceptable water quality, both near the point of discharge and in critical use areas (e.g. bathing waters) nearby. There are two main facets to consider [2] (1) initial dilution, and (2) secondary dispersion and decay.

Initial dilution

Initial dilution is the dilution of a buoyant effluent as it rises from its discharge point near the sea bed to form a 'field' of dilute effluent at the sea surface. It is important for two main reasons: (1) to produce an aesthetically acceptable surface field; and (2) to prevent the formation of a stable surface field which may inhibit secondary dispersion.

For an effluent subject to preliminary treatment only it is generally accepted that a minimum initial dilution of 50–100 is required to achieve these goals. For a primary or secondary treated effluent it would be tempting to design to a lower figure on the basis that aesthetic problems should be much reduced. Indeed, primary treatment should remove a large proportion of the greasy material responsible for surface slicks. The benefits of accepting a lower initial dilution would be simpler diffuser designs with a lower risk of saline intrusion and sedimentation.

However, regardless of the level of pretreatment, reducing the minimum initial dilution below 50–100 will increase the risk of forming a stable surface field. This will inhibit the subsequent mixing of the effluent and carry the risk of high surface concentrations being carried into sensitive use areas, particularly if outfall lengths are reduced.

It is suggested that a minimum initial dilution of 50 should be retained as a design criteria regardless of the level of land-based treatment used.

Secondary dispersion

It is standard practice [2,3] nowadays to use validated mathematical models to simulate the dispersion, dilution and decay of an effluent as it is advected away from an outfall location towards a sensitive use area, such as a bathing water. The model is used to design the outfall, in terms of length and location, such that environmental standards (e.g. EC Bathing Water Directive [4], bacterial standards) are met with a suitable safety margin.

Bacterial concentrations in a secondary treated effluent are likely to be one to two orders of magnitude less than in a preliminary treated effluent. As such there should be scope for shorter outfalls where secondary treatment is used. However, there is no general rule for quantifying this benefit – it will be highly site-specific. In some open coastline situations where tidal currents are consistently parallel to the shore it should be possible to make significant savings in outfall length, provided initial dilution requirements do not become critical. In a headland/bay situation with more complex currents, a certain minimum outfall length may be necessary to ensure that the effluent is not caught up in the bay circulation. In such a situation, secondary treatment may not allow a shorter outfall to be used.

No significant reduction in bacterial concentrations can be expected for a primary treated effluent. As a result outfall design will remain the same.

These considerations suggest that soundly-based hydraulic and environmental modelling will continue to be essential for the effective design of outfalls, even when these are preceded by a high level of land-based treatment.

A further reason for not attempting to make significant reductions in the length of outfalls is the inevitable public desire to move towards tighter standards.

One change which could have a major impact on outfall length is the use of an acceptable disinfection process to kill bacteria. There are, of course, arguments about the effectiveness or otherwise of various disinfection processes, particularly in relation to viruses. Better and more concentrated research may, in fact, demonstrate that an improved design approach is possible.

CATCHMENT PLANNING

The implications of the UWWTD could simply be seen in terms of individual design problems as discussed in the preceding sections. However, on a broader level the proposed changes also point to an increased need for looking at wastewater disposal options on a catchment-wide basis.

Previously, coastal schemes were often looked at on a more or less individual basis. Preliminary treatment plus a long sea outfall would usually be the cheapest option; it would generate no sludge, the headworks could be sited near the community and the environmental impact would be relatively localized near the outfall location.

In the future, meeting the terms of the UWWTD may often mean pumping wastewater flows away from the coastal town to a suitable inland site for treatment. This increases the attractiveness of integration with neighbouring inland towns

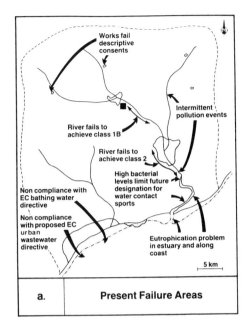

a.   **Present Failure Areas**

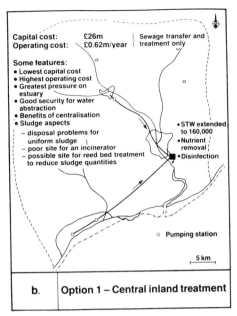

b.   **Option 1 – Central inland treatment**

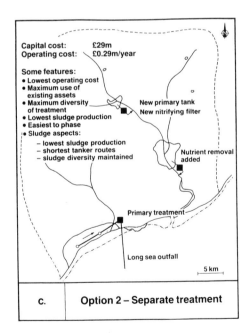

c.   **Option 2 – Separate treatment**

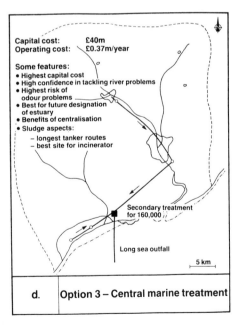

d.   **Option 3 – Central marine treatment**

Fig. 1. Example of comparison of waste disposal options for a catchment.

and of effluent discharge to suitable rivers rather than directly to sea via an outfall. The extra sludge production may increase competition for suitable agricultural or landfill disposal sites and may point to the need for a complete reappraisal of an existing regional sludge disposal strategy. These are complex planning issues requiring careful environmental consideration.

In particular, the extra pressure on rivers as suitable disposal routes for effluents will need to be examined with care. The large assimilative capacity of the sea means that it is usually possible to design marine treatment schemes to meet environmental standards with wide safety margins. This is not the case with many rivers. The potential for eutrophication or deoxygenation problems can be considerably increased in the lower reaches of a river as a result of an extra effluent load. In addition, the river will carry an increased bacteriological load which could adversely affect the quality of coastal bathing waters near the mouth of the river. However, disinfection may have a part to play here.

Fig. 1 shows a comparison between three strategic wastewater disposal options for a catchment involving both coastal and inland towns. All three options meet the required standards but many other issues (noted on the figure), in particular overall robustness and flexibility, require consideration when selecting the Best Practicable Environmental Option (BPEO).

It is suggested that a BPEO type approach to the strategic planning of wastewater disposal, at a catchment-wide level, is likely to become increasingly important in the future. New environmental legislation, such as the UWWTD is one factor in this move towards more integrated planning. Another is the new regulatory framework in England and Wales, involving both environmental and financial regulation of sewerage operations. The National Rivers Authority will want to be confident that all interrelated environmental impacts are being addressed and that maximum flexibility and minimum risk are built into schemes so that the environment is properly safeguarded. Equally the Office of Water Services will want to be confident that capital works programmes do achieve value for money and that the water consumers are safeguarded.

Many marine treatment schemes will need to be investigated and promoted as part of an overall catchment plan to establish the best environmental and cost effective solution.

However, it should be recognized that in such situations there is a real danger that the quest for a 'perfect' solution can produce unacceptable delays in the planning process. There is a large amount of work to be done and limited time available, therefore realistic judgements will be necessary. Planning does not produce tangible works – construction does.

CONCLUSIONS

The main implications of the UWWTD for marine treatment are seen to be as follows.

1. Tackling the challenging problems of building and operating treatment plants in coastal urban situations will require an innovative approach to process design and plant layout. Flexibility for future expansion must always be maintained.
2. Sound environmental and hydraulic modelling will continue to be required for

outfall design. A saving in outfall length, but not in diffuser design, will be possible in some situations.

3. Many marine treatment schemes will need to be integrated into catchment-wide wastewater disposal plans for which overall cost effectiveness and environmental robustness are clearly demonstrated.

4. Marine treatment will continue to provide a valid cost effective option for wastewater disposal with each case being treated on its merits.

REFERENCES

[1] Council of the European Communities (1991) Directive 91/271/EEC concerning urban waste water treatment, *Off. J. Eur. Com* L135/40.

[2] WRc (1990) *Design guide for marine treatment schemes*, WRc Report UM1009.

[3] Crawshaw, D. H., and Head, P. C. (1989) Fylde Coast bathing water improvements – environmental investigations for the design of sea outfalls. *Long sea outfalls*, Proceedings of ICE conference at Glasgow, October 1988, Thomas Telford, London, pp. 75–88.

[4] Council of European Communities (1976) Directive 76/160/EEC concerning the quality of bathing water, *Off. J. Eur. Com* L31/1.

38

Recycling: an environmentally acceptable alternative?

N. P. Board, MSc[†]

INTRODUCTION

Recycling of materials, particularly those comparatively easily sorted at source, for example paper, glass and metals, has now become an area of increased public awareness. Indeed the protection of the environment has become an important marketing tool for many products. The uptake of Chlorofluorocarbon (CFC) free packaging and aerosols, lead-free petrol and low phosphate content detergents are all examples of how the public are prepared to change habits to protect the environment. Similarly the purchase of products which are produced from recycled wastes is now an acceptable activity for the majority, rather than an expression of support for minority causes. The reduction in apparent quality of some recycled products when compared to their virgin counterparts (e.g. loss of brightness in some paper products) is now considered a small price to pay for environmentally friendly products.

This chapter does not provide a detailed reference of pollution loads arising from all recycling activities, nor does it contest the global benefits of recycling and the concomitant improved utiliazion of limited resources. Its purpose is to provoke thought and comment on the local effects of recycling and, as an example, focuses upon the production of paper products from recycled waste.

[†] Environmental Scientist, Water Division, Travers Morgan Environment.

DOMESTIC WASTE ARISINGS

Recycling of waste products from domestic refuse is the subject of many detailed studies in the UK at present. The Recycling City scheme in Sheffield, for instance, will assess the success of dedicated collection systems for various waste types. Historically, however, recycling of wastes in the UK has been consumer led. Examples of successful recycling initiatives are often those associated with charitable bodies who benefit directly from the collection and recovery of waste, perhaps demonstrating that, historically, the UK public have believed that recycling far from being driven from environmental concerns was in fact practised to help local groups obtain much needed funds.

The value of domestic waste in terms of potential revenue has been assessed by Barton [1] who concluded that of the 20 million tonnes of domestic waste produced per year (equivalent to 600 kg/household year) only 42% was suitable for immediate recycling, the remaining 58% requiring either disposal by conventional means of processing into some other saleable form, for example refuse – derived fuel, compost etc. The revenue potential of the 8.4 million tonnes per annum of recycable materials that could be recovered was calculated as approximately £10.25 per tonne (equivalent to £6.15 per household per year). This comparatively low revenue value is surely insufficient to encourage the majority of households to practice recycling from a fund-raising view alone.

It is interesting to note that 18% (or £1.10 per year per household) of the revenue available from recyclable waste was due to the paper content of the waste. Recycling of paper has perhaps been the one route of dedicated waste collection with which the public is most familiar. However, the recovery of paper by charitable groups has been the victim of its own success with many voluntary groups no longer finding it worthwhile to recover newsprint/magazines following the significant drop in funds raised as a result of the 50% fall in merchant prices some 18 months ago.

COMPARISON OF PRODUCTION PROCESSES

To assess the environmental implications of recycling it is necessary to understand the production processes involved in producing a product from both virgin raw materials and recycled waste. Such an assessment of each process should include detailed analysis of energy requirements and should also assess the type and amount of discharges made to the environment. In essence therefore the comparison will necessitate an audit of each production process. The data from such an exercise should be normalized prior to comparison thus producing an environmental index against which other distinct production methods can be compared. The principles described represent those which must be addressed in a formal cradle to grave assessment of any product. This assessment must be conducted albeit within clearly defined limits but their extent must however not be prejudicial to the outcome of the survey. In the case of such a study of paper production processes, one could argue that the assessment must start at forest management and proceed until final packaging of the product is complete.

The production of paper is a complex process requiring various specialized processes to yield wood pulp, to convert that pulp into fibre with suitable papermaking properties and to provide a unique balance of chemical additives to produce the desired end product. Each of these processes is complex and is described briefly below.

In assessing pulping processes one immediately identifies the importance of defining the limits of the cradle to grave assessment. The British Paper and Board Industry Federation [2] estimate that in the UK approximately 47% of the 5 million tonnes of paper produced annually results from virgin fibre. Of that 23% is made from fibre produced wholly within the UK. The capacity for home produced pulp production is constantly changing, however it is understood that there are presently no chemical pulp mills in the UK and further, none is planned. Is it possible therefore to assess the environmental benefits on a local scale of producing paper from recycled fibre when the alternative technology is not practised in this country? If one accepts that the comparison cannot be local, one must still address the types of production processes used.

Pulp falls into one of three main production categories, these may be summarized as follows.

Mechanical pulp

This is a cost-effective method of utilizing timber. The entire log with the exception of the bark is chipped and then ground to produce papermaking fibres. The production process is energy intensive but produces relatively simple effluents as no chemicals are used. The bark, removed before processing, may be used as a fuel in energy production.

Chemical pulp

Bark is again removed before processing. Wood chips are digested with chemicals to remove the 'woody' constituents, such as sap, lignin etc. The fibre is then washed prior to use. The chemical pulp often does not have the desired optical properties and will therefore be bleached prior to use. Similar bleaching procedures may be practised on both mechanical and combination pulps. Energy requirements are low compared with mechanical pulps and recovery of pulping chemicals is often practised to aid process economics (recycling?). In addition, energy may be generated following processing of pulping residues. Discharges from chemical pulp mills may have a significant localized environmental effect. Discharges will contain high concentrations of 'conventional' pollutants, for example biochemical oxygen demand, chemical oxygen demand, total suspended solids. In addition these discharges also contain elevated concentrations of complex organic materials, for example lignin, chlorophenols, vanillins etc. Numerous adverse environmental effects from pulping liquors have been recorded.

Combination pulps

As the name suggests, these are formed using a combination of mechanical and chemical methods. To the pulp producer the advantages of high yield associated with mechanical pulps is maintained, whilst improved chemical characteristics are also achieved. Environmentally the processing remains energy intensive and also results in the discharge of some pulping chemicals.

RECYCLED FIBRE

The alternative to virgin fibre use is of course the use of recycled fibre. This can in its own right take many guises. It is well known that the presence of recycled fibre within a product has been identified as representing a marketing gain. Such claims may be spurious. The re-use of so called 'mill broke' within the mill makes economic sense. 'Mill broke' is the paper that is produced but does not form part of the final product (due to quality limitations, it resulting from off cuts or machine trims etc.) but is that truly recycled? Similarly, is mill broke sold to a wastepaper merchant and bought by a separate mill as wastepaper an example of recycling? Clearly certain paper manufacturers believe so but by no means all of them. The recycled paper that perhaps everyone can identify with is post-consumer waste or that resulting from collection of, for example, newspapers, which is processed by the papermaker before being incorporated into the final product.

The reuse of fibre in the paper mill requires processing of the fibre source prior to formation. The degree of processing and therefore the local environmental effects that may result are dependent both upon the wastepaper type and the product into which it is to be formed. The production of high grade papers, such as printing and writing papers, from recycled fibre is not common. Whilst these products are available it is estimated [2] that only 8% of the product category (some 1.25 million tonnes per year) arise from recycling. Conversely packaging case materials (the familiar brown/dark grey cardboard) are produced almost exclusively from recycled papers. These lower product grades, whilst having to be produced to meet detailed product specifications, are not required to have the same high quality as certain printing papers. The exact processing of recycled papers within the paper mill will vary, however the following processes may be undertaken resulting in emission of the substances identified.

Fibre redispersion

This is undertaken in a hydropulper. Heat energy (steam) may be used to enhance dispersion of the fibres. Elevated temperatures may cause rapid dissolution of organic materials (e.g. starches) which will contribute to the pollution load. Chemical agents may also be added to enhance the breakdown of complex chemical additives present in the paper.

Contaminant removal

Processes to remove large contaminants, for example rags, wire etc., may be relatively crude yet highly efficient. Recycling of lower quality mixed wastepaper grades may require removal of traces of plastics, polystyrenes, grits and other contraries. In addition the reuse of self-adhesive paper products, such as envelopes, may necessitate the use of complex physical and chemical procedures to reduce the presence of sticky contaminants which may have an adverse effect upon ultimate paper product quality.

De-inking

De-inking of waste streams is often necessary for the production of both newsprint and certain tissue products. The process requires the separation of ink particles from the fibre. This process, often augmented by chemical additives, results in a secondary fibre which has improved optical properties to the papermaker. This process may however contribute significantly to the solid waste emissions from a mill. The Research Association for the Paper and Board Printing and Packaging Industries, Pira, estimates that in total, the UK paper industry produces over 250 000 tonnes per annum of mill sludge for disposal, of which de-inking waste contributed approximately 12% [3].

AQUEOUS ENVIRONMENTAL EMISSIONS

Clearly the production of paper from either virgin raw materials or recycled fibre will result in environmental emissions. These emissions are normally controlled by legislation thus providing a framework within which the pulp/paper mill will operate. The environmental performance of the UK paper industry [4] in general is beyond the scope of this discussion, however improvements in effluent treatment facilities during the 1980s have been significant, prompted by both legislative controls and an increased public awareness of environmental issues.

Data collected from various surveys of the UK paper industry indicate that secondary biological treatment of wastewaters increased significantly during the 1980s, with over 30% of UK production capacity discharging biologically treated wastewaters by the end of 1987 (this represents 21 mills with biological treatment in 1987 compared with 9 mills in 1980). It is understood that since this survey several new installations have been made across the industry thus significantly increasing the production capacity that is now discharging biologically treated wastewaters.

Published data regarding comparable pollution loads from waste-based versus virgin fibre production of paper products is not readily available. It is believed that this lack of data is due to the observation that many mills making a particular product type, for example tissues or newsprint, use both fibre sources in their production methods. Further, discharge of wastewaters from different production units is usually made to a common effluent treatment plant. Comparison of final effluent discharges (available from public records) may not therefore be representative. It is known that the concentrations of pollutants (principally dissolved organics) discharges from waste-based processes are significantly higher than those from virgin fibre mills. The nature of these organics also differ due to the presence of differing chemical additives used during production of both the original paper and the waste-based product. These elevated concentrations result from the more efficient use of water in many waste-based mills. If, for instance, one compares typical water consumption from a printing and writing grade usually made from virgin or high quality mill broke waste (typically 50–60 m^3/tonne) with that for processed producing for instance newsprint containing significant quantities of recycled fibre (typically 30–50 m^3/tonne), there are significant differences which may account for the elevated concentrations of dissolved organics present. The pollution load (quoted as kg/COD tonne) may however not be as significantly different.

CONCLUSIONS

The global benefits of a recycling philosophy cannot in themselves be denied. These benefits may be more immediately seen for products which are produced from limited resources, for example aluminium cans etc. Problems of resource exhaustion are clearly less acute for paper products. The pulp industry has a vested interest in maintaining an adequate supply of timber for the future. To this end well-managed forests are maintained by all the major suppliers of paper pulp. Indeed for every tree harvested at least two seedlings are planted. This increases to three seedlings for certain Nordic countries [2].

Recycling is important both in environmental and economic terms for the producer. Efficient use of resources is a familiar topic but, from this overview of those within the paper industry disbenefits with which perhaps the consumer is not aware are: (1) the complex sorting procedures necessary to recover recyclable waste; (2) the complex production processes involved in cleaning recovered material; and (3) the fact that recycling of wastes yields further wastes to be treated/recovered and that these waste materials may impact upon the local environment.

The need for recycling is clear, the benefits undoubted, but public awareness in reality may be limited. By improving consumer awareness the benefits of separation at source may be realized and the value of simple measures in the home appreciated so that they are seen not to be arduous or to be beneficial only to the recycler, but more importantly to be beneficial to the environment.

REFERENCES

[1] Barton, J. R. (1990) *Wastes Manage.*, **80**, No.2, 92–106.
[2] British Paper and Board Industry Federation. News Sheets, Pulp and Paper Information Centre.
[3] Pira, Private Communication.
[4] Board, N. P. (1987) MSc Thesis, *Papermill effluent treatment, current UK practice.*

Index